MODERN EXPERIMENTAL
STRESS ANALYSIS

MODERN EXPERIMENTAL STRESS ANALYSIS
completing the solution of partially specified problems

James F. Doyle
Purdue University, Lafayette, USA

John Wiley & Sons, Ltd

Other Wiley Editorial Offices

John Wiley & Sons Inc., 111 River Street, Hoboken, NJ 07030, USA

Jossey-Bass, 989 Market Street, San Francisco, CA 94103-1741, USA

Wiley-VCH Verlag GmbH, Boschstr. 12, D-69469 Weinheim, Germany

John Wiley & Sons Australia Ltd, 33 Park Road, Milton, Queensland 4064, Australia

John Wiley & Sons (Asia) Pte Ltd, 2 Clementi Loop #02-01, Jin Xing Distripark, Singapore 129809

John Wiley & Sons Canada Ltd, 22 Worcester Road, Etobicoke, Ontario, Canada M9W 1L1

Wiley also publishes its books in a variety of electronic formats. Some content that appears
in print may not be available in electronic books.

British Library Cataloguing in Publication Data

A catalogue record for this book is available from the British Library

ISBN 0-470-86156-8

Produced from LaTeX files supplied by the author and processed by Laserwords Private Limited, Chennai, India
Printed and bound in Great Britain by Antony Rowe Ltd, Chippenham, Wiltshire
This book is printed on acid-free paper responsibly manufactured from sustainable forestry
in which at least two trees are planted for each one used for paper production.

Over the past quarter century and more,
I have benefitted immeasurably
from the knowledge and wisdom
of my colleague, mentor, and friend,
Professor C-T. Sun;
it is with humbled pride that I dedicate this book to him.

Contents

Preface

This book is based on the assertion that, in modern stress analysis, constructing the model is constructing the solution—that the model is the solution. But all model representations of real structures must be incomplete; after all, we cannot be completely aware of every material property, every aspect of the loading, and every condition of the environment, for any particular structure. Therefore, as a corollary to the assertion, we posit that a very important role of modern experimental stress analysis is to aid in completing the construction of the model.

What has brought us to this point? On the one hand, there is the phenomenal growth of finite element methods (FEM); because of the quality and versatility of the commercial packages, it seems as though all analyses are now done with FEM. In companies doing product development and in engineering schools, there has been a corresponding diminishing of experimental methods and experimental stress analysis (ESA) in particular. On the other hand, the nature of the problems has changed. In product development, there was a time when ESA provided the solution directly, for example, the stress at a point or the failure load. In research, there was a time when ESA gave insight into the phenomenon, for example, dynamic crack initiation and arrest. What they both had in common is that they attempted to give "the answer"; in short, we identified an unknown and designed an experiment to measure it. Modern problems are far more complex, and the solutions required are not amenable to simple or discrete answers.

In truth, experimental engineers have always been involved in model building, but the nature of the model has changed. It was once sufficient to make a table, listing dimensions and material properties, and so on, or make a graph of the relationship between quantities, and these were the models. In some cases, a scaled physical construction was the model. Nowadays the model is the FEM model, because, like its physical counterpart, it is a dynamic model in the sense that if stresses or strains or displacements are required, these are computed on the fly for different loads; it is not just a database of numbers or graphs. Actually, it is even more than this; it is a disciplined way of organizing our current knowledge about the structure or component. Once the model is in order or complete, it can be used to provide any desired information like no enormous data bank could ever do; it can be used, in Hamilton's words, "to utter its revelations of the future". It is this predictive and prognostic capability that the current generation of models afford us and that traditional experimental stress analysis is incapable of giving.

Many groups can be engaged in constructing the model; there is always a need for new elements, new algorithms, new constitutive relations, or indeed even new computer hardware/software such as virtual reality caves for accessing and displaying complex models. Experimental stress analysts also have a vital role to play in this.

That the model is the focal point of modern ESA has a number of significant implications. First, collecting data can never be an end in itself. While there are obviously some problems that can be "solved" using experimental methods alone, this is not the norm. Invariably, the data will be used to infer indirectly (or inversely as we will call it) something unknown about the system. Typically, they are situations in which only some aspects of the system are known (geometry, material properties, for example), while other aspects are unknown (loads, boundary conditions, behavior of a nonlinear joint, for example) and we attempt to use measurements to determine the unknowns. These are what we call partially specified problems. The difficulty with partially specified problems is that, far from having no solution, they have great many solutions. The question for us revolves around what supplementary information to use and how to incorporate it in the solution procedure. Which brings us to the second implication. The engineering point to be made is that every experiment or every experimental stress analysis is ultimately incomplete; there will always be some unknowns, and at some stage, the question of coping with missing information must be addressed. Some experimental purists may argue that the proper thing to do is to go and collect more data, "redo the experiment," or design a better experiment. But we reiterate the point that every experiment (which deals with a real structure) is ultimately incomplete, and we must develop methods of coping with the missing information. This is not a statistical issue, where, if the experiment is repeated enough times, the uncertainty is removed or at least characterized. We are talking about experimental problems that inherently are missing enough information for a direct solution.

A final point: a very exciting development coming from current technologies is the possibility of using very many sensors for monitoring, evaluation, and control of engineering systems. Where once systems were limited to a handful of sensors, now we can envisage using thousands (if not even more) of sensors. The shear number of sensors opens up possibilities of doing new things undreamt of before, and doing things in new ways undreamt of before. Using these sensors intelligently to extract most information falls into the category of the inverse problems we are attempting to address. That we can use these sensors in combination with FEM-based models and procedures for real-time analyses of operating structures or post analyses of failed structures is an incredibly exciting possibility.

With some luck, it is hoped that the range of topics covered here will help in the realization of this new potential in experimental mechanics in general and experimental stress analysis in particular.

Lafayette, Indiana   *James F. Doyle*
December, 2003

Notation

Roman letters:

a	radius, plate width, crack length
A	cross-sectional area
b, b_i	thickness, depth, plate length, body force
c_0	longitudinal wave speed, $\sqrt{EA/\rho A}$
$C, [\, C \,]$	damping, capacitance, damping matrix
D	plate stiffness, $Eh^3/12(1 - \nu^2)$
$\hat{e}_i$	unit vectors, direction triads
e_{ij}	dielectric tensor
$E, \hat{E}$	Young's modulus, voltage
EI	beam flexural stiffness
E_{ij}	Lagrangian strain tensor
f	frequency in Hertz
f_σ	photoelastic material fringe value
$F, \{F\}$	member axial force, element nodal force
$g_i(x)$	element shape functions
$G, \hat{G}$	shear modulus, frequency response function, strain energy release rate
h	beam or rod height, plate thickness
h_i	area coordinates
H, H_t, H_s	regularizations
i	complex $\sqrt{-1}$, counter
I	intensity, second moment of area, $I = bh^3/12$ for rectangle
$K, [\, k \,], [\, K \,]$	stiffness, stiffness matrices
K_I, K_1, K_2	stress intensity factor
L	length
M, M_x	moment
$M, [\, m \,], [\, M \,]$	mass, mass matrices
n	index of refraction
N, N_i	fringe order, plate shape functions
$P(t), \hat{P}, \{P\}$	applied force history
$\tilde{P}$	force scale, parameter scale
q	distributed load
$Q, [\, Q \,]$	selector relating experimental and analytical data points
r, R	radial coordinate, radius, electrical resistance
$[\, R \,]$	rotation matrix
S_g	strain gage factor
t, t_i	time, traction vector

T	time window, kinetic energy, temperature
$[\,T\,]$	transformation matrix
$u(t)$	response; velocity, strain, etc.
$u,\ v,\ w$	displacements
V	member shear force, volume, voltage
W	beam width, work
$x^o,\ y^o,\ z^o$	original rectilinear coordinates
$x,\ y,\ z$	deformed rectilinear coordinates

Greek letters:

α	coefficient of thermal expansion
β_{ij}	direction cosines
γ	regularization parameter
δ	small quantity, variation
δ_{ij}	Kronecker delta, same as unit matrix
Δ	determinant, increment
ϵ	small quantity, strain
ϵ_{ijk}	permutation symbol
η	viscosity, damping
θ	angular coordinate
κ	plate curvature
λ	Lamê coefficient, eigenvalue, wavelength
μ	shear modulus, complex frequency
ν	Poisson's ratio
$\rho^o,\ \rho$	mass density
$\sigma,\ \epsilon$	stress, strain
ϕ	phase
$\phi, \{\phi\}, [\ \Phi\ ]$	unit force function, matrix of unit loads
$\psi, \{\psi\}, [\ \Psi\ ]$	sensitivity response functions
χ^2	data error norm
ω	angular frequency

Special Symbols and Letters:

$\mathcal{A}$	data error functional
$\mathcal{B}$	regularization functional
$\mathcal{F}$	equilibrium path
$\mathcal{T}$	kinetic energy
$\mathcal{U}$	strain energy
$\mathcal{V}$	potential energy
∇^2	differential operator, $\dfrac{\partial^2}{\partial x^2} + \dfrac{\partial^2}{\partial y^2}$
$[\ \ ]$	square matrix, rectangular array
$\{\ \}$	vector
o	original configuration
$*$	complex conjugate
$-$	bar, local coordinates
$\cdot$	dot, time derivative
$\,\hat{}\,$	hat, frequency dependent, vector
$\,\tilde{}\,$	scale component
$\prime$	prime, derivative with respect to argument

Abbreviations:

CST	constant strain triangle element
DoF	degree of freedom
DKT	discrete Kirchhoff triangle element
FEM	finite element method
MRT	membrane with rotation triangle element
SRM	sensitivity response method

Introduction

One of the frustrations of an experimental stress analyst is the lack of a universal experimental procedure that solves all problems.

A.S. KOBAYASHI [108]

At present, the finite element method represents a most general analysis tool and is used in practically all fields of engineering analysis.

K-J. BATHE [18]

There are two main types of stress analyses. The first is conceptual, where the structure does not yet exist and the analyst is given reasonable leeway to define geometry, material, loads, and so on. The preeminent way of doing this nowadays is with the finite element method (FEM). The second analysis is where the structure (or a prototype) exists, and it is this particular structure that must be analyzed. Situations involving real structures and components are, by their very nature, only partially specified. After all, the analyst cannot be completely aware of every material property, every aspect of the loading, and every condition of the environment for this particular structure. And yet the results could be profoundly affected by any one of these (and other) factors. These problems are usually handled by an *ad hoc* combination of experimental and analytical methods—experiments are used to measure some of the unknowns, and guesses/assumptions are used to fill in the remaining unknowns. The central role of modern experimental stress analysis is to help complete, through measurement and testing, the construction of an analytical model for the problem. The central concern in this book is to establish formal methods for achieving this.

Partially Specified Problems and Experimental Methods

Experimental methods do not provide a complete stress analysis solution without additional processing of the data and/or assumptions about the structural system. Figure I.1 shows experimental whole-field data for some sample stress analysis problems—these

Modern Experimental Stress Analysis: completing the solution of partially specified problems. James Doyle
© 2004 John Wiley & Sons, Ltd ISBN 0-470-86156-8

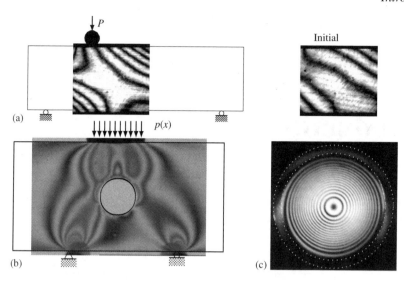

Figure I.1: Example of whole-field experimental data. (a) Moiré in-plane u-displacement fringes for a point-loaded plate (the inset is the initial fringe pattern). (b) Photoelastic stress-difference fringes for a plate with a hole. (c) Double exposure holographic out-of-plane displacement fringes for a circular plate with uniform pressure.

example problems were chosen because they represent a range of difficulties often encountered when doing experimental stress analysis using whole-field optical methods. (Further details of the experimental methods can be found in References [43, 48] and will be elaborated in Chapter 2.) The photoelastic data of Figure I.1(b) can directly give the stresses along a free edge; however, because of edge effects, machining effects, and loss of contrast, the quality of photoelastic data is poorest along the edge, precisely where we need good data. Furthermore, a good deal of additional data collection and processing is required if the stresses away from the free edge is of interest (this would be the case in contact and thermal problems). By contrast, the Moiré methods give objective displacement information over the whole field but suffer the drawback that the fringe data must be spatially differentiated to give the strains and, subsequently, the stresses. It is clear from Figure I.1(a) that the fringes are too sparse to allow for differentiation; this is especially true if the stresses at the load application point are of interest. Also, the Moiré methods invariably have an initial fringe pattern that must be subtracted from the loaded pattern, which leads to further deterioration of the computed strains. Double exposure holography directly gives the deformed pattern but is so sensitive that fringe contrast is easily lost (as is seen in Figure I.1(c)) and fringe localization can become a problem. The strains in this case are obtained by double spatial differentiation of the measured data on the assumption that the plate is correctly described by classical thin plate theory—otherwise it is uncertain as to how the strains are to be obtained.

Conceivably, we can overcome the limitations of each of these methods in special circumstances; in this book, however, we propose to tackle each difficulty directly and in a consistent manner across the different experimental methods and the different types of problems. That is, given a limited amount of sometimes sparse (both spatially and temporally) and sometimes (visually) poor data, determine the complete stress and strain state of the structure. (Within this context, data from the ubiquitous strain gage are viewed as extreme examples of spatially sparse data.) The framework for delivering the complete solution is through the construction of a complete analytical model of the problem.

To begin the connection between analysis (or model building) and experiment, consider a simple situation using the finite element method for analysis of a linear static problem; the problem is mathematically represented by

$$[K]\{u\} = \{P\}$$

where $[K]$ is the stiffness of the structure, $\{P\}$ is the vector of applied loads, and $\{u\}$ is the unknown vector of nodal displacements. (The notation is that of References [67, 71] and will be elaborated in Chapter 1.) The solution of these problems can be put in the generic form

$$\{u\} = [K^{-1}]\{P\} \qquad \text{or} \qquad \{\text{response}\} = [\text{system}]\{\text{input}\}$$

We describe *forward* problems as those where both the system and the input are known and the response is the only unknown. A finite element analysis of a problem where the geometry and material properties (the system) are known, where the loads (the input) are known, and we wish to determine the displacement (the response) is an example of a forward problem. An *inverse* problem is one where we know something of the response (usually by measurement) and wish to infer something either of the system or the input. A simple example is the measurement of load (input) and strain (response) on a uniaxial specimen to infer the Young's modulus (system). In fact, all experimental problems can be thought of as inverse problems because we begin with response information and wish to infer something about the system or the input.

A fully specified forward problem is one in which all the materials, geometries, boundary conditions, loads, and so on are known and the displacements, stresses, and so on are required. A partially specified problem is one where some input information is missing. A common practice in finite element analyses is to try to make all problems fully specified by invoking reasonable modeling assumptions. Consider the following set of unknowns and the type of assumptions that could be made:

- **Dimension** (e.g., a thin-film deposit): Assume a standard deposit thickness and get (from a handbook or previous experience) a mean value. Report the results for a range of values about this mean.

- **Boundary condition** (e.g., a built-in beam or a loose bolt): Assume an elastic support. Report the results for a range of support values, from very stiff to somewhat flexible.

- **Loading** (e.g., a projectile impacting a target): Assume an interaction model (Hertzian, plastic). Report the results for a range of values over the modeling assumptions and parameters.

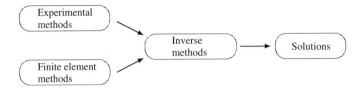

Figure I.2: Finite element methods are combined with experimental methods to effect solutions for an extended range of stress analysis problems.

- **Model** (e.g., a sloshing fuel tank): Model its mass and maybe stiffness but assume that its dynamics has no effect. Since the unknown was not parameterized, its effect cannot be gauged.

Note in each case that because assumptions are used, the results must be reported over the possible latitude of the assumption. This adds considerably to the total cost of the analysis. Additionally, and ultimately more importantly, the use of assumptions makes the results of the analyses uncertain and even unreliable.

On the face of it, the most direct way of narrowing the uncertainty in the results is to simply measure the unknown. In the case of a dimension, this is probably straightforward, but what about the force due to the impact of hale on an aircraft wing? This is not something that can be measured directly. How to use indirect measurements to find the solution of partially specified problems is an important concern of this book. The solution is shown schematically in Figure I.2 where inverse methods are used as the formal mechanisms for combining measurements and analysis.

Inverse problems are very difficult to solve since they are notoriously *ill-conditioned* (small changes in data result in large changes of response); to effect a robust solution we must incorporate (in addition to the measurements and our knowledge of the system) some extra information either about the underlying system or about the nature of the response functions, or both. The question then revolves around what supplementary information should be used and how it should be incorporated in the solution procedure. We will include this supplementary information through the formal mechanism of regularization.

Choosing the Parameterization of the Unknowns

Some of the types of unknowns that arise in modeling real structures are loads, material properties, boundary conditions, and the presence of flaws. The first is considered a force identification problem, while the latter three fall into the category of parameter identification. In our approach, however, force identification is not restricted to just the applied loads but can also be part of the parameter identification process; that is, unknown forces can be used as a particular (unknown) parameterization of the problem. To motivate this, consider again the static response of a linear complex structure to some loading governed by the equation

$$[K]\{u\} = \{P\}$$

Within a structural context, therefore, fracture, erosion, bolt loosening, and so on must appear as a change of stiffness. The changed condition is represented as a perturbation of the stiffness matrix from the original state as $[K^o] \Rightarrow [K^o] + [\Delta K]$. Substitute this into the governing equation and rearrange as

$$[K^o]\{u\} = \{P\} - [\Delta K]\{u\} = \{P\} + \{P_u\}$$

The vector $\{P_u\}$ clearly has information about the change of the structure; therefore, if somehow we can determine $\{P_u\}$ then it should be a direct process to extract the desired structural information.

Actually, the situation is more subtle than what is presented. The fact that the parameters appear implicitly has a profound effect on the solution. For example, consider the simple case of a beam with an unknown thickness distribution along a portion of its span as shown in Figure I.3. Suppose the beam segment is uniform so that there is only one unknown, the thickness h; nonetheless, the number of unknown forces is at least equal to the number of nodes spanned since

$$\{P_u\} = [\Delta K]\{u\} = \frac{\Delta h}{\bar{h}} \left[\bar{h} \frac{\partial K^o}{\partial h} \right] \{u\}$$

where $[\ K\ ]$ is assumed to be an implicit function of h. (For the purpose of this discussion, it is assumed that $[\bar{h} \partial K^o / \partial h]$ can be computed; however, it will be a very important consideration in the actual implementation of our method.) If we have enough sensors to determine $\{P_u\}$, then $\{u\}$ is also known, and Δh can indeed be obtained directly.

The more likely situation is that we do not have a sufficient number of sensors to determine all the components of $\{P_u\}$, then $\{u\}$ is also partially unknown and the force deflection relation has too many unknowns to be solved. In this case, the solution must be achieved by iteration, that is, begin by making a guess h^o for the unknown thickness and solve

$$[K^o]\{u^o\} = \{P\}$$

The equilibrium relation can then be written as

$$[K^o]\{u\} = \{P\} - \frac{\Delta h}{\bar{h}} \left[\bar{h} \frac{\partial K^o}{\partial h} \right] \{u^o\} = \{P\} - \Delta \tilde{P}\{\phi^o\}, \qquad \{\phi^o\} \equiv \left[\bar{h} \frac{\partial K^o}{\partial h} \right] \{u^o\}$$

We identify $\{\phi^o\}$ as a known force distribution and its contribution is scaled by the unknown $\Delta \tilde{P}$. Thus, the collection of unknown forces $\{P_u\}$ has been made more explicit

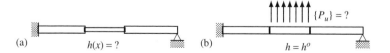

Figure I.3: Two ways of conceiving a partially specified problem. (a) Beam with single unknown thickness in middle span. (b) Beam with known thickness in middle span but many unknown forces.

and the number of unknowns has been reduced to the single unknown scale $\Delta\tilde{P}$; *a priori* knowledge of the underlying unknown (a single thickness) allowed a reduction in the number of unknown forces.

Continuing with the iterative solution, our objective is to determine $\Delta\tilde{P}$, given some response measurements u_i. Solve the two FEM problems

$$[K^o]\{\psi_P\} = \{P\}, \qquad [K^o]\{\psi_K\} = -\{\phi^o\}$$

to construct an estimate of the displacement as

$$\{u\} = \{\psi_P\} + \Delta\tilde{P}\{\psi_K\}$$

At each measurement point compare

$$u_i \qquad \longleftrightarrow \qquad \psi_{Pi} + \Delta\tilde{P}\psi_{Ki}$$

and establish an overdetermined system to determine $\Delta\tilde{P}$. Once done, an improved h can be formed from $h = h^o + \Delta\tilde{P}\overline{h}$ (which would also give an improved force distribution $\{\phi^o\}$) and the process is repeated until convergence. In this arrangement, the key unknown, $\Delta\tilde{P}$, is associated with a known (through iteration) force distribution, and the key computational steps are determining the sensitivity response $\{\psi_K\}$ and solving the system of overdetermined equations. In essence, this is the inverse algorithm to be developed in the subsequent chapters.

There are additional reasons for choosing forces as the set of unknowns. Often in experimental situations, it is impossible, inefficient, or impractical to model the entire structure. Indeed, in many cases, data can only be collected from a limited region. Furthermore, the probability of achieving a good inverse solution is enhanced if the number of unknowns can be reduced. All these situations would be taken care of if we could just isolate the subregion of interest from the rest of the structure: the aircraft wing detached from the fuselage, the loose joint separate from the beams, the center span independent of the rest of the bridge.

The basic idea of our approach to this concern is shown schematically in Figure I.4. The original problem has, in addition to the primary unknowns of interest, additional unknowns associated with the rest of the structure. Through a free-body cut, these are

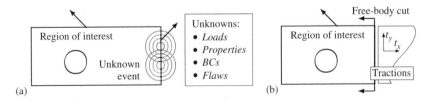

Figure I.4: Schematic of the subdomain concept. (a) The complete system comprises the unknowns of interest plus remote unknowns of no immediate interest. (b) The analyzed system comprises the unknowns of interest plus the unknown boundary tractions.

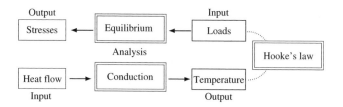

Figure I.5: Concatenated modeling. The conduction modeling is used to generate temperatures that are inputs to the equilibrium modeling.

removed from the problem so that the subdomain has the primary unknowns of interest plus a set of additional unknown tractions that are equivalent to the remainder of the structure. While these unknown tractions appear as applied loads on the subdomain, they are, essentially, a parametric representation of the effect the rest of the structure has on the subregion of interest.

Consider now the problem of thermoelastic stresses: a temperature gradient in a component can generate significant stresses even though there are no applied loads. If the problem is parameterized with the temperatures as the unknowns, there never would be a sufficient number of sensors to determine the great many unknowns. What we can do instead is use a second modeling (in this case a conduction model) to parameterize the temperature distribution in terms of a limited number of boundary temperatures or heat sources. The idea is shown in Figure I.5 as a concatenation of modelings: the stresses are related to the applied loads through the equilibrium model, the applied loads are related to the temperature distribution through Hooke's law, and the temperature distribution is related to the parameterized heat sources through the conduction model. This basic idea can be extended to a wide range of problems including wind and blast loading, radiation loading, and residual stresses. It can also be extended to complex interacting systems such as loads on turbine blades in an engine.

What these examples and discussions show is that a crucial step in the analysis is parameterizing the unknowns of the problem; it also shows that we have tremendous flexibility as to how this can be done. The specific choices made depends on the problem type, the number of sensors available, and the sophistication of the modeling capability; throughout this book we will try to demonstrate a wide variety of cases.

Clearly, the key to our approach is to take the parameterized unknowns (be they representing applied loads or representing structural unknowns) as the fundamental set of unknowns associated with our partially specified FEM model of the system. With these unknowns determined, the finite element modeling can be updated as appropriate, which then allows all other information to be obtained as a postprocessing operation.

Central Role of the Model

What emerges from the foregoing discussion is the central role played by models—the FEM model, the loading model, and so on. As we see it, there are three phases or stages in which the model is used:

(1) Identification—completely specifies the current structure, typically by "tweaking" the nominal properties and loads.

(2) Detection—monitors the structure to detect changes (damage, bolt loosening) or events (sudden loads due to impact). Events could also be long-term loading such as engine loads.

(3) Prediction—actively uses the model to infer additional information. A simple example: given the displacements, predict the stresses. This could also use additional sophisticated models for life prediction and/or reliability.

The schematic relationship among these is shown in Figure I.6. Because of the wide range of possible models needed, it would not make sense to embed these models in the inverse methods. We will instead try to develop inverse methods that use FEM and the like as external processes in a distributed computing sense. In this way, the power and versatility of stand-alone commercial packages can be harnessed.

The components of the unknown parameters $\{\tilde{P}\}$ could constitute a large set of unknowns. One of the important goals of this book, therefore, is to develop methods for force and parameter identification that are robust enough to be applied to complex structures, that can determine multiple isolated as well as distributed unknowns, and that are applicable to dynamic as well as static and nonlinear as well as linear problems. Furthermore, the data used is to be in the form of space-discrete recorded time traces and/or time-discrete spatially distributed images.

A word about an important and related topic not included in the book. In a general sense, the main ideas addressed here have counterparts in the developed area of experimental modal analysis [76] and the developing area of *Model Updating*. Reference [82] is an excellent introduction to this latter area and Reference [126] gives a comprehensive survey (243 citations) of its literature. In this method, vibration testing is used to assemble data from which an improved analytical model can be derived, that is, with an analytical model already in hand, the experimental data are used to "tweak" or update the model so as to give much finer correspondence with the physical behavior of the structure.

Vibration testing and the associated model updating are not covered in this book for a number of reasons. First, it is already covered in great depth in numerous publications. Second, the area has evolved a number of methodologies optimized for vibration-type problems, and these do not lend themselves (conveniently) to other types of problems. For example, methods of modal analysis are not well suited to analyzing wave propagation–type problems, and needless to say, nor for static problems. Another example, accelerometers, are the mainstay of experimental modal analysis, but strain gages are the preeminent tool in experimental stress analysis. Whole-field optical methods such as photoelasticity and Moiré are rarely used as data sources in model updating. Thirdly, modal

Figure I.6: The three stages of model use: identification, detection, prediction.

methods are rooted in the linear behavior of structures and as such are too restrictive when extended to nonlinear problems.

The primary tools we will develop are those for force identification and parameter identification. Actually, by using the idea of sensitivity responses (which is made possible only because models are part of the formulation), we can unify both tools with a common underlying foundation; we will refer to the approach as the sensitivity response method (SRM).

Outline of the Book

The finite element method is the definite method of choice for doing general-purpose stress analysis. This is covered extensively in a number of books, and therefore Chapter 1 considers only those aspects that directly impinge on the methods to be developed in the subsequent chapters. Inverse methods are inseparable from experiments, and therefore the second chapter reviews the two main types of experimental data. It first looks at point sensors typified by strain gages, accelerometers, and force transducers. Then it considers space distributed data as obtained from optical methods typified by photoelasticity, Moiré and holography. Again, this material is covered extensively in a number of books, and therefore the chapter concentrates on establishing a fair sense of the quality of data that can be measured and the effort needed to do so. Chapter 3 introduces the basic ideas of inverse theory and in particular the concept of regularization. A number of examples of parameter identification are used to flesh out the issues involved and the requirements needed for robust inverse methods.

The following four chapters then show applications to various problems in stress analysis and experimental mechanics. Chapter 4 considers static problems; both point sensors and whole-field data are used. Chapters 5 and 6 deal with transient problems; the former uses space-discrete data while the latter uses space-distributed data to obtain histories. Chapter 7 gives basic considerations of nonlinear problems. In each of these chapters, emphasis is placed on developing algorithms that use FEM as a process external to the inverse programming, and thus achieve our goal of leveraging the power of commercial FEM packages to increase the range of problems that can be solved.

The types of problems covered in this book are static and transient dynamic, both linear and nonlinear, for a variety of structures and components. The example studies fall into basically three categories. The first group illustrates aspects of the theory and/or algorithms; for these we use synthetic data from simple structures because it is easier to establish performance metrics. The second group includes experimental studies in which the emphasis is on the practical difficulties. The third group also uses synthetic data, but illustrates problems designed to explore the future possibilities of the methods presented.

In each study, we try to give enough detail so that the results can be duplicated. In this connection, we need to say a word about units: throughout this book, a dual set of units will be used with SI being the primary set. A dual set arises because experimental work always seems to deal with an inconsistent set of units. This inconsistency comes about because of such things as legacy equipment or simply because that is the way

Table I.1:   Conversions between sets of units.

Length	1 in. = 25.4 mm	1 m = 39.4 in.
Force	1 lb = 4.45 N	1 N = 0.225 lb
Pressure	1 psi = 6.90 kPa	1 kPa = 0.145 psi
Mass	1 lbm = 0.454 kg	1 kg = 2.2 lbm
Density	1 lbm/in.^{3} = 27.7 M/m^3	1 M/m^3 = 0.0361 lbm/in.^{3}

things were manufactured. Table I.1 gives the common conversions used when one set of units are converted to another.

1

Finite Element Methods

This chapter reviews some basic concepts in the mechanics of deformable bodies that are needed as foundations for the later chapters; much of the content is abstracted from References [56, 71]. First, we introduce the concept of deformation and strain, followed by stress and the equations of motion. To complete the mechanics formulation of problems, we also describe the constitutive (or material) behavior.

The chapter then reformulates the governing equations in terms of a variational principle; in this, equilibrium is seen as the achievement of a stationary value of the total potential energy. This approach lends itself well to approximate computer methods. In particular, we introduce the finite element method (FEM).

For illustrative purposes, the discussion of the finite element method will be limited to a few standard elements, some applications of which are shown in Figure 1.1.

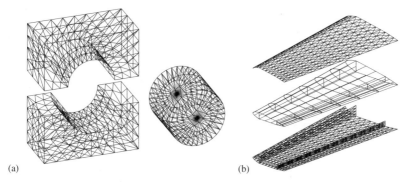

(a)   (b)

Figure 1.1: Exploded views of some FEM applications. (a) Solid bearing block and shaft modeled with tetrahedral elements (The interior edges are removed for clearer viewing). (b) Wing section modeled as a thin-walled structure discretized as collections of triangular shell elements and frame elements.

Modern Experimental Stress Analysis: completing the solution of partially specified problems. James Doyle
© 2004 John Wiley & Sons, Ltd   ISBN 0-470-86156-8

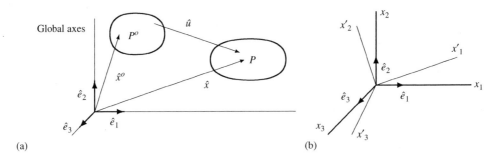

Figure 1.2: Coordinate descriptions. (a) Displacement from an undeformed to a deformed configuration. (b) Base vectors and rotated coordinate system.

More details, elaborations, and examples of FEM in general situations can be found in References [18, 45, 46, 67, 71].

1.1   Deformation and Strain

A deformation is a comparison of two states. In the mechanics of deformable bodies, we are particularly interested in the relative deformation of neighboring points because this is related to the straining of a body.

Motion and Coordinate Descriptions

Set up a common global coordinate system as shown in Figure 1.2(a) and associate x_i^o with the undeformed configuration and x_i with the deformed configuration, that is,

$$\text{initial position:} \quad \hat{x}^o = \sum_i x_i^o \hat{e}_i \qquad \text{final position:} \quad \hat{x} = \sum_i x_i \hat{e}_i$$

where both vectors are referred to the common set of unit vectors $\hat{e}_i$. The variables x_i^o and x_i are called the *Lagrangian* and *Eulerian* variables, respectively. A displacement is the shortest distance traveled when a particle moves from one location to another; that is,

$$\hat{u} = \hat{r} - \hat{r}^o = \sum_i x_i \hat{e}_i - \sum_i x_i^o \hat{e}_i \qquad \text{or} \qquad u_i = x_i - x_i^o$$

and is illustrated in Figure 1.2(a). A motion is expressed as

$$x_i = x_i(x_1^o, x_2^o, x_3^o, t) \tag{1.1}$$

which says that the trajectory of any point is uniquely identified through its original position. In this Lagrangian description, all quantities (stress, strain, and so on) are expressed in terms of the initial position coordinates and time.

The quantities we deal with (such as x_1^o, x_2^o, x_3^o, of the initial position) have components differentiated by their subscripts. These quantities are examples of what we will begin calling *tensors*. A given tensor quantity will have different values for the components in different coordinate systems. Thus, it is important to know how these components change under a coordinate transformation.

Consider two Cartesian coordinate systems (x_1, x_2, x_3) and (x'_1, x'_2, x'_3) as shown in Figure 1.2(b). A *base* vector is a unit vector parallel to a coordinate axis. Let $\hat{e}_1, \hat{e}_2, \hat{e}_3$ be the base vectors for the (x_1, x_2, x_3) coordinate system, and $\hat{e}'_1, \hat{e}'_2, \hat{e}'_3$ the base vectors for the (x'_1, x'_2, x'_3) system as also shown in the figure. The transformation of components in one coordinate system into components in another is given by

$$x'_i = \sum_j \beta_{ij} x_j, \qquad x_i = \sum_j \beta_{ji} x'_j, \qquad \beta_{ij} \equiv \hat{e}'_i \cdot \hat{e}_j$$

where $[\beta_{ij}]$ is the matrix of *direction cosines*. This matrix is orthogonal, meaning that its transpose is its inverse.

A system of quantities is called by different tensor names depending on how the components of the system are defined in the variables x_1, x_2, x_3 and how they are transformed when the variables x_1, x_2, x_3 are changed to x'_1, x'_2, x'_3. A system is called a *scalar* field if it has only a single component ϕ in the variables x_i and a single component ϕ' in the variables x'_i and if ϕ and ϕ' are numerically equal at the corresponding points, that is,

$$\phi(x_1, x_2, x_3) = \phi'(x'_1, x'_2, x'_3)$$

Temperature, volume, and mass density are examples of scalar fields. A system is called a *vector* field or a tensor field of order one if it has three components V_i in the variables x'_i and the components are related by the transformation law

$$V'_i = \sum_k \beta_{ik} V_k, \qquad V_i = \sum_k \beta_{ki} V'_k$$

As already shown, quantities such as position and displacement are first-order tensors; so also are force, velocity, and area. The tensor field of order two is a system which has nine components T_{ij} in the variables x_1, x_2, x_3, and nine components T'_{ij} in the variables x'_1, x'_2, x'_3, and the components are related by

$$T'_{ij} = \sum_{m,n} \beta_{im} \beta_{jn} T_{mn}, \qquad T_{ij} = \sum_{m,n} \beta_{mi} \beta_{nj} T'_{mn}$$

As shown shortly, strain and stress are examples of second-order tensor fields. A special second-order tensor is the Kronecker delta, δ_{ij}, which has the same components (arranged similar to those of the unit matrix) and are the same in all coordinate systems. Tensors such as these are called isotropic tensors and are not to be confused with scalar tensors.

This sequence of defining tensors is easily extended to higher order tensors [56].

Strain Measures

As a body deforms, various points in it will translate and rotate. Strain is a measure of the "stretching" of the material points within a body; it is a measure of the relative displacement without rigid-body motion and is an essential ingredient for the description of the constitutive behavior of materials. The easiest way to distinguish between deformation and the local rigid-body motion is to consider the change in distance between two neighboring material particles. We will use this to establish our strain measures.

There are many measures of strain in existence and we review a few of them here. Assume that a line segment of original length L_o is changed to length L, then some of the common measures of strain are the following:

$$\text{Engineering:} \quad \epsilon = \frac{\text{change in length}}{\text{original length}} = \frac{\Delta L}{L_o}$$

$$\text{True:} \quad \epsilon^{\mathrm{T}} = \frac{\text{change in length}}{\text{final (current) length}} = \frac{\Delta L}{L} = \frac{\Delta L}{L_o + \Delta L}$$

$$\text{Logarithmic:} \quad \epsilon^{\mathrm{N}} = \int_{L_o}^{L} \text{true strain} = \int_{L_o}^{L} \frac{\mathrm{d}L}{L} = \log_n\left(\frac{L}{L_o}\right)$$

The relations among the measures are

$$\epsilon^{\mathrm{T}} = \frac{\epsilon}{1+\epsilon}, \qquad \epsilon^{\mathrm{N}} = \log_n(1+\epsilon)$$

An essential requirement of a strain measure is that it allow the final length to be calculated knowing the original length. This is true of each of the above since

$$\text{Engineering:} \quad L = L_o + \Delta L = L_o + L_o\epsilon = L_o(1+\epsilon)$$

$$\text{True:} \quad L = L_o + \Delta L = L_o + \frac{\epsilon^{\mathrm{T}}L_o}{(1-\epsilon^{\mathrm{T}})} = \frac{L_o}{(1-\epsilon^{\mathrm{T}})}$$

$$\text{Logarithmic:} \quad L = L_o\exp(\epsilon^{\mathrm{N}})$$

The measures give different numerical values for the strain but all are equivalent in that they allow ΔL (or L) to be calculated using L_o. Because the measures are equivalent, it is a matter of convenience as to which measure is to be chosen in an analysis.

The difficulty with these strain measures is that they do not transform conveniently from one coordinate system to another. This poses a problem in developing a three-dimensional theory because the quantities involved should transform as tensors of the appropriate order. We now review a strain measure that has appropriate transformation properties.

Let two material points before deformation have the coordinates (x_i^o) and $(x_i^o + \mathrm{d}x_i^o)$; and after deformation have the coordinates (x_i) and $(x_i + \mathrm{d}x_i)$. The initial distance between these neighboring points is given by

$$\mathrm{d}S_o^2 = \sum_i \mathrm{d}x_i^o \mathrm{d}x_i^o = (\mathrm{d}x_1^o)^2 + (\mathrm{d}x_2^o)^2 + (\mathrm{d}x_3^o)^2$$

and the final distance between the points by

$$\mathrm{d}S^2 = \sum_i \mathrm{d}x_i \mathrm{d}x_i = \sum_{i,j,m} \frac{\partial x_m}{\partial x_i^o} \frac{\partial x_m}{\partial x_j^o} \mathrm{d}x_i^o \mathrm{d}x_j^o$$

where $\partial x_m / \partial x_i^o$ is called the *deformation gradient* and, as per the motion of Equation (1.1), the deformed coordinates are considered a function of the original coordinates. Only in the event of stretching or straining is dS^2 different from dS_o^2, that is,

$$dS^2 - dS_o^2 = dS^2 - \sum_i dx_i^o dx_i^o = \sum_{i,j,m} \left[\frac{\partial x_m}{\partial x_i^o} \frac{\partial x_m}{\partial x_j^o} - \delta_{ij} \right] dx_i^o dx_j^o$$

is a measure of the relative displacements. It is insensitive to rotation as can be easily demonstrated by considering a rigid-body motion. These equations can be written as

$$dS^2 - dS_o^2 = \sum_{i,j} 2E_{ij} dx_i^o dx_j^o , \qquad E_{ij} \equiv \frac{1}{2} \sum_m \left[\frac{\partial x_m}{\partial x_i^o} \frac{\partial x_m}{\partial x_j^o} - \delta_{ij} \right] \qquad (1.2)$$

by introducing the strain measure E_{ij}. It is easy to observe that E_{ij} is a symmetric tensor of the second order. It is called the Lagrangian strain tensor.

Sometimes it is convenient to deal with displacements and displacement gradients instead of the deformation gradient. These are obtained by using the relations

$$x_m = x_m^o + u_m , \qquad \frac{\partial x_m}{\partial x_i^o} = \frac{\partial u_m}{\partial x_i^o} + \delta_{im}$$

The Lagrangian strain tensor is written in terms of the displacement by

$$E_{ij} = \frac{1}{2} \left[\frac{\partial u_i}{\partial x_j^o} + \frac{\partial u_j}{\partial x_i^o} + \sum_m \frac{\partial u_m}{\partial x_i^o} \frac{\partial u_m}{\partial x_j^o} \right]$$

Typical expressions for E_{ij} in unabridged notations are

$$E_{11} = \frac{\partial u_1}{\partial x_1^o} + \frac{1}{2} \left[\left(\frac{\partial u_1}{\partial x_1^o} \right)^2 + \left(\frac{\partial u_2}{\partial x_1^o} \right)^2 + \left(\frac{\partial u_3}{\partial x_1^o} \right)^2 \right]$$

$$E_{22} = \frac{\partial u_2}{\partial x_2^o} + \frac{1}{2} \left[\left(\frac{\partial u_2}{\partial x_2^o} \right)^2 + \left(\frac{\partial u_2}{\partial x_2^o} \right)^2 + \left(\frac{\partial u_3}{\partial x_2^o} \right)^2 \right]$$

$$E_{12} = \frac{1}{2} \left(\frac{\partial u_1}{\partial x_2^o} + \frac{\partial u_2}{\partial x_1^o} \right) + \frac{1}{2} \left[\frac{\partial u_1}{\partial x_1^o} \frac{\partial u_1}{\partial x_2^o} + \frac{\partial u_2}{\partial x_1^o} \frac{\partial u_2}{\partial x_2^o} + \frac{\partial u_3}{\partial x_1^o} \frac{\partial u_3}{\partial x_2^o} \right] \qquad (1.3)$$

Note the presence of the nonlinear terms.

Principal Strains and Strain Invariants

Consider the line segment before deformation as a vector $d\hat{S}^o = \sum_i dx_i^o \hat{e}_i$. We can then interpret Equation (1.2) as a vector product relation

$$dS^2 - dS_o^2 = 2 \sum_j V_j dx_j^o = 2 \hat{V} \cdot d\hat{S}^o , \qquad V_j \equiv \sum_i E_{ij} dx_i^o \qquad (1.4)$$

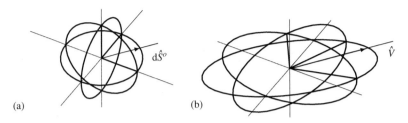

Figure 1.3: The ellipsoid of strain. (a) Traces of the vector $d\hat{S}^o$. (b) Traces of the vector $\hat{V}$.

Consider different initial vectors $d\hat{S}^o$ each of the same size but of different orientations; they will correspond to different vectors $\hat{V}$. Figure 1.3 shows a collection of such vectors where $d\hat{S}^o$ traces out the coordinate circles of a sphere. Note that $\hat{V}$ traces an ellipse and many such traces would form an ellipsoid, that is, the sphere traced by $d\hat{S}^o$ is transformed into an ellipsoid traced by $\hat{V}$.

In general, the vectors $d\hat{S}^o$ and $\hat{V}$ are not parallel. An interesting question to ask, however, is the following: are there any initial vectors that have the same orientation before and after transformation? The answer will give an insight into the properties of all second- order tensors, not just strain but also stress and moments of inertia, to name two more.

To generalize the results, we will consider the symmetric second-order tensor T_{ij}, and unit initial vectors $n_i = dx_i^o/dS^o$. Assume that there is an $\hat{n}$ such that it is proportional to $\hat{V}$, that is,

$$V_i = \sum_j T_{ij} n_j = \lambda n_i \qquad \text{or} \qquad \sum_j \left[T_{ij} - \lambda \delta_{ij} \right] n_j = 0$$

This is given in expanded matrix form as

$$\begin{bmatrix} T_{11} - \lambda & T_{12} & T_{13} \\ T_{12} & T_{22} - \lambda & T_{23} \\ T_{13} & T_{23} & T_{33} - \lambda \end{bmatrix} \begin{Bmatrix} n_1 \\ n_2 \\ n_3 \end{Bmatrix} = 0$$

This is an eigenvalue problem and has a nontrivial solution for n_i only if the determinant of the coefficient matrix vanishes. Expanding the determinantal equation, we obtain the characteristic equation

$$\lambda^3 - I_1 \lambda^2 + I_2 \lambda - I_3 = 0$$

where the invariants I_1, I_2, I_3 are defined as

$$I_1 = \sum_i T_{ii} = T_{11} + T_{22} + T_{33}$$

$$I_2 = \sum_{i,j} \frac{1}{2}[T_{ii}T_{jj} - T_{ij}T_{ji}] = T_{11}T_{22} + T_{22}T_{33} + T_{33}T_{11} - T_{12}^2 - T_{23}^2 - T_{13}^2$$

$$I_3 = \det[T_{ij}] = T_{11}T_{22}T_{33} + 2T_{12}T_{23}T_{13} - T_{11}T_{23}^2 - T_{22}T_{13}^2 - T_{33}T_{12}^2 \qquad (1.5)$$

The characteristic equation yields three roots or possible values for λ: $\lambda^{(1)}$, $\lambda^{(2)}$, $\lambda^{(3)}$. These are called the *eigenvalues* or *principal* values. For each principal value there is a corresponding solution for $\hat{n}$: $\hat{n}^{(1)}$, $\hat{n}^{(2)}$, $\hat{n}^{(3)}$. The three $\hat{n}$s are called the *eigenvectors* or *principal directions*.

The principal values can be computed from the invariants [143] by first defining

$$Q = \tfrac{1}{9}[I_1^2 - 3I_2], \qquad R = \tfrac{1}{54}[-2I_1^3 + 9I_1 I_2 - 27I_3], \qquad \theta = \cos^{-1}[R/\sqrt{Q^3}]$$

then

$$\lambda_1 = \tfrac{1}{3}I_1 - 2\sqrt{Q}\cos\left(\tfrac{1}{3}\theta\right), \qquad \lambda_2, \lambda_3 = \tfrac{1}{3}I_1 - 2\sqrt{Q}\cos\left(\tfrac{1}{3}(\theta \pm 2\pi)\right) \qquad (1.6)$$

For two-dimensional problems where $T_{13} = 0$ and $T_{23} = 0$, this simplifies to

$$\lambda_1, \lambda_2 = \tfrac{1}{2}[T_{11} + T_{22}] \pm \tfrac{1}{2}\sqrt{[T_{11} - T_{22}]^2 + 4T_{12}^2}, \qquad \lambda_3 = T_{33} \qquad (1.7)$$

These can easily be obtained from the Mohr's circle [48].

It is now of interest to know what the components T_{ij} are when transformed to the coordinate system defined by the principal directions. Let the transformed coordinate system be defined by the triad

$$\hat{e}_1' = \hat{n}^{(1)}, \qquad \hat{e}_2' = \hat{n}^{(2)}, \qquad \hat{e}_3' = \hat{n}^{(3)}$$

Then the transformation matrix of direction cosines is given by

$$\beta_{ij} = \hat{e}_i' \cdot \hat{e}_j = \hat{n}^{(i)} \cdot \hat{e}_j = \sum_k n_k^{(i)} \hat{e}_k \cdot \hat{e}_j = n_j^{(i)} \qquad \text{and} \qquad \beta_{ki} = n_i^{(k)}$$

From the eigenvalue problem, we have for the kth eigenmode

$$\sum_j T_{ij} n_j^{(k)} = \lambda^{(k)} n_i^{(k)} \qquad \text{or} \qquad \sum_j T_{ij} \beta_{kj} = \lambda^{(k)} \beta_{ki}$$

Multiply both sides by β_{li} and sum over i; recognizing the left-hand side as the transform of T_{ij} and the right-hand side as the Kronecker delta (because of the orthogonality of β_{ij}) gives

$$T'_{lk} = \lambda^{(k)} \delta_{kl} \qquad \text{or} \qquad \begin{bmatrix} T'_{11} & T'_{12} & T'_{13} \\ & T'_{22} & T'_{23} \\ sym & & T'_{33} \end{bmatrix} = \begin{bmatrix} \lambda^{(1)} & 0 & 0 \\ 0 & \lambda^{(2)} & 0 \\ 0 & 0 & \lambda^{(3)} \end{bmatrix}$$

that is, T_{ij} has a diagonal form when transformed to the principal coordinates. We can further show [56] that the $\lambda^{(i)}$s are the extremums of the associated ellipsoid surface.

The properties just established are valid for all symmetric second-order tensors.

Infinitesimal Strain and Rotation

The full nonlinear deformation analysis of problems is quite difficult and so simplifications are often sought. Three situations for the straining of a block are shown in

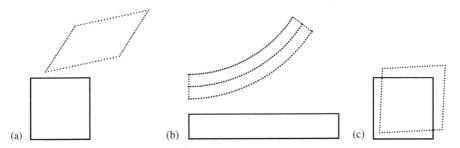

Figure 1.4: Combinations of displacements and strains. (a) Large displacements, rotations, and strains. (b) Large displacements and rotations but small strains. (c) Small displacements, rotations, and strains.

Figure 1.4. The general case is that of large displacements, large rotations, and large strains. In the chapters dealing with the linear theory, the displacements, rotations, and strains are small, and Case (c) prevails. Nonlinear analysis of thin-walled structures such as shells is usually restricted to Case (b), where the deflections and rotations can be large but the strains are small. This is a reasonable approximation because structural materials do not exhibit large strains without yielding and structures are designed to operate without yielding.

If the displacement gradients are small, that is,

$$\left| \frac{\partial u_i}{\partial x_j^o} \right| \ll 1$$

then the nonlinear product terms in the strain tensor definition can be neglected. The result is

$$E_{ij} \simeq \epsilon_{ij} = \frac{1}{2} \left(\frac{\partial u_i}{\partial x_j^o} + \frac{\partial u_j}{\partial x_i^o} \right)$$

where ϵ_{ij} is the infinitesimal strain tensor. This assumption also leads to the conclusion that the components E_{ij} are small as compared with unity.

If, in addition to the above, the following condition on the size of the displacements exists

$$\left| \frac{u_i}{L} \right| \ll 1$$

where L is the smallest dimension of the body, then

$$x_i \simeq x_i^o$$

and the distinction between the Lagrangian and Eulerian variables vanishes. Henceforth, when we use the small-strain approximations, ϵ_{ij} will be used to denote the strain tensor, and x_i will denote both Lagrangian and Eulerian variables.

1.2 Tractions and Stresses

The kinetics of rigid bodies are described in terms of forces; the equivalent concept for continuous media is stress (loosely defined as force over unit area). Actions can be exerted on a continuum through either contact forces or forces contained in the mass. The contact force is often referred to as a surface force or traction as its action occurs on a surface. We are primarily concerned with contact forces but acting inside the body.

Cauchy Stress Principle

Consider a small surface element of area ΔA on an imagined exposed surface A in the deformed configuration as depicted in Figure 1.5. There must be resultant forces and moments acting on ΔA to make it equipollent to the effect of the rest of the material, that is, when the pieces are put back together, these forces cancel each other. Let these forces be thought of as contact forces and so give rise to contact stresses (even though they are inside the body). Cauchy formalized this by introducing his concept of traction vector.

Let $\hat{n}$ be the unit vector that is perpendicular to the surface element ΔA and let $\Delta \hat{F}$ be the resultant force exerted from the other part of the surface element with the negative normal vector. We assume that as ΔA becomes vanishingly small, the ratio $\Delta \hat{F}/\Delta A$ approaches a definite limit $\mathrm{d}\hat{F}/\mathrm{d}A$. The vector obtained in the limiting process

$$\lim_{\Delta A \to 0} \frac{\Delta \hat{F}}{\Delta A} = \frac{\mathrm{d}\hat{F}}{\mathrm{d}A} \equiv \hat{t}^{(\hat{n})}$$

is called the *traction vector*. This vector represents the force per unit area acting on the surface, and its limit exists because the material is assumed continuous. The superscript $\hat{n}$ is a reminder that the traction is dependent on the orientation of the exposed area.

To give explicit representation of the traction vector, consider its components on the three faces of a cube as shown in Figure 1.6(a). Because this description is somewhat cumbersome, we simplify the notation by introducing the components

$$\sigma_{ij} \equiv t_j^{(\hat{e}_i)} \qquad \text{e.g.} \qquad \sigma_{11} \equiv t_1^{(\hat{e}_1)}, \quad \sigma_{13} \equiv t_3^{(\hat{e}_1)}, \quad \sigma_{31} \equiv t_1^{(\hat{e}_3)}, \quad \ldots$$

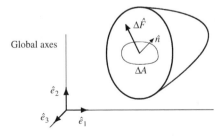

Figure 1.5: Exposed forces on an arbitrary section cut.

where i refers to the face and j to the component. The normal projections of $\hat{t}^{(\hat{n})}$ on these special faces are the normal stress components σ_{11}, σ_{22}, σ_{33}, while projections perpendicular to $\hat{n}$ are shear stress components σ_{12}, σ_{13}; σ_{21}, σ_{23}; σ_{31}, σ_{32}.

It is important to realize that while $\hat{t}$ resembles the elementary idea of stress as force over area, it is not stress; $\hat{t}$ transforms as a vector and has only three components. The tensor σ_{ij} is our definition of stress; it has nine components with units of force over area, but at this stage we do not know how these components transform.

Tractions on Arbitrary Planes

The traction vector $\hat{t}^{(\hat{n})}$ acting on an area $dA\hat{n}$ depends on the normal $\hat{n}$ of the area. The particular relation can be obtained by considering a traction on an arbitrary surface of the tetrahedron shown in Figure 1.7 formed from the stressed cube of Figure 1.6. The vector acting on the inclined surface ABC is $\hat{t}$ and the unit normal vector $\hat{n}$. The equilibrium of the tetrahedron requires that the resultant force acting on it must vanish.

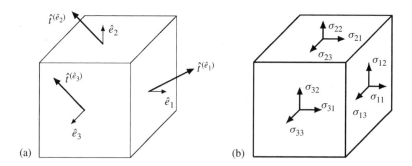

Figure 1.6: Stressed cube. (a) Traction components. (b) Stress components.

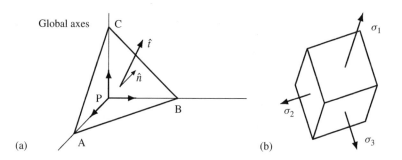

Figure 1.7: Tractions on particular planes. (a) Tetrahedron with exposed traction. (b) Cube with principal stresses.

From the equation for the balance of forces in the x_1-direction for the tetrahedron, and letting $h \to 0$, we obtain

$$t_1 = \sigma_{11}n_1 + \sigma_{21}n_2 + \sigma_{31}n_3 = \sum_j \sigma_{j1}n_j$$

Similar equations can be derived from the consideration of the balance of forces in the x_2- and x_3-directions. These three equations can be written collectively as

$$t_i = \sum_j \sigma_{ji}n_j \qquad (1.8)$$

This compact relation says that we need only know nine numbers $[\sigma_{ij}]$ to be able to determine the traction vector on any area passing through a point. These elements are called the Cauchy stress components and form the Cauchy stress tensor. It is a second-order tensor because t_i and n_j transform as first-order tensors. Later, through consideration of equilibrium, we will establish that it is symmetric.

The traction vector $\hat{t}$ acting on a surface depends on the direction $\hat{n}$ and is usually not parallel to $\hat{n}$. Consider the normal component acting on the face

$$\sigma_n = \hat{t} \cdot \hat{n} = \sum_i t_i n_i = \sum_{i,j} \sigma_{ij} n_i n_j$$

If we make the associations

$$t_i \longrightarrow V_i, \qquad n_i \longrightarrow dx_i^o, \qquad \sigma_{ij} \longrightarrow E_{ij}$$

then we may infer that the behavior of the normal stress component is the same as that for the stretching of Equation (1.4). In particular, the stress tensor $[\sigma_{ij}]$ has principal values (called *principal stresses*) and a corresponding set of principal directions. The principal stresses are shown in Figure 1.7(b); note that the shear stresses are zero on a principal cube. It is usual to order the principal stress according to

$$\sigma_3 < \sigma_2 < \sigma_1$$

These values can be computed according to the formulas of Equation (1.6).

Stress Referred to the Undeformed Configuration

Stress is most naturally established in the deformed configuration, but we have chosen to use the Lagrangian variables (i.e., the undeformed configuration) for the description of a body with finite deformation. For consistency, we need to introduce a measure of stress referred to the undeformed configuration. To help appreciate the new definitions of stress to be introduced, it is worthwhile to keep the following in mind:

- The traction vector is first defined in terms of a force divided by area.

- The stress tensor is then defined according to a transformation relation for the traction and area normal.

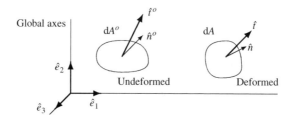

Figure 1.8: Traction vectors in the undeformed and deformed configurations.

To refer our description of tractions to the surface before deformation, we must define a traction vector $\hat{t}^o$ acting on an area dA^o as indicated in Figure 1.8. The introduction of such a vector is somewhat arbitrary, so we first reconsider the Cauchy stress so as to motivate the developments.

In the deformed state, on every plane surface passing through a point, there is a traction vector $\hat{t}_i$ defined in terms of the deformed surface area; that is, letting the traction vector be $\hat{t}$ and the total resultant force acting on dA be $d\hat{F}$, then

$$\hat{t} \equiv \frac{d\hat{F}}{dA} \qquad \text{or} \qquad t_i \equiv \frac{dF_i}{dA}$$

Let all the traction vectors and unit normals in the deformed body form two respective vector spaces. Then, the Cauchy stress tensor σ_{ij} was shown to be the transformation between these two vector spaces, that is,

$$t_i = \sum_j \sigma_{ji} n_j$$

Defined in this manner, the Cauchy stress tensor is an abstract quantity; however, on special plane surfaces such as those with unit normals parallel to $\hat{e}_1$, $\hat{e}_2$, and $\hat{e}_3$, respectively, the nine components of $[\,\sigma_{ij}\,]$ can be related to the traction vector and thus have physical meaning; that is, the meaning of σ_{ij} is the components of stress derived from the force vector dF_i divided by the deformed area. This, in elementary terms, is called *true stress*.

We will now do a parallel development for the undeformed configuration. Let the resultant force $d\hat{F}^o$, referred to the undeformed configuration, be given by a transformation of the force $d\hat{F}$ acting on the deformed area. One possibility is to take $dF_i^o = dF_i$, and this gives rise to the so-called Lagrange stress tensor, which in simple terms would correspond to "force divided by original area." Instead, let

$$dF_i^o = \sum_j \frac{\partial x_i^o}{\partial x_j} dF_j$$

which follows the analogous rule for the deformation of line segments. The reason for this choice will become apparent later when we consider the equations of motion. It is important to realize that this is not a rotation transformation but that the force components are being "deformed". The Kirchhoff traction vector is defined as

$$t_i^o \equiv \frac{dF_i^o}{dA^o} = \sum_j \frac{\partial x_i^o}{\partial x_j} \frac{dF_j}{dA^o}$$

This leads to the definition of the Kirchhoff stress tensor σ_{ij}^K:

$$t_i^o = \sum_j \sigma_{ji}^K n_j^o$$

The meaning of σ_{pq}^K components of stress derived from the transformed components of the force vector, divided by the original area. There is no elementary equivalence to this stress. This stress is often referred to as the second Piola–Kirchhoff stress tensor; we will abbreviate it simply as the Kirchhoff stress tensor.

It can be shown [56] that the relation between the Kirchhoff and Cauchy stress tensors is

$$\sigma_{ji}^K = \sum_{m,n} \frac{\rho^o}{\rho} \frac{\partial x_i^o}{\partial x_m} \frac{\partial x_j^o}{\partial x_n} \sigma_{mn}, \qquad \sigma_{ji} = \sum_{m,n} \frac{\rho}{\rho^o} \frac{\partial x_i}{\partial x_m^o} \frac{\partial x_j}{\partial x_n^o} \sigma_{mn}^K \qquad (1.9)$$

where ρ is the mass density. We will show that the Cauchy stress tensor is symmetric, hence these relations show that the Kirchhoff stress tensor is also a symmetric tensor.

1.3 Governing Equations of Motion

Newton's laws for the equation of motion of a rigid body will be used to establish the equations of motion of a deformable body. It will turn out, however, that they are not the most suitable form, and we look at other formulations. In particular, we look at the forms arising from the principle of virtual work and leading to stationary principles such as Hamilton's principle and Lagrange's equation.

Strong and Weak Formulations

Consider an arbitrary volume V taken from the deformed body as shown in Figure 1.9; it has tractions $\hat{t}$ on the boundary surface A, and body force per unit mass $\hat{b}$. Newton's laws of motion become

$$\int_A \hat{t} \, dA + \int_V \rho \hat{b} \, dV = \int_V \rho \ddot{u} \, dV$$

$$\int_A \hat{x} \times \hat{t} \, dA + \int_V \hat{x} \times \hat{b} \rho \, dV = \int_V \hat{x} \times \ddot{u} \rho \, dV$$

where ρ is the mass density. These are the equations of motion in terms of the traction. We now state the equations of motion in terms of the stress. In doing this, there is a choice between using the deformed state and the undeformed state.

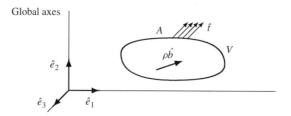

Figure 1.9: Arbitrary small volume taken from the deformed configuration.

Reference [56] shows how the stress-traction relation, $t_i = \sum_j \sigma_{ji} n_j$, combined with the integral theorem can be used to recast the first equilibrium equation into the form

$$\frac{\partial \sigma_{11}}{\partial x_1} + \frac{\partial \sigma_{21}}{\partial x_2} + \frac{\partial \sigma_{31}}{\partial x_3} + \rho b_1 = \rho \ddot{u}_1$$

$$\frac{\partial \sigma_{12}}{\partial x_1} + \frac{\partial \sigma_{22}}{\partial x_2} + \frac{\partial \sigma_{32}}{\partial x_3} + \rho b_2 = \rho \ddot{u}_2$$

$$\frac{\partial \sigma_{13}}{\partial x_1} + \frac{\partial \sigma_{23}}{\partial x_2} + \frac{\partial \sigma_{33}}{\partial x_3} + \rho b_3 = \rho \ddot{u}_3 \qquad (1.10)$$

and that the second becomes $\sigma_{ij} = \sigma_{ji}$, showing the symmetry of the stress tensor. The statement of a problem in terms of a set of differential equations plus the associated boundary conditions is called the *strong formulation* of the problem. The equations of motion in terms of the Kirchhoff stress are more complicated than those using the Cauchy stress because they explicitly include the deformed state. We will not state them here because the finite element formulation can be done more simply in terms of a variational form.

Let $u_i(x_i^o)$ be the displacement field, which satisfies the equilibrium equations in V. On the surface A, the surface traction t_i is prescribed on A_t and the displacement on A_u. Consider a variation of displacement δu_i (we will sometimes call this the virtual displacement), then

$$\bar{u}_i = u_i + \delta u_i$$

where u_i satisfy the equilibrium equations and the given boundary conditions. Thus, δu_i must vanish over A_u but be arbitrary over A_t. Let δW_e be the virtual work done by the body force b_i and traction t_i; that is,

$$\delta W_e = \sum_i \int_V \rho b_i \delta u_i \, dV + \sum_i \int_{A_t} t_i \delta u_i \, dA + \sum_i \int_{A_u} t_i \delta u_i \, dA \qquad (1.11)$$

After some manipulation and substitution for the traction vector, we can write the virtual work form of equilibrium as [71]

$$\delta W = \delta W_e - \sum_{m,n} \int_V \sigma_{mn} \delta \epsilon_{mn} \, dV = \delta W_e - \sum_{p,q} \int_{V^o} \sigma_{pq}^K \delta E_{pq} \, dV^o = 0 \qquad (1.12)$$

Hence, the Cauchy stress/small-strain combination is energetically equivalent to the Kirchhoff stress/Lagrangian strain combination. In contrast to the differential equations of motion, there are no added complications using the undeformed state as reference. This formulation of the problem is known as the *weak formulation*.

The virtual work formulation is completely general, but there are further developments that are more convenient to use in some circumstances. We now look at some of these developments.

Stationary Principles

The internal virtual work is associated with the straining of the body and therefore we will use the representation

$$\delta \mathcal{U} = \sum_{i,j} \int_V \sigma_{ij} \delta \epsilon_{ij} \, \mathrm{d}V = \sum_{i,j} \int_{V^o} \sigma_{ij}^K \delta E_{ij} \, \mathrm{d}V^o$$

and call $\mathcal{U}$ the strain energy of the body.

A system is *conservative* if the work done in moving the system around a closed path is zero. We say that the external force system is conservative if it can be obtained from a potential function. For example, for a set of discrete forces, we have

$$P_i = -\frac{\partial \mathcal{V}}{\partial u_i} \qquad \text{or} \qquad \mathcal{V} = -\sum_i P_i u_i$$

where u_i is the displacement associated with the load P_i, and $\mathcal{V}$ is the potential. The negative sign in the definition of $\mathcal{V}$ is arbitrary, but choosing it so gives us the interpretation of $\mathcal{V}$ as the capacity (or potential) to do work. The external work term now becomes

$$\delta W_e = \sum_i P_i \delta u_i = -\sum_i \frac{\partial \mathcal{V}}{\partial u_i} \delta u_i = -\delta \mathcal{V}$$

We get almost identical representations for conservative body forces and conservative traction distributions. The principle of virtual work can be rewritten as

$$\delta \mathcal{U} + \delta \mathcal{V} = 0 \qquad \text{or} \qquad \delta \Pi \equiv \delta [\mathcal{U} + \mathcal{V}] = 0 \qquad (1.13)$$

The term inside the brackets is called the *total potential energy*. This relation is called the *principle of stationary potential energy*. We may now restate the principle of virtual work as follows: *For a conservative system to be in equilibrium, the first-order variation in the total potential energy must vanish for every independent admissible virtual displacement.* Another way of stating this is as follows: among all the displacement states of a conservative system that satisfy compatibility and the boundary constraints, those that also satisfy equilibrium make the potential energy stationary. In comparison to the conservation of energy theorem, this is much richer, because instead of one equation it leads to as many equations as there are degrees of freedom (independent displacements).

To apply the idea of virtual work to dynamic problems, we need to account for the presence of inertia forces, and the fact that all quantities are functions of time. The inertia leads to the concept of *kinetic energy*, defined as

$$\mathcal{T} \equiv \frac{1}{2}\sum_i \int_V \rho \dot{u}_i \dot{u}_i \, dV \qquad \text{such that} \qquad \delta\mathcal{T} \equiv \sum_i \int_V \rho \dot{u}_i \, \delta\dot{u}_i \, dV$$

Hamilton refined the concept that a motion can be viewed as a path in configuration space; he showed [89] that, for a system with given configurations at times t_1 and t_2, of all the possible configurations between these two times the actual that occurs satisfies a stationary principle. Hamilton integrated the equation over time and since by stipulation, the configuration has no variations at the extreme times, the virtual work relation becomes

$$\int_{t_1}^{t_2} [\delta W^s + \delta W^b + \delta\mathcal{T} - \delta\mathcal{U}] \, dt = 0 \tag{1.14}$$

This equation is generally known as the *extended Hamilton's principle*. In the special case when the applied loads, both body forces and surface tractions, can be derived from a scalar potential function $\mathcal{V}$, the variations become complete variations and we can write

$$\delta \int_{t_1}^{t_2} [\mathcal{T} - (\mathcal{U} + \mathcal{V})] \, dt = 0 \tag{1.15}$$

This equation is the one usually referred to as *Hamilton's principle*.

When we apply these stationary principles, we need to identify two classes of boundary conditions, called *essential* and *natural* boundary conditions, respectively. The essential boundary conditions are also called *geometric* boundary conditions because they correspond to prescribed displacements and rotations; these geometric conditions must be rigorously imposed. The natural boundary conditions are associated with the applied loads and are implicitly contained in the variational principle.

Lagrange's Equations

Hamilton's principle provides a complete formulation of a dynamical problem; however, to obtain solutions to some problems, the Hamilton integral formulation must be converted into one or more differential equations of motion. For a computer solution, these must be further reduced to equations using discrete unknowns; that is, we introduce some generalized coordinates (or degrees of freedom with the constrained degrees removed). At present, we will not be explicit about which coordinates we are considering but accept that we can write any function (the displacement, say) as

$$u(x, y, z) = u(u_1 g_1(x, y, z), u_2 g_2(x, y, z), \ldots, u_N g_N(x, y, z))$$

where u_i are the generalized coordinates and $g_i(x, y, z)$ are prescribed (known) functions of (x, y, z). The generalized coordinates are obtained by the imposition of *holonomic* constraints—the constraints are geometric of the form $f_i(u_1, u_2, \ldots, u_N, t) = 0$ and do not depend on the velocities.

Hamilton's extended principle now takes the form

$$\int_{t_1}^{t_2} \sum_{j=1}^{N} \left\{ -\frac{d}{dt}\left(\frac{\partial T}{\partial \dot{u}_j}\right) + \frac{\partial T}{\partial u_j} - \frac{\partial (U + V)}{\partial u_j} + Q_j \right\} \delta u_j \, dt = 0$$

where Q_j are a set of nonconservative forces. Because the virtual displacements δu_j are independent and arbitrary, and because the time limits are arbitrary, each integrand is zero. This leads to the *Lagrange's equation of motion*:

$$\mathcal{F}_i \equiv \frac{d}{dt}\left(\frac{\partial T}{\partial \dot{u}_i}\right) - \frac{\partial T}{\partial u_i} + \frac{\partial}{\partial u_i}(U + V) - Q_i = 0 \qquad (1.16)$$

for $i = 1, 2, \ldots, N$. The expression, $\mathcal{F}_i = 0$, is our statement of (dynamic) equilibrium. It is apparent from the Lagrange's equation that, if the system is not in motion, then we recover the principle of stationary potential energy expressed in terms of generalized coordinates.

It must be emphasized that the transition from Hamilton's principle to Lagrange's equation was possible only by identifying u_i as generalized coordinates; that is, Hamilton's principle holds true for constrained as well as generalized coordinates, but Lagrange's equation is valid only for the latter. A nice historical discussion of Hamilton's principle and Lagrange's equation is given in Reference [176].

Consider the Lagrange's equation when the motions are small. Specifically, consider small motions about an equilibrium position defined by $u_i = 0$ for all i. Perform a Taylor series expansion on the strain energy function to get

$$U(u_1, u_2, \ldots) = U(0) + \sum_i \frac{\partial U}{\partial u_i}\bigg|_0 u_i + \frac{1}{2}\sum_{i,j} \frac{\partial^2 U}{\partial u_i \partial u_j}\bigg|_0 u_i u_j + \cdots$$

The first term in this expansion is irrelevant and the second term is zero, since, by assumption, the origin is an equilibrium position. We therefore have the representation of the strain energy as

$$U(u_1, u_2, \ldots) \approx \frac{1}{2}\sum_{i,j} K_{ij} u_i u_j, \qquad K_{ij} \equiv \frac{\partial^2 U}{\partial u_i \partial u_j}\bigg|_0$$

We can do a similar expansion for the kinetic energy; in this case, however, we also assume that the system is linear in such a way that T is a function only of the velocities $\dot{u}_j$. This leads to

$$T(\dot{u}_1, \dot{u}_2, \ldots) \approx \frac{1}{2}\sum_{i,j} M_{ij} \dot{u}_i \dot{u}_j, \qquad M_{ij} \equiv \frac{\partial^2 T}{\partial \dot{u}_i \partial \dot{u}_j}\bigg|_0$$

The potential of the conservative forces also has an expansion similar to that for U, but we retain only the linear terms in u_j such that

$$V = -\sum_j P_j u_j, \qquad P_j \equiv -\frac{\partial V}{\partial u_j}\bigg|_0$$

Finally, assume that the nonconservative forces are of the viscous type such that the virtual work is

$$\delta W^d = Q^d \delta u = -c\dot{u}\,\delta u$$

This suggests the introduction of a function analogous to the potential for the conservative forces

$$Q_j^d = -\frac{\partial \mathcal{D}}{\partial \dot{u}_j} \qquad \text{where} \qquad \mathcal{D} = \mathcal{D}(\dot{u}_1, \dot{u}_2, \ldots, \dot{u}_N)$$

For small motions, this gives

$$\mathcal{D}(\dot{u}_1, \dot{u}_2, \ldots, \dot{u}_N) \approx \frac{1}{2}\sum_{i,j} C_{ij}\dot{u}_i\dot{u}_j, \qquad C_{ij} \equiv \frac{\partial^2 \mathcal{D}}{\partial \dot{u}_i \partial \dot{u}_j}\bigg|_0$$

The function $\mathcal{D}$ is called the *Rayleigh dissipation function*.

Substitute these forms for $\mathcal{U}$, $\mathcal{V}$, $\mathcal{T}$, and $\mathcal{D}$ into Lagrange's equation to get

$$\sum_j \{K_{ij}u_j + C_{ij}\dot{u}_j + M_{ij}\ddot{u}_j\} = P_i, \qquad i = 1, 2, \ldots, N$$

This is put in the familiar matrix form as

$$[\,K\,]\{u\} + [\,C\,]\{\dot{u}\} + [\,M\,]\{\ddot{u}\} = \{P\} \tag{1.17}$$

By comparison with a single spring–mass system, we have the meaning of $[\,K\,]$, $[\,M\,]$, and $[\,C\,]$ as the (generalized) structural stiffness, mass, and damping matrices, respectively. As yet, we have not said how the actual coefficients can be obtained or the actual meaning of the generalized coordinates; this is the subject of a later section where we consider some specific finite elements.

In this system of equations, $\{u\}$ is the vector of all the free degrees of freedom and is of size $\{M_u \times 1\}$, $[\,K\,]$ is the $[M_u \times M_u]$ stiffness matrix, $[\,C\,]$ is the $[M_u \times M_u]$ damping matrix, $\{P\}$ is the $\{M_u \times 1\}$ vector of all applied loads (some of which are zero).

1.4   Material Behavior

The concepts of stress, on the one hand, and strain, on the other, were developed independently of each other and apart from the assumption of a continuum, the development placed no restrictions on the material; that is, the concepts developed so far apply whether the material is elastic or plastic, isotropic or anisotropic. Indeed, they apply even if the material is a fluid. This section makes the material behavior explicit.

Types of Materials

There is a wide range of materials available, but engineering structures are made from relatively few. (Reference [86] gives an enjoyable account of a variety of structures and the types of materials used for their construction.) Most structures are designed to sustain only elastic loads and are fabricated from isotropic and homogeneous materials (steel and aluminum, for instance); and therefore a discussion of these materials will suffice for most situations (see Table 1.1 and Table 1.2). Analysis of structural components fabricated with composite materials, on the other hand, requires the use of anisotropic elasticity theory. Analysis of problems in metal-forming and ductile fracture are based on the inelastic and plastic responses of materials, particularly those under large deformation. Polymeric materials require knowledge of their time-dependent stress relaxation and creep properties. We consider these latter behaviors only briefly.

While the behavior of a real material is very complicated, most structural materials can be divided into a limited number of classes; four of the main classes will be considered briefly here. To simplify the following relations, only small deformations will be considered.

We begin the discussion by considering the ubiquitous stress/strain diagram of a tensile test. Load (P) and extension (ΔL) are the typical data collected in the performance of a tensile test; when these are plotted as stress (P/A) against strain ($\Delta L/L$) they may look like one of the plots in Figure 1.10. A number of terms are used to describe these plots. The *Proportionality Limit* is the last point where stress and strain are linearly related (Point a). The *Elastic Limit* is the last point from which, after removal of load, there is no permanent strain (Point b). The *Yield Point* is technically the same as the elastic limit but is usually associated with the gross onset of permanent strain. Many materials do not exhibit a clearly defined yield point and so this point is often taken to correspond to a certain offset of strain. Point c is the yield point for 0.2% (.002 strain) offset. The *Yield Strength* is the stress at the yield point, σ_Y. The *Tensile Strength*, or sometimes called the Ultimate Tensile Strength (UTS), is the maximum stress reached during loading (Point d). The *Rupture Strain* is the maximum strain reached during loading (Point ×).

The unloading curve is usually parallel to the elastic loading curve and results in a permanent set or a plastic strain at complete unloading. The difference in yield stresses between the virgin and plastically deformed material is caused by its strain-hardening response.

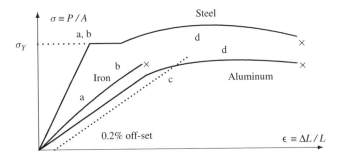

Figure 1.10: Some stress/strain diagrams.

Table 1.1: Typical structural material properties in SI units.

Material	ρ [kg/m^3]	E [GPa]	G [GPa]	σ_Y [MPa]	σ_U [MPa]	K_{1c} [MPa$\sqrt{m}$]	α [$\mu\epsilon/°$C]
Aluminum:							
2024	2768	73	27.6	276	414	26	22.5
7075		71	27.6	483	538	24	22.5
Titanium:							
6AL-4 V	4429	110	41	924	924	115	9.4
Steel:							
4340	7832	200	27.6	1170	1240	–	11.9
AMS6520	183	27.6	1720	1730	–	11.9	
Stainless		200	27.6	1117	1241	(–)	11.9
300 M		200	27.6	165	193	(–)	11.9
Plastic:							
PMMA	1107	4.1	–	76	–	–	90
Epoxy		3.5	–	–	–	–	90
Wood:							
Spruce	443	9.0	(–)	65	–	–	–
Fir		11	–	47	55	–	–

Table 1.2: Typical structural material properties in common units.

Material	ρ [lb/in^3]	E [msi]	G [msi]	σ_Y [ksi]	σ_U [ksi]	K_{1c} [ksi$\sqrt{in}$]	α [$\mu\epsilon/°$F]
Aluminum							
2024	0.10	10.6	4.00	40	60	29.0	12.5
7075		10.3	3.9	70	78	22.0	12.5
Titanium							
6AL-4 V	0.16	16.0	6.0	130	134	105	5.2
Steel							
4340	0.283	29.0	12.0	170	180	–	6.6
AMS6520		26.5	12.0	250	252	–	6.6
17-7 PH		29.0	12.0	162	180	40.0	6.6
300 M		29.0	12.0	240	280	47.0	6.6
Plastic							
PMMA	0.04	0.6	–	11	–	–	50
Epoxy		0.475	–	–	–	–	50
Wood							
Spruce	0.016	1.3	–	9.4	–	–	–
Fir		1.6	–	6.8	8	–	–

If there is a one to one relation between the stress and strain, and, on unloading, all the strain is instantaneously recovered, then the material is said to be elastic. Most structural materials in common use are adequately described by this type of material. In linear elasticity, the deformation is a linear function of the stress

$$\sigma = E\epsilon$$

where E is called the Young's modulus.

For some materials, it is found that beyond a certain stress level large deformations occur for small increments in load, and furthermore, much of the deformation is not recovered when the load is removed. On the load/unload cycle, if $\sigma > \sigma_Y$ (σ_Y is the yield stress), this material cannot recover the deformation caused after yielding. This remaining deformation is called the permanent or plastic strain. The total strain is considered as composed of elastic and plastic parts

$$\epsilon = \epsilon^e + \epsilon^p$$

and constitutive relations are written for the separate parts. An explicit constitutive relation is considered later in this chapter.

The mechanical properties of all solids are affected to varying degrees by the temperature and rate of deformation. Although such effects are not measurable (for typical structural materials) at ordinary temperatures, they become noticeable at high temperatures relative to the glassy transition temperature for polymers or the melting temperature for metals. Above the glassy transition temperature, many amorphous polymers flow like a Newtonian fluid and are referred to as *viscoelastic* materials; that is, they exhibit a combination of solid and fluid effects. Viscoelastic materials are time-dependent and the relation must be written in terms of the time derivatives; the standard linear solid, for example, is described by

$$\frac{d\epsilon}{dt} + \frac{E_2}{\eta}\epsilon = \frac{1}{E}\frac{d\sigma}{dt} + \frac{1}{\eta}\left[1 + \frac{E_2}{E_1}\right]\sigma$$

In addition to the time dependence, there is also a dependence on more material coefficients. This material exhibits a residual strain after the load is removed, but over time it recovers this strain to leave the material in its original state.

The structural properties of some base materials can be improved by combining them with other materials to form composite materials. A historical survey is given in Reference [86]. For a sandwich material such as plywood, the cross-grain weakness of the wood is improved by alternating directions of the lamina. Concrete is an example of particulate materials with cement as the bonding agent. Reinforced concrete uses steel bars to improve the tensile strength of the base concrete. Sheets of glass are stiff but prone to brittle failure because of small defects. Glass-fiber-reinforced composites combine the stiffness of glass in the form of fibers with the bonding of a matrix material such as epoxy; in this way, a defect in individual fibers does affect the overall behavior. It is possible to describe these structured materials through their constituent behavior, but for structural analyses this is very rarely done because it is computationally prohibitive. It is more usual to treat these materials as homogeneous materials and establish average properties based on a representative volume. The variety of behaviors from these structured materials is much richer than that of traditional homogeneous materials and not easily summarized in a brief discussion; by way of example, we will consider their anisotropic properties.

Failure Criteria

In broad terms, failure refers to any action that leads to an inability on the part of a structure to function as intended. Common modes of failure include permanent deformation (yielding), fracture, buckling, creep, and fatigue. The successful use of a material in any application requires assurance that it will function safely. Therefore, the design process must involve steps where the predicted in-service stresses, strains, and deformations are limited to appropriate levels using failure criteria based on experimental data.

The basic assumption underlying all failure criteria is that failure is predicted to occur at a particular point in a material only when the value of a certain measure of stress reaches a critical value. The critical level of the selected measure is obtained experimentally, usually by a uniaxial test similar to that of Figure 1.10; that is, the goal of such criteria is to predict failure in multiaxial states of stress using uniaxial failure stress as the only input parameter. We discuss some of the more common criteria below.

The stress state of a body is characterized in terms of the principal stresses because they are the extremum values. In the following, we assume the principal stresses are ordered according to $\sigma_3 < \sigma_2 < \sigma_1$.

The maximum principal stress theory predicts that failure will occur at a point in a material when the maximum principal normal stress becomes equal to or exceeds the uniaxial failure stress for that material; that is, this criterion predicts failure to occur at a point when

$$\sigma_1 \geq \sigma_f \qquad \text{or} \qquad \sigma_3 \leq -\sigma_f$$

where σ_f is the magnitude of the uniaxial failure stress in tension and an equal but opposite failure stress is assumed in compression. This theory provides a generally poor prediction of yield onset for most metals and is not typically used for materials that behave in a ductile fashion. It has, however, been applied successfully to predict fracture of some brittle materials in multiaxial stress states.

The maximum shear stress theory states that the yield is predicted to occur at a point in a material when the absolute maximum shear stress at that point becomes equal to or exceeds the magnitude of the maximum shear stress at yield in a uniaxial tensile test. On the basis of the stress transformation equations, this critical yielding value of the maximum shear stress during a uniaxial test is $\tau_Y = \frac{1}{2}\sigma_Y$. Thus, yield failure is predicted to occur in a multiaxial state of stress if

$$\tau_{\max} \geq \tfrac{1}{2}\sigma_Y$$

In terms of the principal stresses, this can be recast into the form

$$\sigma_1 - \sigma_3 \geq \sigma_Y$$

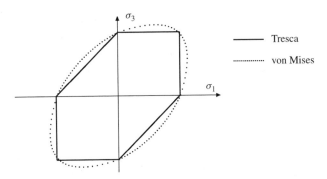

Figure 1.11: Tresca and von Mises yield criteria.

This criterion is shown in Figure 1.11. Experimental evidence for various states of stress attest that the maximum shear stress criterion is a good theory for predicting yield failure of ductile metals.

The total strain energy of an isotropic linear elastic material is often divided into two parts: the dilatation energy, which is associated with change of volume under a mean hydrostatic pressure, and the distortion energy, which is associated with change in shape. The energy of distortion theory states that yield is predicted to occur at a point in a material when the distortion energy at that point becomes equal to or exceeds the magnitude of the distortion energy at yield in a uniaxial tensile test of the same material. The critical yielding value of the distortion energy in a uniaxial test is

$$\mathcal{U}_d = \frac{1+\nu}{6E}[2\sigma_Y^2]$$

This leads to the statement that yield will occur if

$$\sigma_{vm} \equiv \sqrt{2[(\sigma_1 - \sigma_2)^2 + (\sigma_2 - \sigma_3)^2 + (\sigma_3 - \sigma_1)^2]} \geq \sigma_Y$$

where σ_{vm} is called the *von Mises stress*. The behavior of this is shown in Figure 1.11. The predictions are close to those of the maximum shear stress criterion, but it has the slight advantage of using a single function for any state of stress. In 3-D complex stress states, the von Mises stress can be computed in terms of the component stresses through

$$\sigma_{vm} = \sqrt{I_1^2 - 3I_2}$$

where I_1 and I_2 are the invariants computed from Equation (1.5).

Elastic Constitutive Relations

Because of our interest in structures, the elastic material is the one of most relevance to us. Furthermore, the situations that arise are usually of the type of Cases (b) and (c) of Figure 1.4 and we utilize this to make approximations.

Consider a small volume of material under the action of applied loads on its surface. Then one way to describe elastic behavior is as follows: *The work done by the applied forces is transformed completely into strain (potential) energy, and this strain energy is completely recoverable.* That the work is transformed into potential energy and that it is completely recoverable means the material system is conservative. Using Lagrangian variables, the increment of work done on the small volume is

$$dW_e = \int_{V^o} \left[\sum_{i,j} \sigma_{ij}^K \, dE_{ij} \right] dV^o$$

The potential is comprised entirely of the strain energy $\mathcal{U}$; the increment of strain energy is

$$d\mathcal{U} = d\mathcal{U}(E_{ij}) = \int_{V^o} \left[\sum_{i,j} \frac{\partial \overline{\mathcal{U}}}{\partial E_{ij}} \, dE_{ij} \right] dV^o$$

where $\overline{\mathcal{U}}$ is the strain energy density. From the hypothesis, we can equate dW_e and $d\mathcal{U}$, and because the volume is arbitrary, the integrands must be equal; hence we have

$$\sigma_{ij}^K = \frac{\partial \overline{\mathcal{U}}}{\partial E_{ij}} \tag{1.18}$$

A material described by this relation is called *hyperelastic*. Note that it is valid for large deformations and for anisotropic materials; however, rather than develop this general case, we will look at each of these separately.

Many structural materials (steel and aluminum, for example) are essentially isotropic in that the stiffness of a sheet is about the same in all directions. Consequently, the strain energy is a function of the strain invariants only; that is, $\overline{\mathcal{U}} = \overline{\mathcal{U}}(I_1, I_2, I_3)$, where the invariants are computed by

$$I_1 = \sum_k E_{kk}, \qquad I_2 = \tfrac{1}{2}I_1^2 - \tfrac{1}{2}\sum_{i,k} E_{ik}E_{ik}, \qquad I_3 = \det[E_{ij}]$$

On substituting these into the above constitutive relation, and rearranging them, we get

$$\sigma_{ij}^K = \beta_o \delta_{ij} + \beta_1 E_{ij} + \beta_2 \sum_p E_{ip}E_{pj} \tag{1.19}$$

which is a nice compact relation. The coefficients β_i are functions of only the invariants [71].

Reinforced materials are likely to have directional properties and are therefore anisotropic. They are also more likely to have small operational strains, and we take advantage of this to effect another set of material approximations. Take the Taylor series expansion of the strain energy density function which, on differentiation, then leads to

$$\sigma_{pq}^K = \frac{\partial \overline{\mathcal{U}}}{\partial E_{pq}} \approx \left[\frac{\partial \overline{\mathcal{U}}}{\partial E_{pq}}\right]_0 + \sum_{r,s} \left[\frac{\partial^2 \overline{\mathcal{U}}}{\partial E_{pq}\partial E_{rs}}\right]_0 E_{rs} + \cdots = \sigma_{pq}^o + \sum_{r,s} D_{pqrs} E_{rs} + \cdots$$

where σ_{pq}^o corresponds to an initial stress. Because of symmetry in σ_{ij}^K and E_{ij}, C_{pqrs} reduces to 36 coefficients. But because of the explicit form of C_{pqrs} in terms of derivatives, we have the further restriction

$$D_{pqrs} = \left.\frac{\partial^2 \overline{\mathcal{U}}}{\partial E_{pq} \partial E_{rs}}\right|_0 = \left.\frac{\partial^2 \overline{\mathcal{U}}}{\partial E_{rs} \partial E_{pq}}\right|_0 = D_{rspq}$$

This additional symmetry reduces the elastic tensor to 21 constants. This is usually considered to be the most general linearly elastic material. We can write this relation in the matrix form

$$\{\sigma\} = [\ D\ ]\{E\}, \quad \{\sigma\} \equiv \{\sigma_{11}^K, \sigma_{22}^K, \sigma_{33}^K, \sigma_{23}^K, \ldots\}, \quad \{\epsilon\} \equiv \{E_{11}, E_{22}, E_{33}, 2E_{23}, \ldots\}$$

where $[\ D\ ]$ is of size $[6 \times 6]$. Because of the symmetry of both the stress and strain, we have $[\ D\ ]^T = [\ D\ ]$. Special materials are reduced forms of this relation.

An *orthotropic* material has three planes of symmetry, and this reduces the number of material coefficients to nine; and the elastic matrix is given by

$$\begin{bmatrix} d_{11} & d_{12} & d_{13} & 0 & 0 & 0 \\ d_{12} & d_{22} & d_{23} & 0 & 0 & 0 \\ d_{13} & d_{23} & d_{33} & 0 & 0 & 0 \\ 0 & 0 & 0 & \frac{1}{2}(d_{11}-d_{12}) & 0 & 0 \\ 0 & 0 & 0 & 0 & \frac{1}{2}(d_{11}-d_{12}) & 0 \\ 0 & 0 & 0 & 0 & 0 & \frac{1}{2}(d_{11}-d_{12}) \end{bmatrix}$$

For a transversely isotropic material, this reduces to five coefficients because $d_{55} = d_{44}$ and $d_{66} = (d_{11} - d_{12})/2$. A thin fiber-reinforced composite sheet is usually considered to be transversely isotropic [100].

For the isotropic case, every plane is a plane of symmetry and every axis is an axis of symmetry. It turns out that there are only two independent elastic constants, and the elastic matrix is given as above but with

$$d_{11} = d_{22} = d_{33} = \lambda + 2\mu, \qquad d_{12} = d_{23} = d_{13} = \lambda$$

The constants λ and μ are called the Lamé constants. The stress/strain relation for isotropic materials (with no initial stress) are usually expressed in the form

$$\sigma_{ij}^K = 2\mu E_{ij} + \lambda \delta_{ij} \sum_k E_{kk}, \qquad 2\mu E_{ij} = \sigma_{ij}^K - \frac{\lambda}{3\lambda + 2\mu}\delta_{ij}\sum_k \sigma_{kk}^K \qquad (1.20)$$

This is called *Hooke's law* and is the linearized version of Equation (1.19). The expanded form of the Hooke's law for strains in terms of stresses is

$$E_{xx} = \frac{1}{E}[\sigma_{xx}^K - \nu(\sigma_{yy}^K + \sigma_{zz}^K)]$$

$$E_{yy} = \frac{1}{E}[\sigma_{yy}^K - \nu(\sigma_{zz}^K + \sigma_{xx}^K)]$$

$$E_{zz} = \frac{1}{E}[\sigma_{zz}^K - v(\sigma_{xx}^K + \sigma_{yy}^K)] \tag{1.21}$$

$$2E_{xy} = \frac{2(1+v)}{E}\sigma_{xy}^K, \quad 2E_{yz} = \frac{2(1+v)}{E}\sigma_{yz}^K, \quad 2E_{xz} = \frac{2(1+v)}{E}\sigma_{xz}^K$$

and for stresses in terms of strains

$$\sigma_{xx}^K = \frac{E}{(1+v)(1-2v)}[(1-v)E_{xx} + v(E_{yy} + E_{zz})]$$

$$\sigma_{yy}^K = \frac{E}{(1+v)(1-2v)}[(1-v)E_{yy} + v(E_{zz} + E_{xx})]$$

$$\sigma_{zz}^K = \frac{E}{(1+v)(1-2v)}[(1-v)E_{zz} + v(E_{xx} + E_{yy})] \tag{1.22}$$

$$\sigma_{xy}^K = \frac{E}{2(1+v)}2E_{xy}, \quad \sigma_{yz}^K = \frac{E}{2(1+v)}2E_{yz}, \quad \sigma_{xz}^K = \frac{E}{2(1+v)}2E_{xz}$$

where E is the Young's modulus and v is the Poisson's ratio related to the Lamé coefficients by

$$E = \frac{\mu(3\lambda + 2\mu)}{\lambda + \mu}, \quad v = \frac{\lambda}{2(\lambda + \mu)}, \quad \lambda = \frac{vE}{(1-2v)(1+v)}, \quad \mu = G = \frac{E}{2(1+v)}$$

The coefficient $\mu = G$ is called the shear modulus.

A special case that arises in the analysis of thin-walled structures is that of *plane stress*. Here, the stress through the thickness of the plate is approximately zero such that $\sigma_{zz}^K \approx 0$, $\sigma_{xz}^K \approx 0$, and $\sigma_{yz}^K \approx 0$. This leads to

$$E_{zz} = \frac{-v}{E}[\sigma_{xx}^K + \sigma_{yy}^K] = \frac{-v}{1-v}[E_{xx} + E_{yy}]$$

Substituting this into the 3-D Hooke's law then gives

$$E_{xx} = \frac{1}{E}[\sigma_{xx}^K - v\sigma_{yy}^K], \qquad \sigma_{xx}^K = \frac{E}{(1-v^2)}[E_{xx} + vE_{yy}]$$

$$E_{yy} = \frac{1}{E}[\sigma_{yy}^K - v\sigma_{xx}^K], \qquad \sigma_{yy}^K = \frac{E}{(1-v^2)}[E_{yy} + vE_{xx}] \tag{1.23}$$

The shear relation is unaffected.

A final point to note is that, except for the isotropic material, the material coefficients are given with respect to a particular coordinate system. Hence, we must transform the coefficients into the new coordinate system when the axes are changed.

Strain Energy for Some Linear Elastic Structures

When the strains are small, we need not distinguish between the undeformed and deformed configurations. Under this circumstance, let the material obey Hooke's law

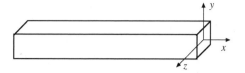

Figure 1.12: A slender structural member in local coordinates.

and be summarized in the matrix forms

$$\{\epsilon\} = [\,C\,]\{\sigma\}, \qquad \{\sigma\} = [\,D\,]\{\epsilon\}, \qquad [\,C\,] = [\,D\,]^{-1}$$

The general expression for the strain energy is

$$\mathcal{U} = \frac{1}{2}\int_V [\sigma_{xx}\epsilon_{xx} + \sigma_{yy}\epsilon_{yy} + \sigma_{xy}\gamma_{xy} + \cdots]\,dV = \frac{1}{2}\int_V \{\sigma\}^{\mathrm{T}}\{\epsilon\}\,dV$$

where $\gamma_{ij} = 2\epsilon_{ij}(i \neq j)$ is called the engineering shear strain. Using Hooke's law, the strain energy can be put in the alternate forms

$$\mathcal{U} = \frac{1}{2}\int_V \{\epsilon\}^{\mathrm{T}}[\,D\,]\{\epsilon\}\,dV = \frac{1}{2}\int_V \{\sigma\}^{\mathrm{T}}[\,C\,][\,\sigma\,]\,dV \qquad (1.24)$$

The above relations will now be particularized to some structural systems of interest by writing the distributions of stress and strain in terms of resultants and displacements, respectively.

For example, for the rod member of Figure 1.12, there is only an axial stress present and it is uniformly distributed on the cross section. Let F be the resultant force; then $\sigma_{xx} = F/A = E\epsilon_{xx}$ and

$$\text{axial:} \qquad \mathcal{U} = \frac{1}{2}\int_0^L \frac{F^2}{EA}\,dx = \frac{1}{2}\int_0^L EA\left[\frac{du}{dx}\right]^2 dx$$

For the beam member in bending, there is only an axial stress, but it is distributed linearly on the cross section in such a way that there is no resultant axial force. Let M be the resultant moment; then $\sigma_{xx} = -My/I = E\epsilon_{xx}$ and

$$\text{bending:} \qquad \mathcal{U} = \frac{1}{2}\int_0^L \frac{M^2}{EI}\,dx = \frac{1}{2}\int_0^L EI\left[\frac{d^2v}{dx^2}\right]^2 dx$$

where I is the second moment of area. Other members are handled similarly [71].

1.5 The Finite Element Method

The discussions of the previous sections show that we can pose the solution of a solid mechanics problem in terms of extremizing a functional; this is known as the *weak* form or *variational* form of the problem. What we wish to pursue in the following is approximation arising from the weak form; specifically, we will approximate the functional itself and use the variational principle to obviate consideration of the natural boundary conditions. This is called the Ritz method. The computational implementation will be in the form of the finite element method.

Ritz Method

In general, a continuously distributed deformable body consists of an infinity of material points and therefore has infinitely many degrees of freedom. The Ritz method is an approximate procedure by which continuous systems are reduced to systems with finite degrees of freedom. The fundamental characteristic of the method is that we operate on the functional corresponding to the problem. To fix ideas, consider the static case where the solution of $\delta\Pi = 0$ with prescribed boundary conditions on u is sought. Let

$$u(x, y, z) = \sum_{i=1}^{\infty} a_i g_i(x, y, z)$$

where $g_i(x, y, z)$ are independent known *trial functions*, and the a_i are unknown multipliers to be determined in the solution. The trial functions satisfy the essential (geometric) boundary conditions but not necessarily the natural boundary conditions. The variational problem states that

$$\Pi(u) = \Pi(a_1, a_2, \ldots) = \text{stationary}$$

Thus, $\Pi(a_1, a_2, \ldots)$ can be regarded as a function of the variables $a_1, a_2, \cdots$. To satisfy $\Pi = $ stationary, we require that

$$\mathcal{F}_1 = \frac{\partial \Pi}{\partial a_1} = 0, \qquad \mathcal{F}_2 = \frac{\partial \Pi}{\partial a_2} = 0, \qquad \cdots$$

These equations are then used to determine the coefficients a_i. Normally, only a finite number of terms is included in the expansion.

An important consideration is the selection of the trial functions $g_i(x, y, z)$; but it must also be kept in mind that these functions need only satisfy the essential boundary conditions and not (necessarily) the natural boundary conditions. For practical analyses, this is a significant point and largely accounts for the effectiveness of the displacement-based finite element analysis procedure.

We will use the simple example of Figure 1.13 to illustrate some aspects of the Ritz method. To begin, we must have a compatible displacement field; this is achieved by imposing the geometric boundary conditions. For this problem, these conditions are $u = 0$ at both $x = 0$ and $x = L$. To satisfy these, assume a polynomial

$$u(x) = a_0 + a_1 x + a_2 x^2 + a_3 x^3 + \cdots$$

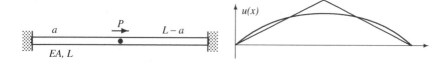

Figure 1.13: Fixed fixed rod with a concentrated load.

Imposing the boundary conditions allows a_0 and a_1 to be determined in terms of the other coefficients. This leads to the displacement representation

$$u(x) = a_2[x^2 - xL] + a_3[x^3 - xL^2] + \cdots = a_2 g_2(x) + a_3 g_3(x) + \cdots$$

that will satisfy the boundary conditions. Indeed, each of the functions separately satisfy the boundary conditions and therefore are individually acceptable Ritz functions.

The total potential energy of the rod problem is

$$\Pi = \frac{1}{2} \int_0^L EA \left(\frac{du}{dx}\right)^2 dx - Pu|_{x=a}$$

Substituting the assumed displacements and invoking the stationarity of Π with respect to the coefficients a_n, we obtain after differentiation

$$\frac{\partial \Pi}{\partial a_2} = \int_0^L EA \left[a_2[2x - L] + a_3[3x^2 - L^2] + \cdots \right] [2x - L] \, dx - P \left[[a^2 - aL] \right] = 0$$

$$\frac{\partial \Pi}{\partial a_3} = \int_0^L EA \left[a_2[2x - L] + a_3[3x^2 - L^2] + \cdots \right] [3x^2 - L^2] \, dx - P \left[[a^3 - aL^2] \right] = 0$$

Performing the required integrations gives

$$\frac{EA}{30} \begin{bmatrix} 10L^3 & 15L^4 & \cdots \\ 15L^4 & 24L^5 & \cdots \\ \vdots & \vdots & \vdots \end{bmatrix} \begin{Bmatrix} a_2 \\ a_3 \\ \vdots \end{Bmatrix} = P \begin{Bmatrix} a^2 - aL \\ a^3 - aL^2 \\ \vdots \end{Bmatrix}$$

It is typical to end up with a set of simultaneous equations. in these problems Note that the system matrix is symmetric; this too is usually the case.

Solving this system when only the a_2 term is used in the expansion gives the approximate solution

$$u(x) = \frac{P}{EAL^3} [a^2 - aL][x^2 - xL]$$

In the special case of $a = L/2$, the maximum deflection (also at $x = L/2$) is

$$u_{max} = \frac{PL}{4EA} \frac{3}{4}, \qquad u_{exact} = \frac{PL}{4EA}$$

This underestimates the maximum as shown in Figure 1.13

We will now redo the problem but this time assume a compatible displacement field that is piecewise linear, that is, we let

$$u(x) = \left[\frac{x}{a}\right]u_o \qquad 0 \le x \le a, \qquad\qquad u(x) = \left[\frac{L-x}{L-a}\right]u_o \qquad a \le x \le L$$

where u_o is the displacement at the load application point. Since the displacement distribution is specified in a piecewise fashion, the integral for the strain energy must be divided up as

$$\mathcal{U} = \frac{1}{2}\int_0^L EA\left(\frac{du}{dx}\right)^2 dx = \frac{1}{2}\int_0^a EA\left[\frac{u_o}{a}\right]^2 dx + \frac{1}{2}\int_a^L EA\left[\frac{-u_o}{L-a}\right]^2 dx$$

This gives, after expanding and integrating,

$$\mathcal{U} = \frac{1}{2}EA\frac{u_o^2 a}{a^2} + \frac{1}{2}EA\frac{u_o^2(L-a)}{(L-a)^2} = \frac{1}{2}EA\frac{u_o^2 L}{a(L-a)}$$

The total potential for the problem is then

$$\Pi = \frac{1}{2}EA\frac{u_o^2 L}{a(L-a)} - Pu_o$$

Minimizing with respect to u_o gives

$$\frac{\partial \Pi}{\partial u_o} = EA\frac{u_o L}{a(L-a)} - P = 0 \qquad \text{or} \qquad u_o = \frac{P}{EA}\frac{a(L-a)}{L}$$

As it happens, this is the exact result.

This simple example is important in demonstrating the relationship between a Ritz analysis and a finite element analysis; that is, the classical Ritz method uses functions that cover the entire region, but as the example shows, we can alternatively use simpler functions, but limited to smaller regions. As discussed next, these smaller regions are actually the finite elements.

The Finite Element Discretization

One disadvantage of the conventional Ritz analysis is that the trial functions are defined over the whole region. This causes a particular difficulty in the selection of appropriate functions; in order to solve accurately for large stress gradients, say, we may need many functions. However, these functions are also defined over the regions in which the stresses vary rather slowly and where not many functions are required. Another difficulty arises when the total region is made up of subregions with different kinds of strain distributions. As an example, consider a building modeled by plates for the floors and beams for the

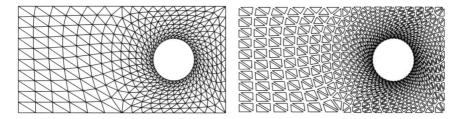

Figure 1.14: Continuous domain discretized as finite elements. Right figure has shrunk elements for easier viewing.

vertical frame. In this situation, the trial functions used for one region (e.g., the floor) are not appropriate for the other region (e.g., the frame), and special displacement continuity conditions and boundary relations must be introduced. We conclude that the conventional Ritz analysis is, in general, not particularly computer-oriented.

We can view the finite element method as an application of the Ritz method where, instead of the trial functions spanning the complete domain, the individual functions span only subdomains (the finite elements) of the complete region and are zero everywhere else. Figure 1.14 shows an example of a bar with a hole modeled as a collection of many triangular regions. The use of relatively many functions in regions of high strain gradients is made possible simply by using many elements as shown around the hole in the figure. The combination of domains with different kinds of strain distributions (e.g., a frame member connected to a plate) may be achieved by using different kinds of elements to idealize the domains.

In order that a finite element solution be a Ritz analysis, it must satisfy the essential boundary conditions. However, in the selection of the displacement functions, no special attention need be given to the natural boundary conditions, because these conditions are imposed with the load vector and are satisfied approximately in the Ritz solution. The accuracy with which these natural boundary conditions are satisfied depends on the specific trial functions employed and on the number of elements used to model the problem. This idea is demonstrated in the convergence studies of the next section.

The basic finite element procedure will now be described with reference to Figure 1.15. First, the structure is discretized into many subregions and for each subregion the displacement field is written in terms of nodal values. The total potential energy (which includes the strain energy and the potential of the nodal forces) is then minimized with respect to the nodal values to give the equilibrium relation

$$\{F_e\} = [\ k\ ]\{u\}$$

where $\{u\}$ is the *vector of nodal displacements*, $\{F_e\}$ is the *vector of element nodal forces*, and $[\ k\ ]$ is called the *element stiffness matrix*. We will also refer to $\{u\}$ as the *nodal degrees of freedom*.

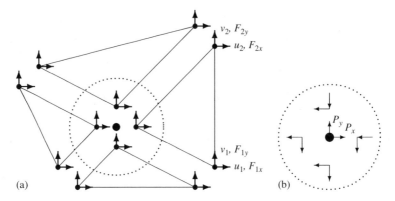

Figure 1.15: An assemblage of elements. (a) Nodal displacements and forces. (b) Equilibrium at common nodes.

When a continuous medium is discretized into a collection of piecewise simple regions, the issue of assemblage then arises. There are two aspects to this. The first is that there must be compatibility between the elements. This is achieved (in simple terms) through having neighboring elements share nodes and have common interpolations along shared interfaces. The second is that there must be equilibrium of the assembled elements. This can be viewed as either imposing equilibrium between the neighboring elements as depicted in Figure 1.15(b) or as adding the strain energies of elements. We will take the latter view.

Introduce the vector of all degrees of freedom (size $\{N \times 1\}$)

$$\{u\} \equiv \{u_1, u_2, u_3, \ldots\}^{\mathrm{T}}$$

The total potential of all the strain energies and applied loads is then

$$\Pi = \mathcal{U} + \mathcal{V} = \sum_m \mathcal{U}_m - \{P\}^{\mathrm{T}}\{u\}$$

The applied loads vector $\{P\}$ is considered to comprise nodal applied loads, nodal loads arising from any load distributions $q(x)$, and inertia loads. Extremizing this with respect to the degrees of freedom then gives the equilibrium condition

$$\mathcal{F} = \frac{\partial \Pi}{\partial u} = 0 = \{P\} - \{F\}, \qquad \{F\} = \sum_m \frac{\partial \mathcal{U}_m}{\partial u} = \sum_m \{F_{\mathrm{e}}\}_m \qquad (1.25)$$

The vector $\{F\}$ is the assemblage of element nodal forces. This relation (referred to as the *loading equation*) must be satisfied throughout the loading history and is valid for nonlinear problems.

We can give an alternate interpretation of the loading equation for linear problems. Augment each element stiffness with zeros to size $[N \times N]$ and call it $[\,K\,]_m = [k_{\mathrm{augment}}]_m$. The total strain energy is

$$\mathcal{U} = \frac{1}{2}\sum_m \{u\}^{\mathrm{T}}[\,K\,]_m\{u\} = \{u\}^{\mathrm{T}}[\,K\,]\{u\}$$

After extremizing, this leads to

$$[K]\{u\} = \{P\}, \qquad [K] \equiv \sum_m [K]_m = \sum_m [k_{\text{augment}}]_m$$

The $[N \times N]$ square matrix $[K]$ is called the *structural stiffness matrix*. Thus, the assemblage process (for linear problems) can be thought of as adding the element stiffnesses suitably augmented (with zeros) to full system size.

Convergence of the Approximate Solutions

Clearly, for the finite element method to be useful, its solutions must be capable of converging to the exact solutions.

As a Ritz procedure, it must satisfy the geometric boundary conditions and interelement compatibility. In addition, to insure convergence there are three more requirements. The essential requirement is that the element must be capable of modeling a constant state of strain and stress in the limit of small element size. For example, consider the following expansions for use in constructing a beam element:

$$v(x) = a_0 + a_1 x + a_2 x^2 + \cdots, \qquad v(x) = a_0 + a_1 x + a_3 x^3 \cdots$$

Both are adequate for ensuring compatibility between elements; however, since the strain in a beam is related to the second derivative of the deflection, the second expansion would not produce a state of constant strain in the limit of small elements (it would produce zero strain at one end of the element).

A subsidiary requirement is that the element should be capable of rigid-body displacements without inducing strains. This is actually an extreme case of the constant strain condition with $\epsilon_{ij} = 0$. The third requirement is that the assumed displacement functions should be complete up to orders required by the first criterion. In the expansions used for constructing a beam element, the first is complete, but the second is not.

We will have more to say about convergence as we construct particular elements for particular types of structures. In general, when the criteria are satisfied, a lower bound on the strain energy is assured, and convergence is monotonic.

1.6   Some Finite Element Discretizations

There are a great variety of elements documented in the literature; we consider just a few of them for illustrative purposes. We begin with the tetrahedron element; this 3-D element satisfies all convergence criteria and therefore it is suitable for computing accurate solutions for all problems. Unfortunately, it is not very practical in many situations because it requires too many elements to achieve the required accuracy. As a consequence, other elements which take advantage of the nature of the structure have been developed—we look at the frame and shell elements, but a great many more can be found in References [18, 45, 46, 71].

Tetrahedral Solid Element

Conceive of the 3-D solid as discretized into many tetrahedrons. Consider one of those tetrahedrons divided into four volumes where the common point is at (x, y, z) as shown in Figure 1.16. Define

$$h_1 = V_1/V, \qquad h_2 = V_2/V, \qquad h_3 = V_3/V, \qquad h_4 = V_4/V, \qquad h_i = h_i(x, y, z)$$

We have the obvious constraint that $h_1 + h_2 + h_3 + h_4 = 1$. The volumes of these tetrahedrons uniquely define the position of the common point.

The position of the point (x, y, z) in the tetrahedron can be written as

$$\begin{Bmatrix} 1 \\ x \\ y \\ z \end{Bmatrix} = \begin{bmatrix} 1 & 1 & 1 & 1 \\ x_1 & x_2 & x_3 & x_4 \\ y_1 & y_2 & y_3 & y_4 \\ z_1 & z_2 & z_3 & z_4 \end{bmatrix} \begin{Bmatrix} h_1 \\ h_2 \\ h_3 \\ h_4 \end{Bmatrix} \tag{1.26}$$

where the subscripts 1, 2, 3 refer to the counterclockwise nodes of the triangle as viewed from Node 4. We can invert this relation, but the expressions are rather lengthy.

In order for the four coordinates $\{h_1, h_2, h_3, h_4\}$ to describe the three coordinates (x, y, z), it must be supplemented by the constraint $h_1 + h_2 + h_3 + h_4 = 1$. We can invoke this constraint explicitly by introducing *natural coordinates* as shown in Figure 1.16(b) and given as

$$h_1 = \xi, \qquad h_2 = \eta, \qquad h_3 = \zeta, \qquad h_4 = 1 - \xi - \eta - \zeta \tag{1.27}$$

These are volumes in (ξ, η, ζ) space. This notation is from Reference [35]. We have for a typical function

$$u(x, y, z) = \sum_i^4 h_i(\xi, \eta, \zeta) u_i$$

where u_i are the nodal values of the function. The x coordinate, for example, is given from Equation (1.26) as

$$x = x_1 h_1 + x_2 h_2 + x_3 h_3 + x_4 h_4 = x_4 + x_{14}\xi + x_{24}\eta + x_{34}\zeta$$

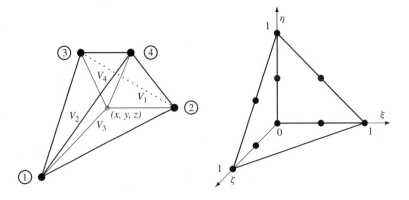

Figure 1.16: Volume and natural coordinates for a tetrahedron.

where $x_{14} = x_1 - x_4$, and so on. The strains are obtained in terms of derivatives of element displacements. Using the natural coordinate system, we get, for example,

$$\frac{\partial}{\partial x} = \frac{\partial}{\partial \xi}\frac{\partial \xi}{\partial x} + \frac{\partial}{\partial \eta}\frac{\partial \eta}{\partial x} + \frac{\partial}{\partial \zeta}\frac{\partial \zeta}{\partial x}$$

But, to evaluate the derivatives of (ξ, η, ζ) with respect to (x, y, z), we need to have the explicit relation between the two sets of variables. We obtain this as

$$\begin{Bmatrix} \dfrac{\partial}{\partial \xi} \\[6pt] \dfrac{\partial}{\partial \eta} \\[6pt] \dfrac{\partial}{\partial \zeta} \end{Bmatrix} = \begin{bmatrix} \dfrac{\partial x}{\partial \xi} & \dfrac{\partial y}{\partial \xi} & \dfrac{\partial z}{\partial \xi} \\[6pt] \dfrac{\partial x}{\partial \eta} & \dfrac{\partial y}{\partial \eta} & \dfrac{\partial z}{\partial \eta} \\[6pt] \dfrac{\partial x}{\partial \zeta} & \dfrac{\partial y}{\partial \zeta} & \dfrac{\partial z}{\partial \zeta} \end{bmatrix} \begin{Bmatrix} \dfrac{\partial}{\partial x} \\[6pt] \dfrac{\partial}{\partial y} \\[6pt] \dfrac{\partial}{\partial z} \end{Bmatrix} \qquad \text{or} \qquad \begin{Bmatrix} \dfrac{\partial}{\partial \xi} \end{Bmatrix} = [J] \begin{Bmatrix} \dfrac{\partial}{\partial x} \end{Bmatrix}$$

where $[J]$ is called the *Jacobian operator* relating the natural coordinates to the local coordinates and having the explicit form

$$[J] = \begin{bmatrix} x_{14} & y_{14} & z_{14} \\ x_{24} & y_{24} & z_{24} \\ x_{34} & y_{34} & z_{34} \end{bmatrix}$$

The inverse relation for the derivatives

$$\begin{Bmatrix} \dfrac{\partial}{\partial x} \end{Bmatrix} = [J^{-1}] \begin{Bmatrix} \dfrac{\partial}{\partial \xi} \end{Bmatrix}$$

requires that $[J^{-1}]$ exists. In most cases, the existence is clear; however, in cases where the element is highly distorted or folds back on itself the Jacobian transformation can become singular. The inverse is given by

$$[J^{-1}] = \frac{1}{\det[J]} \begin{bmatrix} y_{24}z_{34} - y_{34}z_{24} & y_{34}z_{14} - y_{14}z_{34} & y_{14}z_{24} - y_{24}z_{14} \\ z_{24}x_{34} - z_{34}x_{24} & z_{34}x_{14} - z_{14}x_{34} & z_{14}x_{24} - z_{24}x_{14} \\ x_{24}y_{34} - x_{34}y_{24} & x_{34}y_{14} - x_{14}y_{34} & x_{14}y_{24} - x_{24}y_{14} \end{bmatrix} = [A]$$

where

$$\det[J] = x_{14}[y_{24}z_{34} - y_{34}z_{24}] + y_{14}[z_{24}x_{34} - z_{34}x_{24}] + z_{14}[x_{24}y_{34} - x_{34}y_{24}] = 6|V|$$

The determinant of the Jacobian is related to the volume of the tetrahedron. Note that $[J]$ and its inverse are constants related just to the coordinates of the tetrahedron.

The six components of strains are related to the displacements by

$$\epsilon_{xx} = \frac{\partial u}{\partial x}, \qquad \epsilon_{yy} = \frac{\partial v}{\partial y}, \qquad \epsilon_{zz} = \frac{\partial w}{\partial z}$$

$$2\epsilon_{yz} = \frac{\partial v}{\partial z} + \frac{\partial w}{\partial y}, \qquad 2\epsilon_{xz} = \frac{\partial u}{\partial z} + \frac{\partial w}{\partial x}, \qquad 2\epsilon_{xy} = \frac{\partial u}{\partial y} + \frac{\partial v}{\partial x}$$

We will place them, in this order, into the vector $\{\epsilon\}$. They are written in terms of the nodal DoF and inverse Jacobian components as

$$\{\epsilon\} = [\,B\,]\{u\}$$

where

$$[\,B\,] = \begin{bmatrix} A_{11} & 0 & 0 & A_{12} & 0 & 0 & A_{13} & 0 & 0 & -\overline{A}_1 & 0 & 0 \\ 0 & A_{21} & 0 & 0 & A_{22} & 0 & 0 & A_{23} & 0 & 0 & -\overline{A}_2 & 0 \\ 0 & 0 & A_{31} & 0 & 0 & A_{32} & 0 & 0 & A_{33} & 0 & 0 & -\overline{A}_3 \\ 0 & A_{31} & A_{21} & 0 & A_{32} & A_{22} & 0 & A_{33} & A_{23} & 0 & -\overline{A}_3 & -\overline{A}_2 \\ A_{31} & 0 & A_{11} & A_{32} & 0 & A_{12} & A_{33} & 0 & A_{13} & -\overline{A}_3 & 0 & -\overline{A}_1 \\ A_{21} & A_{11} & 0 & A_{22} & A_{12} & 0 & A_{23} & A_{13} & 0 & -\overline{A}_2 & -\overline{A}_1 & 0 \end{bmatrix}$$

and $\overline{A}_i = \sum_i A_{ij}$. The stresses are related to the strains by Hooke's law

$$\{\sigma\} = [\,D\,]\{\epsilon\} = [\,D\,][\,B\,]\{u\}$$

with $\{\sigma\}$ organized the same as $\{\epsilon\}$.

The strain energy of the element is

$$\mathcal{U} = \frac{1}{2}\int_V \{\sigma\}^{\mathrm{T}}\{\epsilon\}\,\mathrm{d}V = \frac{1}{2}\int_V \{u\}^{\mathrm{T}}[\,B\,]^{\mathrm{T}}[\,D\,][\,B\,]\{u\}\,\mathrm{d}V$$

Noting that none of the quantities inside the integral depends on the position coordinates, we have for the total potential

$$\Pi = \tfrac{1}{2}\{u\}^{\mathrm{T}}[\,B\,]^{\mathrm{T}}[\,D\,][\,B\,]\{u\}V - \{F\}^{\mathrm{T}}\{u\}$$

The associated equilibrium equation is

$$\{F\} = [\,k\,]\{u\}, \qquad [\,k\,] \equiv [\,B\,]^{\mathrm{T}}[\,D\,][\,B\,]V \qquad (1.28)$$

The $[12 \times 12]$ square matrix $[\,k\,]$ is called the stiffness matrix for the *Constant Strain Tetrahedron* (TET4) element. The coordinate system used to formulate the element (i.e., the local coordinate system) is also the global coordinate system; hence we do not need to do any rotation of the element stiffness before assemblage. The structural stiffness matrix is simply

$$[\,K\,] = \sum_m [\,k\,]_m$$

where the element stiffnesses are suitably augmented to conform to the global system.

One of the most important characteristics of an element is its *convergence* properties, that is, it should converge to the exact result in the limit of small element size. When looking at convergence, it is necessary to change the mesh in a systematic way, preferably each time halving the element size in each dimension. The test problem we will use is that of the annulus shown in Figure 1.17 and loaded with an internal uniformly distributed traction. The mesh shown is made of $[32 \times 4 \times 1]$ modules where each module is a

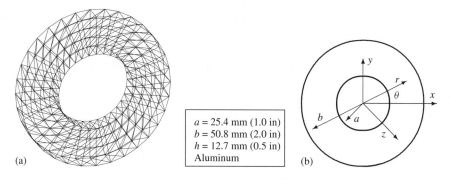

$a = 25.4$ mm (1.0 in)
$b = 50.8$ mm (2.0 in)
$h = 12.7$ mm (0.5 in)
Aluminum

(a) (b)

Figure 1.17: Annulus modeled with tetrahedral elements. (a) Properties of mesh with [32 × 4 × 1] modules. (b) Cylindrical coordinates.

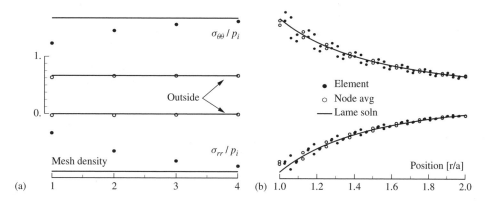

Figure 1.18: Annulus convergence results. (a) Boundary stresses for different mesh densities. (b) Stress distributions for mesh density 3.

hexahedron block made of six elements; the other meshes are similar except that the number of hoop and radial modules was changed by a factor of two each time.

Figure 1.18 shows the convergence behavior for the stresses on the inside and outside of the annulus. As expected, there is monotonic convergence from below. The mesh in Figure 1.17(a) is labeled as a density of "2"; the highest density tested was [128 × 16 × 1] modules, which have 12,288 elements. It is easy to see how the number of elements can escalate as the mesh is refined.

It is interesting that the outside stresses converge to the exact solution much more rapidly than the inside ones. We can get a better insight into what is happening by looking at the distributions.

The exact radial and hoop stresses are given by the Lamé solution for thick cylinders [56]

$$\sigma_{rr} = \frac{p_i}{(1 - a^2/b^2)}\left[\frac{a^2}{b^2} - \frac{a^2}{r^2}\right], \qquad \sigma_{\theta\theta} = \frac{p_i}{(1 - a^2/b^2)}\left[\frac{a^2}{b^2} + \frac{a^2}{r^2}\right]$$

where p_i is the internal pressure. These distributions of stress are shown as the full line in Figure 1.18(b). The tetrahedral element has a constant stress distribution and these are shown as the full dots in Figure 1.18(b) plotted against the element centroidal radial position (they do not all lie on the same radial line). Even for mesh density 3, the values are spread about the exact solution, with the bigger spread occurring at the inside radius.

Typically, in an FEM analysis, the results are reported with respect to the nodal values. Therefore, some rule is needed to convert the element centroidal values into nodal values. A popular rule is to use nodal averages, that is, for a given node, the nodal value is the average value of all the elements attached to that node. These are shown in Figure 1.18(b) plotted as the empty circles, and for the most part they are very close to the exact solution. This rule is least accurate when the elements form a boundary with a stress gradient. In those cases, more sophisticated rules can be used (Reference [35] gives a nice scheme using least squares and the element interpolation functions) or the mesh can be designed to have a higher density in the vicinity of boundaries with stress gradients.

The tetrahedron element satisfies all convergence criteria, and therefore it is suitable for obtaining accurate solutions to all problems. Unfortunately, as just seen, it is usually not a very practical element because it requires too many elements to obtain sufficient accuracy. We now look at some other elements that *a priori* have built into them some structural simplifications.

Beam and Frame Element Stiffness

Consider a straight homogeneous beam element of length L as shown in Figure 1.19. Assume that there are no external loads applied between the two ends of Node 1 and Node 2 (i.e., $q(x) = 0$). At each node, there are two essential beam actions, namely, the bending moment and shear force. The corresponding nodal degrees of freedom are the rotation ϕ (or the slope of the deflection curve at the node) and the vertical displacement v. The positive directions of nodal forces and moments and the corresponding displacements and rotations are shown in the figure.

In this simple structural model, the 3-D distributions of displacement are reduced to the single displacement distribution $v(x)$ for the transverse deflection. We wish to write

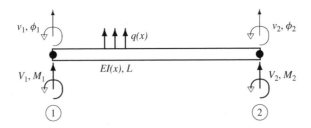

Figure 1.19: Sign convention for beam element.

this distribution in terms of the nodal displacements. There are four nodal values, so the simplest representation is

$$v(x) = a_0 + a_1 x + a_2 x^2 + a_3 x^3$$

where a_0, a_1, a_2, and a_3 are constants. By using the end conditions

$$v(0) = v_1, \qquad \frac{dv(0)}{dx} = \phi_1; \qquad v(L) = v_2, \qquad \frac{dv(L)}{dx} = \phi_2$$

we can rewrite the constants in terms of the nodal displacements v_1 and v_2, and the nodal rotations ϕ_1 and ϕ_2. Substitution of these into the expression for the deflection leads us to

$$v(x) = \left[1 - 3\left(\frac{x}{L}\right)^2 + 2\left(\frac{x}{L}\right)^3 \right] v_1 + \left(\frac{x}{L}\right)\left[1 - 2\left(\frac{x}{L}\right) + \left(\frac{x}{L}\right)^2 \right] L\phi_1$$

$$+ \left(\frac{x}{L}\right)^2\left[3 - 2\left(\frac{x}{L}\right) \right] v_2 + \left(\frac{x}{L}\right)^2\left[-1 + \left(\frac{x}{L}\right) \right] L\phi_2$$

$$\equiv g_1(x)v_1 + g_2(x)L\phi_1 + g_3(x)v_2 + g_4(x)L\phi_2 \qquad (1.29)$$

The functions $g_n(x)$ are called the *beam shape functions*. The complete description of the element is captured in the four nodal degrees of freedom v_1, ϕ_1, v_2, and ϕ_2 (since the shape functions are known explicitly). For example, the curvature is obtained (in terms of the nodal degrees of freedom) simply by differentiation; that is,

$$\frac{d^2 v}{dx^2} = [g_1''(x)v_1 + g_2''(x)L\phi_1 + g_3''(x)v_2 + g_4''(x)L\phi_2]$$

$$= \frac{1}{L^3}[(-6L + 12x)v_1 + (-4L + 6x)L\phi_1 + (6L - 12x)v_2 + (-2L + 6x)L\phi_2]$$

If, in any problem, these degrees of freedom can be determined, then the solution has been obtained.

The strain energy stored in the beam element is (neglecting shear)

$$\mathcal{U} = \frac{1}{2}\int_0^L EI\left[\frac{d^2 v}{dx^2}\right]^2 dx = \frac{1}{2}\int_0^L EI[g_1''(x)v_1 + g_2''(x)L\phi_1 + g_3''(x)v_2 + g_4''(x)L\phi_2]^2 dx$$

The total potential for the problem is therefore

$$\Pi = \frac{1}{2}\int_0^L EI[g_1''(x)v_1 + g_2''(x)L\phi_1 + \cdots]^2 dx - V_1 v_1 - M_1\phi_1 - V_2 v_2 - M_2\phi_2$$

The element stiffness relation is obtained by extremizing the total potential to give, for example,

$$\mathcal{F}_{v_1} = \frac{\partial\Pi}{\partial v_1} = 0 = \int_0^L EIg_1''(x)g_1''(x)\,dx - V_1$$

In general, the elements of the stiffness matrix are

$$k_{ij} = \frac{\partial^2\Pi}{\partial u_i\,\partial u_j} = \int_0^L EIg_i''(x)g_j''(x)\,dx$$

In the present form, these relations are applicable to nonuniform sections. Note that it is the symmetry of the terms $g_i''(x)g_j''(x)$ that insures the symmetry of the stiffness matrix.

Carrying out these integrations under the condition of constant EI gives the beam element stiffness matrix. Then, the nodal loads-displacement relations can be rearranged into the form

$$
\begin{Bmatrix} V_1 \\ M_1 \\ V_2 \\ M_2 \end{Bmatrix} = \{F_e\} = [\ k\ ]\{u\} = \frac{EI}{L^3} \begin{bmatrix} 12 & 6L & -12 & 6L \\ 6L & 4L^2 & -6L & 2L^2 \\ -12 & -6L & 12 & -6L \\ 6L & 2L^2 & -6L & 4L^2 \end{bmatrix} \begin{Bmatrix} v_1 \\ \phi_1 \\ v_2 \\ \phi_2 \end{Bmatrix} \tag{1.30}
$$

where $[\ k\ ]$ is the *beam element stiffness matrix*. Note that this stiffness matrix is symmetric. Also note that the nodal loads satisfy the equilibrium conditions for the free-body diagram of the element.

A typical member of a 3-D space frame can be thought of as having four actions: two bending actions, a twisting action, and an axial action. Thus, the displacement of each node is described by three translational and three rotational components of displacement, giving six degrees of freedom at each unrestrained node. Corresponding to these degrees of freedom are six nodal loads. The notations we will use for the displacement and force vectors at each node are, respectively,

$$
\{u\} = \{u, v, w, \phi_x, \phi_y, \phi_z\}^{\mathrm{T}}, \qquad \{F_e\} = \{F_x, F_y, F_z, M_x, M_y, M_z\}^{\mathrm{T}}
$$

On the local level, for each member, the forces are related to the displacements by the partitioned matrices

$$
\{\overline{F}_e\} = \begin{bmatrix} \overline{k}_{11} & \overline{k}_{12} & \overline{k}_{13} & \overline{k}_{14} \\ \overline{k}_{21} & \overline{k}_{22} & \overline{k}_{23} & \overline{k}_{24} \\ \overline{k}_{31} & \overline{k}_{32} & \overline{k}_{33} & \overline{k}_{34} \\ \overline{k}_{41} & \overline{k}_{42} & \overline{k}_{43} & \overline{k}_{44} \end{bmatrix} \{\overline{u}\} = [\ \overline{k}\ ]\{\overline{u}\}
$$

By way of example, the upper left quadrant is

$$
\begin{bmatrix} \overline{k}_{11} & \overline{k}_{12} \\ \overline{k}_{21} & \overline{k}_{22} \end{bmatrix} = \frac{+1}{L} \begin{bmatrix} EA & 0 & 0 & 0 & 0 & 0 \\ 0 & 12EI_z/L^2 & 0 & 0 & 0 & 6EI_z/L \\ 0 & 0 & 12EI_y/L^2 & 0 & -6EI_y/L & 0 \\ 0 & 0 & 0 & GI_x & 0 & 0 \\ 0 & 0 & -6EI_y/L & 0 & 4EI_y & 0 \\ 0 & 6EI_z/L & 0 & 0 & 0 & 4EI_z \end{bmatrix}
$$

where EA and GI_x are associated with the axial and torsional behaviors respectively. The other quadrants are similar. For example, the first quadrant of $[\ \overline{k}\ ]$ is the same as the fourth quadrant except for the signs of the off-diagonal terms.

The transformation of the components of a vector $\{v\}$ from the local to the global axes is given by

$$
\{\overline{v}\} = [\ R\ ]\{v\}
$$

where [R] is the transformation matrix. The same matrix will transform the vectors of nodal forces and displacements. To see this, note that the element nodal displacement vector is composed of four separate vectors, namely,

$$\{u\} = \{\{u_1, v_1, w_1\}; \{\phi_{x1}, \phi_{y1}, \phi_{z1}\}; \{u_2, v_2, w_2\}; \{\phi_{x2}, \phi_{y2}, \phi_{z2}\}\}$$

Each of these are separately transformed by the [3 × 3] rotation matrix [R]. Hence, the complete transformation is

$$\{\overline{F}_e\} = [\ T\ ]\{F_e\}, \qquad \{\overline{u}\} = [\ T\ ]\{u\}, \qquad [\ T\ ] \equiv \begin{bmatrix} R & 0 & 0 & 0 \\ 0 & R & 0 & 0 \\ 0 & 0 & R & 0 \\ 0 & 0 & 0 & R \end{bmatrix}$$

where [**T**] is a [12 × 12] matrix. Substituting for the barred vectors in the element stiffness relation allows us to obtain the global stiffness as

$$[\ k\ ] = [\ T\ ]^T [\ \overline{k}\ ][\ T\ ]$$

For assemblage, each member stiffness is rotated to the global coordinate system and then augmented to the system size. The structural stiffness matrix is then

$$[\ K\ ] = \sum_m [\ T\ ]_m^T [\ \overline{k}\ ]_m [\ T\ ]_m$$

where the summation is over each member. It is important to realize that the transformation occurs before the assembly. The element stiffnesses just derived are exact for frames with uniform cross-sectional properties and therefore there need not be a discussion of convergence properties. It must be realized, however, that the beam model itself is an approximation to the actual 3-D case, and this discrepancy may need to be taken into account. We give an example of such a situation in Section 5.3.

Triangular Plate and Shell Elements

Plates and shells have one of their dimensions (the thickness) considerably smaller that the other two dimensions. In a manner similar to what is done for rods and beams, the behavior through the thickness is replaced in terms of resultants. We illustrate this for the membrane and flexural behavior.

Consider a triangle divided into three areas where the common apcx is at location (x, y) as shown in Figure 1.20(a). Define

$$h_1 = A_1/A, \qquad h_2 = A_2/A, \qquad h_3 = A_3/A, \qquad h_i = h_i(x, y)$$

with the obvious constraint that $h_1 + h_2 + h_3 = 1$. The areas of these triangles define uniquely the position of the common apex.

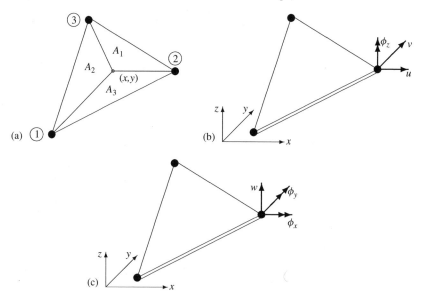

Figure 1.20: Triangular discretizations with nodal degrees of freedom. (a) Area coordinates for a triangle. (b) Membrane DoF. (c) Bending DoF.

The position of a point (x, y) in the triangle can be written as

$$\begin{Bmatrix} 1 \\ x \\ y \end{Bmatrix} = \begin{bmatrix} 1 & 1 & 1 \\ x_1 & x_2 & x_3 \\ y_1 & y_2 & y_3 \end{bmatrix} \begin{Bmatrix} h_1 \\ h_2 \\ h_3 \end{Bmatrix} \tag{1.31}$$

where the subscripts $1, 2, 3$ refer to the counterclockwise nodes of the triangle. We can invert this to get the expressions for the area coordinates

$$\begin{Bmatrix} h_1 \\ h_2 \\ h_3 \end{Bmatrix} = \frac{1}{2A} \begin{bmatrix} x_2 y_3 - x_3 y_2 & y_{23} & x_{32} \\ x_3 y_1 - x_1 y_3 & y_{31} & x_{13} \\ x_1 y_2 - x_2 y_1 & y_{12} & x_{21} \end{bmatrix} \begin{Bmatrix} 1 \\ x \\ y \end{Bmatrix} = \frac{1}{2A} \begin{bmatrix} a_1 & b_1 & c_1 \\ a_2 & b_2 & c_2 \\ a_3 & b_3 & c_3 \end{bmatrix} \begin{Bmatrix} 1 \\ x \\ y \end{Bmatrix} \tag{1.32}$$

with $2A \equiv x_{21} y_{31} - x_{31} y_{21}$, $x_{ij} \equiv x_i - x_j$, and so on. From this, it is apparent that functions of (x, y) can equally well be written as functions of (h_1, h_2, h_3); that is, any function of interest can be written as

$$u(x, y) = \sum_i^3 h_i(x, y) u_i$$

where u_i are the nodal values.

Perhaps the simplest continuum element to formulate is that of the constant strain triangle (CST) element. We therefore begin with a discussion of this element. The basic assumption in the formulation is that the displacements are described by

$$u(x, y) = \sum_i^3 h_i(x, y) u_i, \qquad v(x, y) = \sum_i^3 h_i(x, y) v_i$$

The displacement gradients are given by

$$\frac{\partial u}{\partial x} = \sum \frac{\partial h_i}{\partial x} u_i = \frac{1}{2A} \sum_i b_i u_i , \qquad \frac{\partial v}{\partial x} = \sum \frac{\partial h_i}{\partial x} v_i = \frac{1}{2A} \sum_i b_i v_i$$

$$\frac{\partial u}{\partial y} = \sum \frac{\partial h_i}{\partial y} u_i = \frac{1}{2A} \sum_i c_i u_i , \qquad \frac{\partial v}{\partial y} = \sum \frac{\partial h_i}{\partial y} v_i = \frac{1}{2A} \sum_i c_i v_i$$

where the coefficients b_i and c_i are understood to be evaluated with respect to the original configuration. The strains are

$$\epsilon_{xx} = \frac{\partial u}{\partial x} , \qquad \epsilon_{yy} = \frac{\partial v}{\partial y} , \qquad 2\epsilon_{xy} = \gamma_{xy} = \frac{\partial u}{\partial y} + \frac{\partial v}{\partial x}$$

which are expressed in matrix form as

$$\left\{ \begin{array}{c} \epsilon_{xx} \\ \epsilon_{yy} \\ 2\epsilon_{xy} \end{array} \right\} = \frac{1}{2A} \begin{bmatrix} b_1 & 0 & b_2 & 0 & b_3 & 0 \\ 0 & c_1 & 0 & c_2 & 0 & c_3 \\ c_1 & b_1 & c_2 & b_2 & c_3 & b_3 \end{bmatrix} \left\{ \begin{array}{c} u_1 \\ v_1 \\ u_2 \\ v_2 \\ u_3 \\ v_3 \end{array} \right\} \qquad \text{or} \quad \{\epsilon\} = [B_L]\{u\}$$

The stresses are related to the strains by the plane stress Hooke's law

$$\left\{ \begin{array}{c} \sigma_{xx} \\ \sigma_{yy} \\ \sigma_{xy} \end{array} \right\} = \frac{E}{1 - v^2} \begin{bmatrix} 1 & v & 0 \\ v & 1 & 0 \\ 0 & 0 & (1-v)/2 \end{bmatrix} \left\{ \begin{array}{c} \epsilon_{xx} \\ \epsilon_{yy} \\ 2\epsilon_{xy} \end{array} \right\} \qquad \text{or} \quad \{\sigma\} = [D]\{\epsilon\}$$

The virtual work for a plate in plane stress is

$$\delta W = \int_V \left[\sigma_{xx} \delta\epsilon_{xx} + \sigma_{yy} \delta\epsilon_{yy} + \sigma_{xy} \delta\gamma_{xy} \right] dV \qquad \text{or} \qquad \delta W = \int_V \{\sigma\}^T \delta\{\epsilon\} dV$$

Substituting for the stresses and strains in terms of the degrees of freedom gives

$$\delta W = \int_V \{u\}^T [B_L]^T [D][B_L] \delta\{u\} dV$$

Noting that none of the quantities inside the integral depends on the position coordinates, we have

$$\delta W = \{u\}^T [B_L]^T [D][B_L] \delta\{u\} V$$

The virtual work of the nodal loads is

$$\delta W = \{F_e\}^T \delta\{u\}$$

From the equivalence of the two, we conclude that the nodal forces are related to the degrees of freedom through

$$\{F_e\} = [k]\{u\}, \qquad [k] \equiv [B_L]^T [D][B_L] V \qquad (1.33)$$

The [6 × 6] square matrix [k] is called the stiffness matrix for the *Constant Strain Triangle* (CST) element.

The CST element has only two displacements in its formulation and is generally overstiff in situations involving rotations. A three-noded triangular element (which we will label as MRT) is discussed in Reference [71] and taken from References [25, 26]. This element is shown to have superior in-plane performance over the CST element. The DoF at each node are $\{u\} = \{u, v, \phi_z\}$. A similar element was developed in Reference [9], and a nice comparison of the performance of these elements is given in Reference [135].

The convergence study is for the deep cantilever beam shown in Figure 1.21; the basic mesh is also shown in the figure—other meshes are similar except that the number of modules per side was changed by factors of two.

The results for the tip deflection at B and stresses at A are shown in Figure 1.22 where they are compared with the exact solution. The normalizations are with respect to the exact solution for v_B and σ_A. Both elements show a nice convergence to the exact solution.

The basic idea of the discrete Kirchhoff triangular (DKT) element is to treat the plate element in flexure as composed of a series of plane stress triangular laminas stacked on top of each other. From the previous discussion, we know that we can describe each lamina adequately using the CST or higher element; hence, it then only remains to impose the

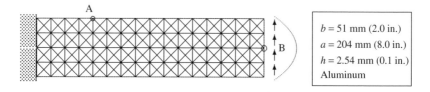

Figure 1.21: Mesh geometry [4 × 16] for a cantilever beam.

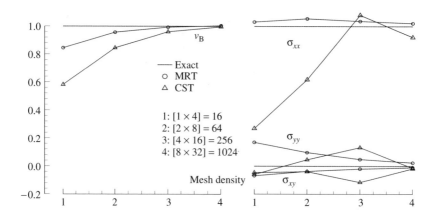

Figure 1.22: Convergence of membrane displacements and stresses.

constraints that the laminas form a structure in flexure. An explicit formulation is given in Reference [20] and coding is given in References [45, 96].

This element, now called the Discrete Kirchhoff Triangular (DKT) element was first introduced by Stricklin, Haisler, Tisdale, and Gunderson in 1968 [162]. It has been widely researched and documented as being one of the more efficient flexural elements. Batoz, Bathe, and Ho [19] performed extensive testing on three different elements including the hybrid stress model (HSM), DKT element, and a selective reduced integration (SRI) element. Comparisons between the different element types were made based on the results from different mesh orientations and different boundary and loading conditions. The authors concluded that the DKT and HSM elements are the most effective elements available for bending analysis of thin plates. Of these two elements, the DKT element was deemed superior to the HSM element based on the comparison between the experimental and theoretical results [19].

We now look at the convergence properties of the DKT element. The problem considered is that of a simply supported plate (with mesh as shown in Figure 1.23) and the transverse load applied uniformly. When looking at convergence, it is necessary to change the mesh in a systematic way. The mesh shown is made of [4 × 8] modules, where each module is made of four elements. The other meshes are variations of this mesh, obtained by uniformly changing the number of modules.

The results are shown in Figure 1.24 where it is compared with the exact solution. The normalizations for the displacements and moments are

$$w_o = \frac{16 q_o b^4}{D \pi^6}, \qquad M_o = \frac{q_o b^2}{4 \pi^2}$$

The DKT element exhibits excellent convergence. It is pleasing to see that it gives good results even for relatively few modules.

The performance of the DKT element for the moments is also very good. These reported moments are nodal averages. Because a lumped load was used in this example, we conclude that a lumped representation is adequate when a suitably refined mesh is used.

In our formulation, the displacement of each node of a space frame or shell is described by three translational and three rotational components of displacement, giving

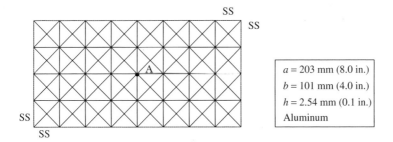

Figure 1.23: Generic [4 × 8] mesh used in the plate flexure convergence study.

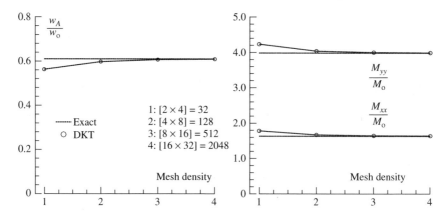

Figure 1.24: Convergence study for displacements and moments.

six degrees of freedom at each unrestrained node. Corresponding to these degrees of freedom are six nodal loads. The notations for the shell are identical to those of the space frame. For each element in local coordinates, the forces are related to the displacements by

$$\{\overline{F}_e\} = \{\overline{u}\} = [\; \overline{k}\;]\{\overline{u}\}$$

where $[\; \overline{k}\;]$ is of size $[18 \times 18]$. It is comprised of a combination of the $[9 \times 9]$ MRT and $[9 \times 9]$ DKT elements. Just as for the frame, we can write the element stiffness of the shell in global coordinates as a transformation of the local stiffness

$$[\; k\;] = [\; T\;]^{\mathrm{T}}[\; \overline{k}\;][\; T\;]$$

Here, however, $[\; T\;]$ is of size $[18 \times 18]$ and its construction is given in Reference [71]. The assemblage procedure is the same as for the frame.

Applied Body and Traction Loads

The applied loads fall into two categories. The first are point forces and moments; these do not need any special treatment since they are already in nodal form. The others arise from distributed loads in the form of body loads such as weight and applied tractions such as pressures; these are the loads of interest here. We want to develop a scheme that converts the distributed loads into equivalent concentrated forces and moments. In this process, this adds another level of approximation to the FEM solution; but keep in mind that any desired degree of accuracy can always be achieved by increasing the number of nodes. (This is at the expense of computing time, but has the advantage of being simple.) The question now is as follows: is there a best way to convert a distributed load into a set of concentrated loads? We will investigate two approximate ways of doing this.

To get the equivalent nodal loads, we will equate the virtual work of the nodal loads to that of the distributed load. The virtual work is the load times the virtual displacement, we will take the displacement as represented (interpolated) by

$$\{\overline{u}(x, y, z)\} = [N(x, y, z)]\{u\}$$

where $\{\overline{u}\}$ is the collection of relevant displacements.

Consider the body forces first. The virtual work is

$$\{P\}^{T}\{\delta u\} = \int_{V} \{\rho\overline{b}(x, y, z)\}^{T}\{\delta\overline{u}(x, y, z)\}\,dV = \int_{V} \{\rho\overline{b}(x, y, z)\}^{T}[N(x, y, z)]\{\delta u\}\,dV$$

This leads to

$$\{P\} = \int_{V} [N(x, y, z)]^{T}\{\rho\overline{b}(x, y, z)\}\,dV$$

If, furthermore, the body force distribution is discretized similar to the displacements, that is,

$$\{\rho\overline{b}(x, y, z)\} = [N(x, y, z)]\{\rho b\}$$

then the equivalent nodal forces are given by

$$\{P\} = \left\{\int_{V} [N(x, y, z)]^{T}[N(x, y, z)]\,dV\right\}\{\rho b\}$$

We will gives a few examples of this.

Consider the beam shown in Figure 1.25. The virtual work of the equivalent system is

$$\delta W_{e} = P_{1}\delta v_{1} + T_{1}\delta\phi_{1} + P_{2}\delta v_{2} + T_{2}\delta\phi_{2}$$

The displacements are interpolated as

$$v(x) = g_{1}(x)v_{1} + g_{2}(x)L\phi_{1} + g_{3}(x)v_{2} + g_{4}(x)L\phi_{2} = [g_{1}, g_{2}L, g_{3}, g_{4}L]\{u\} = [\,N\,]\{u\}$$

If the body force per unit length is uniform such that $\rho b = q_{0} = $ constant, then the integrations lead to

$$\{P\} = \{P_{1}, T_{1}, P_{2}, T_{2}\}^{T} = \tfrac{1}{2}q_{0}AL\{1, \tfrac{1}{6}L, 1, -\tfrac{1}{6}L\}^{T}$$

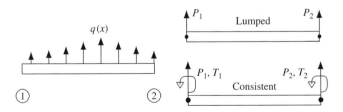

Figure 1.25: Replacement of an arbitrary load distribution by equivalent nodal values.

It is interesting to note the presence of the moments. Consider the triangular membrane element with displacements represented by

$$\left\{ \begin{array}{c} u(x,y) \\ v(x,y) \end{array} \right\} = \frac{1}{2A} \left[\begin{array}{cccccc} h_1 & 0 & h_2 & 0 & h_3 & 0 \\ 0 & h_1 & 0 & h_2 & 0 & h_3 \end{array} \right] \left\{ \begin{array}{c} u_1 \\ \vdots \\ v_3 \end{array} \right\} = [\,N\,]\{u\}$$

Assuming $\rho b_i = q_{i0} = $ constant leads to

$$\{P\} = \{P_{x1}, P_{y1}, \ldots, P_{y3}\}^{\mathrm{T}} = \tfrac{1}{3} Ah\{q_{x0}, q_{y0}; q_{x0}, q_{y0}; q_{x0}, q_{y0}\}^{\mathrm{T}}$$

The other elements are treated in a similar fashion. The next section, when dealing with the mass matrix, shows examples of using linear interpolations on the body forces.

Traction distributions are treated similar to the body forces except that the equivalent loads act only on the loaded section of the element. The virtual work is

$$\{P\}^{\mathrm{T}}\{\delta u\} = \int_A \{\bar{t}(x,y,z)\}^{\mathrm{T}}\{\delta \bar{u}(x,y,z)\}\, \mathrm{d}A = \int_A \{\bar{t}(x,y,z)\}^{\mathrm{T}}[N(x,y,z)]\{\delta u\}\, \mathrm{d}A$$

where A is the loaded surface. This leads to

$$\{P\} = \int_A [N(x,y,z)]^{\mathrm{T}}\{\bar{t}(x,y,z)\}\, \mathrm{d}A$$

If, furthermore, the body force distribution is discretized similar to the displacements, then the equivalent nodal forces are given by

$$\{P\} = \left\{ \int_A [N(x,y,z)]^{\mathrm{T}}[N(x,y,z)]\, \mathrm{d}A \right\} \{t\}$$

We will give just a single example of this.

Consider the triangle shown in Figure 1.26. Let the tractions at the end points be t_{x1}, t_{y1} and so on, and let them be linearly interpolated with the same interpolation functions

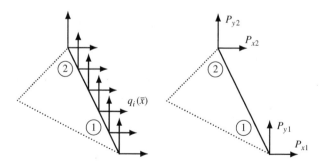

Figure 1.26: Replacement of an arbitrary traction load by equivalent nodal values.

as for the element. Then, the nodal forces are

$$\begin{Bmatrix} P_{x1} \\ P_{y1} \\ P_{x2} \\ P_{y2} \end{Bmatrix} = \frac{hL_{12}}{6} \begin{bmatrix} 2 & 0 & 1 & 0 \\ 0 & 2 & 0 & 1 \\ 1 & 0 & 2 & 0 \\ 0 & 1 & 0 & 2 \end{bmatrix} \begin{Bmatrix} t_{x1} \\ t_{y1} \\ t_{x2} \\ t_{y2} \end{Bmatrix}$$

We see that these replace the load distribution with nodal values that are weighted averages of the distribution. Because the weighting functions are the shape functions for the element displacements this representation is called the *consistent* load representation. Notice that the consistent loads in the case of the beam have moments even though the applied distributed load does not.

The consistent loads are a statically equivalent load system. This can be shown in general, but we demonstrate it with the following special cases. For a uniform distributed load q_0 acting on a beam, for example,

$$P_1 = \tfrac{1}{2} q_0 L, \qquad P_2 = \tfrac{1}{2} q_0 L$$
$$T_1 = +\tfrac{1}{12} q_0 L^2, \quad T_2 = -\tfrac{1}{12} q_0 L^2$$

The resultant vertical force is $P_1 + P_2 = q_0 L$, which is in agreement with the total distributed load. The resultant moment about the first node is $T_1 + T_2 + P_2 L = \tfrac{1}{2} q_0 L^2$ again in agreement with the value from the distributed load.

An alternative loading scheme is to simply lump the loads at the nodes, that is, for the beam, let

$$P_i = \tfrac{1}{2} \int_L q \, dx, \qquad T_i = 0$$

A triangular plate element with a distributed pressure would put one-third of the resultant on each node. As shown in the example of Figure 1.23 for the rectangular plate, this is usually adequate when reasonable mesh refinements are used; the difference in the performance is only noticeable for coarse meshes. There is also another reason for preferring the lumped approach: during nonlinear deformations, the applied moments are nonconservative loads and therefore would require special treatment.

1.7 Dynamic Considerations

We showed earlier that the linearized version of Lagrange's equations leads to

$$[M]\{\ddot{u}\} + [C]\{\dot{u}\} + [K]\{u\} = \{P\} \qquad (1.34)$$

The computer solution of the structural equations of motion is discussed in more detail in later chapters, but it is of value now to consider some of the major problem types originating from this set. The above equations of motion are to be interpreted as a system of differential equations in time for the unknown nodal displacements $\{u\}$, subject to the known forcing histories $\{P\}$, and a set of boundary and initial conditions. The categorizing of problems is generally associated with the type of applied loading. But first we will establish the mass and damping matrices.

Mass and Damping Matrices

By D'Alembert's principle, we can consider the body force loads as comprising the applied force loads and the inertia loads, that is

$$\rho b_i \Rightarrow \rho b_i - \rho \ddot{u}_i - \eta \dot{u}_i$$

Thus, we can do a similar treatment as used for the body forces above. This leads to

$$-\sum_i \int_{V^o} (\rho \ddot{u}_i + \eta \dot{u}_i) \delta u_i \, dV^o \Rightarrow -[\, m\,]\{\ddot{u}\} = -\left[\int_{V^o} \rho [N]^T [N] \, dV^o \right]\{\ddot{u}\}$$

$$\Rightarrow -[\, c\,]\{\dot{u}\} = -\left[\int_{V^o} \eta [N]^T [N] \, dV^o \right]\{\dot{u}\}$$

where $[\, m\,]$ is called the element mass matrix and $[\, c\,]$ is called the element damping matrix. Note that the integrations are over the original geometry. When the shape functions $[\, N\,]$ are the same as used in the stiffness formulation, the mass and damping matrices are called *consistent*.

The assemblage process for the mass and damping matrices is done in exactly the same manner as for the linear elastic stiffness. As a result, the mass and damping matrices will exhibit all the symmetry and bandedness properties of the stiffness matrix. When the structural joints have concentrated masses, we need only amend the structural mass matrix as follows

$$[\, M\,] = \sum_i [m^{(i)}] + \lceil M_c \rfloor$$

where $\lceil M_c \rfloor$ is the collection of joint concentrated masses. This is a diagonal matrix.

Note that the resulting matrix forms for $[\, m\,]$ and $[\, c\,]$ are identical, that is,

$$[\, c\,] = \frac{\eta}{\rho}[\, m\,]$$

This is an example of the damping matrix being proportional to the mass matrix on an element level. For proportional damping at the structural level, we assume

$$[\, C\,] = \alpha[\, M\,] + \beta[\, K\,]$$

where α and β are constants chosen to best represent the physical situation. Note that this relation is not likely to hold for structures composed of different materials. However, for lightly damped structures it can be a useful approximation.

We will establish the mass matrix for the beam. The accelerations are represented by

$$\{\ddot{v}\} = [\, g_1 \;\; g_2 L \;\; g_3 \;\; g_4 L\,] \begin{Bmatrix} \ddot{v}_1 \\ \vdots \\ \ddot{\phi}_2 \end{Bmatrix} \qquad \text{or} \qquad \{\ddot{u}\} = [\, N\,]\{\ddot{u}\}$$

with [N] being a [1 × 4] matrix. The consistent mass matrix for a uniform beam is

$$m_{ij} = \int_0^L \rho A g_i(x) g_j(x)\, dx \qquad \text{or} \qquad [\ m\ ] = \frac{\rho AL}{420} \begin{bmatrix} 156 & 22L & 54 & -13L \\ 22L & 4L^2 & 13L & -3L^2 \\ 54 & 13L & 156 & -22L \\ -13L & -3L^2 & -22L & 4L^2 \end{bmatrix}$$

The mass matrix of the frame is a composition of that of the rod and beam suitably augmented, for example, to [6 × 6] for a plane frame.

For the membrane behavior, we use the shape functions associated with the constant strain triangle. The accelerations are represented as

$$\begin{Bmatrix} \ddot{u} \\ \ddot{v} \end{Bmatrix} = \begin{bmatrix} h_1 & 0 & h_2 & 0 & h_3 & 0 \\ 0 & h_1 & 0 & h_2 & 0 & h_3 \end{bmatrix} \begin{Bmatrix} \ddot{u}_1 \\ \ddot{v}_1 \\ \vdots \\ \ddot{v}_3 \end{Bmatrix} \qquad \text{or} \qquad \{\ddot{u}\} = [\ N\ ]\{\ddot{u}\}$$

Then the mass matrix is

$$[\ m\ ] = \int_{V^o} \rho [\ N\ ]^T [\ N\ ]\, dV^o \Longrightarrow [\ m\ ] = \frac{\rho Ah}{12} \begin{bmatrix} 2 & 0 & 1 & 0 & 1 & 0 \\ 0 & 2 & 0 & 1 & 0 & 1 \\ 1 & 0 & 2 & 0 & 1 & 0 \\ 0 & 1 & 0 & 2 & 0 & 1 \\ 1 & 0 & 1 & 0 & 2 & 0 \\ 0 & 1 & 0 & 1 & 0 & 2 \end{bmatrix}$$

For the bending behavior, let the displacements be represented by

$$\{w(x, y)\} = [\ N\ ]\{u\}, \qquad \{u\} = \{w_1, \phi_{x1}, \phi_{y1}; \dot{w}_2, \phi_{x2}, \phi_{y2}; w_3, \phi_{x3}, \phi_{y3}\}$$

where the shape functions $N(x, y)$ are appropriately selected. Again, we get

$$[\ m\ ] = \int \rho h [\ N\ ]^T [\ N\ ]\, dA$$

The expressions are too lengthy to write here. We will generally find it more beneficial, anyway, to use the lumped mass matrix.

It is useful to realize that because the mass matrix does not involve derivatives of the shape function, we can be more lax about the choice of shape function than for the stiffness matrix. In fact, in many applications, we will find it preferable to use a lumped mass (and damping) approximation where the only nonzero terms are on the diagonal. We show some examples here.

The simplest mass model is to consider only the translational inertias, which are obtained simply by dividing the total mass by the number of nodes and placing this value of mass at the node. Thus, the diagonal terms for the 3-D frame and plate are

$$\lceil\ m\ \rfloor = \tfrac{1}{2}\rho AL \lceil 1, 1, 1, 0, 0, 0; 1, 1, 1, 0, 0, 0 \rfloor$$

$$\lceil\ m\ \rfloor = \tfrac{1}{3}\rho Ah \lceil 1, 1, 1, 0, 0, 0; 1, 1, 1, 0, 0, 0; 1, 1, 1, 0, 0, 0 \rfloor$$

respectively. These neglect the rotational inertias of the flexural actions. Generally, these contributions are negligible and the above are quite accurate especially when the elements are small. There is, however, a very important circumstance when a zero diagonal mass is unacceptable and reasonable nonzero values are needed. Later in this section, we develop an explicit numerical integration scheme where the time step depends on the highest resonant frequencies of the structure; these frequencies in turn are dictated by the rotational inertias—a zero value would imply an infinitesimally small Δt.

First consider the frame. We will use the diagonal terms of the consistent matrix to form an estimate of the diagonal matrix. Note that the translation diagonal terms add up to only $\rho AL312/420$. Hence, by scaling each diagonal term by $420/312$ we get

$$\lceil m \rfloor = \tfrac{1}{2}\rho AL\lceil 1, 1, 1, 2\beta, \beta, \beta; 1, 1, 1, 2\beta, \beta, \beta \rfloor, \qquad \beta = \alpha L^2/40$$

where α is typically taken as unity. This scheme has the merit of correctly giving the translational inertias. We treat the plate in an analogous manner as

$$\lceil m \rfloor = \tfrac{1}{3}\rho Ah\lceil 1, 1, 1, \beta, \beta, 20\beta; 1, 1, 1, \beta, \beta, 20\beta; 1, 1, 1, \beta, \beta, 20\beta \rfloor$$

with $\beta = \alpha L^2/40$. We estimate the effective length $L \approx \sqrt{A/\pi}$ as basically the radius of a disk of the same area as the triangle. Again, α is typically taken as unity.

Types of Linear Dynamic Problems

For *transient* dynamic problems, the applied load $P(t)$ are general functions of time. Typically, the equations require some numerical scheme for integration over time and become computationally intensive in two respects. First, a substantial increase in the number of elements must be used in order to model the mass and stiffness distributions accurately. The other is that the complete system of equations must be solved at each time increment. When the applied load is short lived, the resulting problem gives rise to *free wave propagation*.

For the special case when the excitation force is time harmonic, that is,

$$\{P\} = \{\hat{P}\}e^{i\omega t} \qquad \text{or} \qquad \left\{ \begin{array}{c} P_1 \\ \vdots \\ P_n \end{array} \right\} = \left\{ \begin{array}{c} \hat{P}_1 \\ \vdots \\ \hat{P}_n \end{array} \right\} e^{i\omega t}$$

where ω is the angular frequency and many of the P_n could be zero, then the response is also harmonic and given by

$$\{u\} = \{\hat{u}\}e^{i\omega t} \qquad \text{or} \qquad \left\{ \begin{array}{c} u_1 \\ \vdots \\ u_n \end{array} \right\} = \left\{ \begin{array}{c} \hat{u}_1 \\ \vdots \\ \hat{u}_n \end{array} \right\} e^{i\omega t}$$

This type of analysis is referred to as *forced frequency* analysis. Substituting these forms into the differential equations gives

$$[K]\{\hat{u}\}e^{i\omega t} + i\omega[C]\{\hat{u}\}e^{i\omega t} - \omega^2[M]\{\hat{u}\}e^{i\omega t} = \{\hat{P}\}e^{i\omega t}$$

or, after canceling through the common time factor,

$$\left[[\ K\ ]+i\omega[\ C\ ]-\omega^2[\ M\ ]\right]\{\hat{u}\}=\{\hat{P}\}\qquad\text{or}\qquad[\ \hat{K}\ ]\{\hat{u}\}=\{\hat{P}\}$$

Thus, the solution can be obtained analogous to the static problem; the difference is that the stiffness matrix is modified by the inertia term $\omega^2[\ M\ ]$ and the complex damping term $i\omega[\ C\ ]$. This is the discrete approximation of the dynamic structural stiffness. It is therefore frequency dependent as well as complex. This system of equations is now recognized as the spectral form of the equations of motion of the structure. One approach, then, to transient problems is to evaluate the above at each frequency and use the FFT [31] for time domain reconstructions. This is feasible, but a more full-fledged spectral approach based on the exact dynamic stiffness is developed in References [70].

A case of very special interest is that of free vibrations. When the damping is zero this case gives the mode shapes that are a very important aspect of modal analysis [76, 82]. For free vibrations of the system, the applied loads $\{P\}$ are zero giving the equations of motion as

$$[[\ K\ ]-\omega^2[\ M\ ]]\{\hat{u}\}=0$$

This is a system of homogeneous equations for the nodal displacements $\{\hat{u}\}$. For a nontrivial solution, the determinant of the matrix of coefficients must be zero. We thus conclude that this is an eigenvalue problem, ω^2 are the eigenvalues, and the corresponding $\{\hat{u}\}$ are the eigenvectors of the problem. Note that the larger the number of elements (for a given structure), the larger the system of equations; consequently, the more eigenvalues we can obtain.

In the remainder of this section, and in some of the forthcoming chapters, we will be primarily concerned with the transient dynamic problem.

Time Integration Methods

This section introduces some time integration methods for finding the dynamic response of structures; that is, the dynamic equilibrium equations are integrated directly in a step-by-step fashion.

To construct the central difference algorithm, we begin with finite difference expressions for the nodal velocities and accelerations at the current time t

$$\{\dot{u}\}_t=\frac{1}{2\Delta t}\{u_{t+\Delta t}-u_{t-\Delta t}\},\qquad\{\ddot{u}\}_t=\frac{1}{\Delta t^2}\{u_{t+\Delta t}-2u_t+u_{t-\Delta t}\}\qquad(1.35)$$

Substitute these into the equations of motion written at time t to get

$$[\ K\ ]\{u\}_t+[\ C\ ]\frac{1}{2\Delta t}\{u_{t+\Delta t}-u_{t-\Delta t}\}+[\ M\ ]\frac{1}{\Delta t^2}\{u_{t+\Delta t}-2u_t+u_{t-\Delta t}\}=\{P\}_t$$

We can rearrange this equation so that only quantities evaluated at time $t + \Delta t$ are on the left-hand side

$$\left[\frac{1}{2\Delta t}C + \frac{1}{\Delta t^2}M\right]\{u\}_{t+\Delta t} = \{P\}_t \tag{1.36}$$

$$-\left[K - \frac{2}{\Delta t^2}M\right]\{u\}_t - \left[\frac{1}{\Delta t^2}M - \frac{1}{2\Delta t}C\right]\{u\}_{t-\Delta t}$$

This scheme is therefore explicit. Note that the stiffness is on the right-hand side in the effective load vector; therefore, these equations cannot recover the static solution in the limit of large Δt. Furthermore, if the mass matrix is not positive definite (that is, if it has some zeros on the diagonal) then the scheme does not work, because the square matrix on the left-hand side is not invertible.

The algorithm for the step-by-step solution operates as follows: we start at $t = 0$; initial conditions prescribe $\{u\}_0$ and $\{\dot{u}\}_0$. From these and the equation of motion we find the acceleration $\{\ddot{u}\}_0$ if it is not prescribed. These equations also yield the displacements $\{u\}_{-\Delta t}$ needed to start the computations; that is, from the differences for the velocity and acceleration we get

$$\{u\}_{-\Delta t} = \{u\}_0 - (\Delta t)\{\dot{u}\}_0 + \tfrac{1}{2}(\Delta t)^2\{\ddot{u}\}_0$$

The set of Equations (1.35) and (1.36) are then used repeatedly; the equation of motion gives $\{u\}_{\Delta t}$, then the difference equations give $\{\ddot{u}\}_{\Delta t}$ and $\{\dot{u}\}_{\Delta t}$, and then the process is repeated.

We now derive a different integration scheme by considering the equations of motion at the time step $t + \Delta t$. Assume that the acceleration is constant over the small time step Δt and is given by its average value, that is,

$$\ddot{u}(t) \approx \tfrac{1}{2}(\ddot{u}_t + \ddot{u}_{t+\Delta t}) = \text{constant} = \alpha$$

Integrate this to give the velocity and displacement as

$$\dot{u}(t) = \dot{u}_t + \alpha t, \qquad u(t) = u_t + \dot{u}_t t + \tfrac{1}{2}\alpha t^2$$

We estimate the average acceleration by evaluating the displacement at time $t + \Delta t$ to give

$$\tfrac{1}{2}(\ddot{u}_t + \ddot{u}_{t+\Delta t}) = \alpha = \frac{2}{\Delta t^2}\{u_{t+\Delta t} - u_t - \dot{u}_t \Delta t\}$$

These equations can be rearranged to give difference formulas for the new acceleration and velocity (at time $t + \Delta t$) in terms of the new displacement as

$$\{\ddot{u}\}_{t+\Delta t} = 2\{\alpha\} - \{\ddot{u}\}_t = \frac{4}{\Delta t^2}\{u_{t+\Delta t} - u_t\} - \frac{4}{\Delta t}\{\dot{u}\}_t - \{\ddot{u}\}_t$$

$$\{\dot{u}\}_{t+\Delta t} = \{\alpha\}\Delta t + \{\dot{u}\}_t = \frac{2}{\Delta t}\{u_{t+\Delta t} - u_t\} - \{\dot{u}\}_t \tag{1.37}$$

Substitute these into the equations of motion at the new time $t + \Delta t$ and shift all terms that have been evaluated at time t to the right-hand side to obtain the scheme

$$\left[K + \frac{2}{\Delta t} C + \frac{4}{\Delta t^2} M \right] \{u\}_{t+\Delta t} = \{P\}_{t+\Delta t} + [\ C\ ] \left\{ \frac{2}{\Delta t} u + \dot{u} \right\}_t$$

$$+ [\ M\ ] \left\{ \frac{4}{\Delta t^2} u + \frac{4}{\Delta t} \dot{u} + \ddot{u} \right\}_t \qquad (1.38)$$

The new displacements are obtained by solving this system of simultaneous equations, which makes it an implicit scheme. This scheme is a special case of Newmark's method [18].

The algorithm operates very similar to the explicit scheme: We start at $t = 0$; initial conditions prescribe $\{u\}_0$ and $\{\dot{u}\}_0$. From these and the equations of motion (written at time $t = 0$) we find $\{\ddot{u}\}_0$ if it is not prescribed. Then the above system of equations are solved for the displacement $\{u\}_{\Delta t}$, from which estimates of the accelerations $\{\ddot{u}\}_{\Delta t}$ and the velocities $\{\dot{u}\}_{\Delta t}$ can also be obtained. These are used to obtain current values of the right-hand side. Then solving the equation of motion again yields $\{u\}_{2\Delta t}$, and so on. The solution procedure for $\{u\}_{t+\Delta t}$ is not trivial, but the coefficient matrix need to be reduced to $[\ U\ ]^T [\ D\ ][\ U\ ]$ form only once if Δt and all of the system matrices do not change during the integration. Note that in the limit of large ΔT we recover the static solution.

Explicit Versus Implicit Schemes

The computational cost for the explicit method is approximately

$$\text{cost} = \tfrac{1}{2} N B_m^2 + [2N(2B_m - 1) + N2B]q$$

where q is the number of time increments, B is the semibandwidth of the stiffness matrix, and B_m is the semibandwidth of the mass (and damping) matrix. When the mass matrix is diagonal this reduces to

$$\text{cost} = 2NBq$$

Thus, there is a considerable reduction when the mass and damping matrices are diagonal. By contrast, the computational cost for the implicit method is approximately

$$\text{cost} = \tfrac{1}{2} N B^2 + [2NB + 2N(2B_m - 1)]q$$

where the first term is the cost of the matrix decomposition. When the mass matrix is diagonal this reduces to

$$\text{cost} = \tfrac{1}{2} N B^2 + 2NBq$$

Except for the cost of the initial decomposition, the total cost is the same as for the central difference method.

Numerical integration schemes are susceptible to (numerical) instabilities, a symptom of which is that the solution diverges at each time step. For the central difference algorithm, we require that [71]

$$\Delta t < \frac{2}{\omega} = \frac{T}{\pi} \tag{1.39}$$

where T is the period associated with the frequency ω. Therefore, for numerical stability, the step size must be less than one-third of the period. Hence, the method is only *conditionally stable*, since it is possible for this criterion to not be satisfied in some circumstances. For the average acceleration method, we get that the system is *unconditionally stable* [71], gives no amplitude decay, but will exhibit a phase shift.

The foregoing analysis shows that the computational cost for the explicit central difference scheme can be less than the implicit average acceleration method. Further, the stability criterion of Equation (1.39) will be automatically satisfied for the explicit scheme because of accuracy considerations; that is, this seems easily achieved since it is generally considered [24, 45] that the step size should be less than one-tenth the period for an accurate solution. Hence, we could conclude that the explicit scheme is preferable of the two.

A very important factor was overlooked in the above discussion: a multiple degree of freedom system will have many (vibrational) modes, and when direct integration is used, this is equivalent to integrating each mode with the same time step Δt. Therefore, the above stability criterion must be applied to the highest modal frequency of the system even if our interest is in the low frequency response. In other words, if we energize the system in such a way as to excite only the lower frequencies, we must nonetheless choose an integration step corresponding to the highest possible mode. The significance of this is that the matrix analyses of structures produce so-called *stiff equations*. In the present context, stiff equations characterize a structure whose highest natural vibration frequencies are much greater than the lowest. Especially stiff structures therefore include those with a very fine mesh, and a structure with near-rigid support members. If the conditionally stable algorithm is used for these structures, Δt must be very small, usually orders of magnitude smaller than for the implicit scheme.

In summary, explicit methods are conditionally stable and therefore require a small Δt but produce equations that are cheap to solve. The implicit methods are (generally) unconditionally stable and therefore allow a large Δt but produce equations that are more expensive to solve. The size of Δt is governed by considerations of accuracy rather than stability; that is, we can adjust the step size appropriate to the excitation force or the number of modes actually excited. The difference factor can be orders of magnitude and will invariably outweigh any disadvantage in having to decompose the system matrices. On the basis of these considerations, the implicit scheme is generally the method of choice for general structural dynamics problems that are linear and where the frequency content is relatively low. The explicit scheme is generally the method of choice for wave-type dynamic problems where the frequency content is relatively high. A schematic of this division is shown in Figure 1.27. Additional considerations come into play when we look at nonlinear systems. More discussions can be found in References [24, 71].

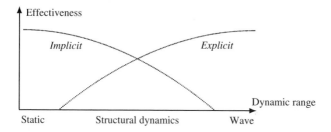

Figure 1.27: Overlap of effectiveness of implicit and explicit integration methods.

1.8 Geometrically Nonlinear Problems

In this section, using some simple examples taken from Reference [71], we lay out the basic difficulties inherent in directly solving nonlinear problems. All nonlinear problems are solved in an incremental/iterative manner with some sort of linearization done at each time or load step; that is, we view the deformation as occurring in a sequence of steps associated with time increments Δt, and at each step it is the increment of displacements that are considered to be the unknowns. We first consider this incremental aspect to the solution.

Incremental Solution Scheme

Consider the loading equation (1.25) at time step t_{n+1}

$$\{\mathcal{F}\}_{n+1} = \{P\}_{n+1} - \{F(u)\}_{n+1} = \{0\}$$

where $\{P\}$ are the applied loads, $\{F(u)\}$ are the deformation dependent element nodal forces, $\{\mathcal{F}\}$ is the vector of out-of-balance forces; all are of size $\{M_u \times 1\}$. For equilibrium, we must have that $\{\mathcal{F}\} = \{0\}$, but as we will see this is not necessarily (numerically) true during an incremental approximation of the solution.

We do not know the displacements $\{u\}_{n+1}$; hence we cannot compute the nodal forces $\{F\}_{n+1}$. As is usual in such nonlinear problems, we linearize about a known state; that is, assume we know everything at time step t_n, then write the Taylor series approximation for the element nodal forces

$$\{F(u)\}_{n+1} \approx \{F(u)\}_n + \left[\frac{\partial F}{\partial u}\right]_n \{\Delta u\} + \cdots = \{F(u)\}_n + [K_T]_n\{\Delta u\} + \cdots \quad (1.40)$$

The square matrix $[K_T]$ is called the *tangent stiffness matrix*. It is banded, symmetric of size $[M_u \times B]$.

We are now in a position to solve for the increments of displacement; rearrange the approximate equilibrium equation into a loading equation as

$$\{P\}_{n+1} - \{F\}_n - [K_T]\{\Delta u\} \approx 0 \implies [K_T]\{\Delta u\} = \{P\}_{n+1} - \{F\}_n$$

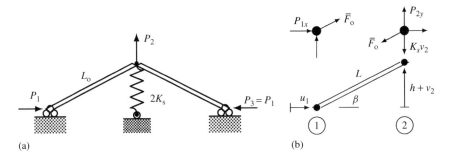

Figure 1.28: Simple pinned truss with a grounded spring. (a) Geometry. (b) Free-body diagrams.

The basic procedure is to compute the increment $\{\Delta u\}$, update the displacement and element nodal forces, and move to the next load level.

To make the ideas explicit, consider the simple truss whose geometry is shown in Figure 1.28. The members are of original length L_o and the unloaded condition has the apex at a height of h. The two ends are on pinned rollers. When h is zero, this problem gives rise to a static instability (buckling); nonzero h acts as a geometric imperfection. This problem is considered in greater depth in Reference [71].

Let the height h be small relative to the member length, and let the deflections be somewhat small; then we have the geometric approximations

$$L_x = L \cos \alpha \approx L_o - u_1, \qquad L_y = L \sin \alpha \approx h + v_2, \qquad u_1 \ll v_2$$

The deformed length of the member is

$$L = \sqrt{(L_o - u_1)^2 + (h + v_2)^2} \approx L_o - u_1 + \frac{h}{L_o} v_2 + \frac{1}{2} L_o \left(\frac{v_2}{L_o} \right)^2$$

The axial force is computed from the axial strain as

$$\overline{F}_o = E A_o \overline{\epsilon} = E A_o \frac{L - L_o}{L_o} = E A_o \left[-\frac{u_1}{L_o} + \frac{h}{L_o} \frac{v_2}{L_o} + \frac{1}{2} \left(\frac{v_2}{L_o} \right)^2 \right]$$

Note that we consider the parameters of the constitutive relation to be unchanged during the deformation.

Look at equilibrium in the deformed configuration as indicated in Figure 1.28; specifically, consider the resultant horizontal force at Node 1 and vertical force at Node 2 giving

$$0 = P_{1x} + \overline{F}_o \cos \beta \approx P_{1x} + \overline{F}_o$$

$$0 = P_{2y} - \overline{F}_o \sin \beta - K_s v_2 \approx P_{2y} - \beta \overline{F}_o - K_s v_2, \qquad \beta \equiv (h + v_2)/L_o$$

Rewrite these in vector form as

$$\begin{Bmatrix} 0 \\ 0 \end{Bmatrix} = \begin{Bmatrix} P_{1x} \\ P_{2y} \end{Bmatrix} + \begin{Bmatrix} \overline{F}_o \\ -\beta \overline{F}_o \end{Bmatrix} - \begin{Bmatrix} 0 \\ K_s v_2 \end{Bmatrix} \qquad \text{or} \qquad \{\mathcal{F}\} = \{P\} - \{F\} = \{0\}$$

We can solve these equations explicitly for the special case $P_{1x} = P$, $P_{2y} = 0$. This leads to $\overline{F}_o = -P$ and the two deflections

$$u_1 = \left[\frac{P}{EA} + \left(\frac{h}{L_o} \right)^2 \frac{P}{K_s L_o - P} + \frac{1}{2} \left(\frac{h}{L_o} \right)^2 \left(\frac{P}{K_s L_o - P} \right)^2 \right] L_o$$

$$v_2 = \left[\frac{h}{L_o} \frac{P}{K_s L_o - P} \right] L_o$$

The load/deflection relations are nonlinear even though the deflections are assumed to be somewhat small. Indeed, when the applied load is close to $K_s L_o$, we get very large deflections. (This is inconsistent with our above stipulation that the deflections are "somewhat small," let us ignore that issue for now and accept the results as indicated.) These results, for different values of h, are shown in Figure 1.29 as the continuous lines. The effect of a decreasing h is to cause the transition to be more abrupt. Also shown are the behaviors for $P > K_s L_o$. These solutions could not be reached using monotonic loading, but they do in fact represent equilibrium states that can cause difficulties for a numerical scheme that seeks the equilibrium path approximately, that is, it is possible to accidentally converge on these spurious equilibrium states.

The tangent stiffness for our truss problem takes the explicit form

$$[K_T]_n = \left[\frac{\partial F}{\partial u} \right]_n = \begin{bmatrix} \dfrac{\partial F_{1x}}{\partial u_1} & \dfrac{\partial F_{1x}}{\partial v_2} \\[2ex] \dfrac{\partial F_{2x}}{\partial u_1} & \dfrac{\partial F_{2x}}{\partial v_2} \end{bmatrix}_n = \begin{bmatrix} \dfrac{\partial \overline{F}_o}{\partial u_1} & -\dfrac{\partial \overline{F}_o}{\partial v_2} \\[2ex] \beta \dfrac{\partial \overline{F}_o}{\partial u_1} & \dfrac{1}{L_o} \overline{F}_o + \beta \dfrac{\partial \overline{F}_o}{\partial u_2} + K_s \end{bmatrix}_n$$

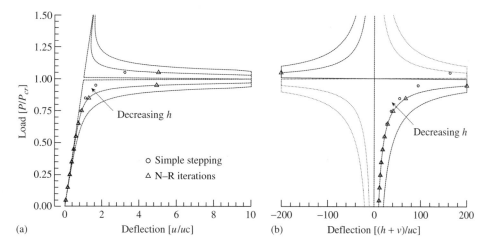

(a)   Deflection [u/uc]

(b)   Deflection [$(h+v)/uc$]

Figure 1.29: Load/deflection behavior for the simple truss. (a) Horizontal displacement u_1. (b) Vertical displacement v_2.

Performing the differentiations

$$\frac{\partial \overline{F}_o}{\partial u_1} = EA\left[-\frac{1}{L_o}\right], \qquad \frac{\partial \overline{F}_o}{\partial v_2} = EA\left[\frac{h}{L_o^2} + \frac{v_2}{L_o^2}\right]$$

then leads to the stiffness

$$[K_T] = \left[\frac{\partial F}{\partial u}\right] = \frac{EA}{L_o}\begin{bmatrix} 1 & -\beta \\ -\beta & \beta^2 + \frac{K_s L_o}{EA} \end{bmatrix} + \frac{\overline{F}_o}{L_o}\begin{bmatrix} 0 & 0 \\ 0 & 1 \end{bmatrix} = [K_E] + [K_G]$$

Note that both matrices are symmetric. The first matrix is the elastic stiffness—the elastic stiffness of a truss member oriented slightly off the horizontal by the angle $\beta = (h + v_2)/L_o$. The second matrix is called the *geometric stiffness matrix* because it arises due to the rotation of the member; note that it depends on the axial load $\overline{F}_o$.

Continuing the special case when $P_{1x} = P$, $P_{2y} = 0$, the system of equations to be solved is

$$\left[\frac{EA}{L_o}\begin{bmatrix} 1 & -\beta \\ -\beta & \beta^2 + \gamma \end{bmatrix} + \frac{\overline{F}_o}{L_o}\begin{bmatrix} 0 & 0 \\ 0 & 1 \end{bmatrix}\right]_n \begin{Bmatrix} \Delta u_1 \\ \Delta v_2 \end{Bmatrix} = \begin{Bmatrix} P \\ 0 \end{Bmatrix}_{n+1} - \begin{Bmatrix} -\overline{F}_o \\ -\beta\overline{F}_o + K_s v_2 \end{Bmatrix}_n$$

with $\gamma = K_s L_o/EA$. A simple solution scheme, therefore, involves computing the increments at each step and updating the displacements as

$$u_{1(n+1)} = u_{1(n)} + \Delta u_1, \qquad v_{2(n+1)} = v_{2(n)} + \Delta v_2$$

The axial force and orientation β also need to be updated as

$$\overline{F}_o|_{n+1} = EA\left[-\frac{u_1}{L_o} + \frac{h}{L_o}\frac{v_2}{L_o} + \frac{1}{2}\left(\frac{v_2}{L_o}\right)^2\right]_{n+1}, \qquad \beta_{n+1} = \frac{h + v_2}{L_o}\Big|_{n+1}$$

Table 1.3 and Figure 1.29 shows the results using this simple stepping scheme where $P_{cr} = K_s L_o$ and $u_{cr} = P_{cr}/EA$.

Table 1.3 also shows the out-of-balance force

$$\{\mathcal{F}\}_{n+1} = \{P\}_{n+1} - \{F\}_{n+1}$$

computed at the end of each step. Clearly, nodal equilibrium is not being satisfied and it deteriorates as the load increases. In order for this simple scheme to give reasonable results, it is necessary that the increments be small. This can be computationally prohibitive for large systems because at each step, the tangent stiffness must be formed and decomposed. A more refined incremental version that uses an iterative scheme to enforce nodal equilibrium will now be developed.

Newton–Raphson Equilibrium Iterations

The increments in displacement are obtained by solving

$$\{K_T\}_n\{\Delta u\} = \{P\}_{n+1} - \{F\}_n$$

Table 1.3: Incremental results using simple stepping.

P/P_{cr}	u_1/u_{cr}	$(h + v_2)/u_{cr}$	P	$\mathcal{F}_{1x}$	$\mathcal{F}_{2y}$
0.0500	0.051000	10.5000	100.00	0.050003	0.989497E−02
0.1500	0.153401	11.6315	300.00	0.256073	0.446776E−01
0.2500	0.257022	13.1329	500.00	0.450836	0.589465E−01
0.3500	0.362372	15.0838	700.00	0.761169	0.759158E−01
0.4500	0.470656	17.7037	900.00	1.37274	0.100333
0.5500	0.584416	21.3961	1100.0	2.72668	0.137041
0.6500	0.709589	26.9600	1300.0	6.19165	0.192208
0.7500	0.862467	36.1926	1500.0	17.0481	0.257334
0.8500	1.09898	53.8742	1700.0	62.5275	0.938249E−01
0.9500	1.66455	94.1817	1900.0	324.939	-4.00430
1.050	3.23104	163.411	2100.0	958.535	-24.0588

from which an estimate of the displacements is obtained as

$$\{u\}_{n+1} \approx \{u\}_n + \{\Delta u\}$$

As just pointed out, if these estimates for the new displacements are substituted into Equation (1.8) then this equation will not be satisfied because the displacements were obtained using only an approximation of the nodal forces given by Equation (1.40). What we can do, however, is repeat the above process at the same applied load level until we get convergence; that is, with i as the iteration counter, we repeat

$$\text{solve:} \quad \{K_T\}_{n+1}^{i-1}\{\Delta u\}^i = \{P\}_{n+1} - \{F\}_{n+1}^{i-1}$$

$$\text{update:} \quad \{u\}_{n+1}^i = \{u\}_{n+1}^{i-1} + \{\Delta u\}^i$$

$$\text{update:} \quad \{K_T\}_{n+1}^i, \quad \{F\}_{n+1}^i$$

until $\{\Delta u\}^i$ becomes less than some tolerance value. The iteration process is started (at each increment) using the starter values

$$\{u\}_{n+1}^0 = \{u\}_n, \quad \{K_T\}_{n+1}^0 = \{K_T\}_n, \quad \{F\}_{n+1}^0 = \{F\}_n$$

This basic algorithm is known as the *full Newton–Raphson method*.

Combined incremental and iterative results are also given in Figure 1.29. We see that it gives the exact solution. Iteration results for a single load step equal to 0.95 P_{cr} are given in Table 1.4 in which the initial guesses correspond to the linear elastic solution. We see that convergence is quite rapid, and the out-of-balance forces go to zero.

It is worth pointing out the converged value above P_{cr} in Figure 1.29; this corresponds to a vertical deflection where the truss has "flipped" over to the negative side. Such a situation would not occur physically, but does occur here due to a combination of linearizing the problem (i.e., the small angle approximation) and the nature of the iteration process (i.e., no restriction is placed on the size of the iterative increments).

Table 1.4:   Newton–Raphson iterations for a load step $0.95\,P_{cr}$.

i	u_1	v_2	$u_1 - u_{1ex}$	$v_2 - v_{2ex}$	$\mathcal{F}_{1x}$	$\mathcal{F}_{2y}$
1	−3.02304	76.0009	−3.12184	72.2009	10.3207E+7	−0.244E+7
2	29.0500	75.9986	28.9512	72.1986	0.21923	−72.3657
3	−25.8441	3.95793	−25.9429	0.157933	0.2594E+7	−107896.
4	0.105242	3.95793	0.64416E-2	0.157926	0.6835E-2	−0.158213
5	0.09867	3.80001	−0.12436E-3	0.8344E-5	12.4694	−0.498780
6	0.09880	3.80000	0.15646E-6	0.4053E-5	0.000000	−0.3948E-5
7	0.09880	3.80000	0.000000	0.000000	−0.4882E-3	0.1878E-4
exact	0.09880	3.80000				

The full Newton–Raphson method has the disadvantage that during each iteration, the tangent stiffness matrix must be formed and decomposed. The cost of this can be quite prohibitive for large systems. Thus, effectively, the computational cost is like that of the incremental solution with many steps. It must be realized, however, that because of the quadratic convergence, six Newton–Raphson iterations, say, is more effective than six load increments.

The *modified Newton–Raphson method* is basically as above except that the tangent stiffness is not updated during the iterations but only after each load increment. This generally requires more iterations and sometimes is less stable, but it is less computationally costly.

Both schemes are illustrated in Figure 1.30 where the starting point is from the zero load state—it is clear why the modified method will take more iterations. The plot for the modified method has the surprising implication that we do not need to know the actual tangent stiffness in order to compute correct results—what must be realized in the incremental/iterative scheme is that we are imposing equilibrium (iteratively) in terms

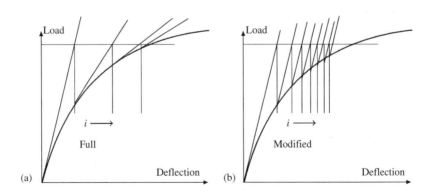

Figure 1.30: Full and modified Newton–Raphson methods for a single load step. (a) The tangent stiffness is updated at each iteration. (b) The tangent stiffness remains the same at each iteration.

of the applied loads and resultant nodal forces; we need good quality *element* stiffness matrices in order to get the good quality element nodal forces, but the assembled tangent stiffness matrix is used only to suggest a direction for the iterative increments; that is, to get the correct converged results we need to have good element stiffness relations, but not necessarily a good assembled tangent stiffness matrix. Clearly, however, a good quality tangent stiffness will give more rapid convergence as well as increase the radius of convergence.

The Corotational Scheme

A particularly effective method for handling the analysis of structures, which undergo large deflections and rotations is the corotational scheme [23, 46, 71]. In this, a local coordinate system moves with each element, and constitutive relations and the like are written with respect to this coordinate system. Consequently, for linear materials, all of the nonlinearities of the problem are shifted into the description of the moving coordinates. We will review just the essentials of the scheme here.

The main objectives are establishing the relation between the local variables and the global variables, from which we can establish the stiffness relations in global variables. To help in the generalization, assume that each element has N nodes with three components of force at each node.

There are $2N$ position variables we are interested in: the global position of each node before deformation ($\hat{x}_{oj}$), and after deformation ($\hat{x}_j$), where the subscript j enumerates the nodes as illustrated in Figure 1.31 for a triangular element. The local positions are given by

$$\hat{\bar{x}}_{oj} = [E_o^T]\{\hat{x}_{oj} - \hat{x}_{o1}\}, \qquad \hat{\bar{x}}_j = [E^T]\{\hat{x}_j - \hat{x}_{o1}\}$$

where $[E_o]$ and $[E]$ are the triads describing the orientation of the element before and after deformation, respectively. The local displacements are defined as

$$\hat{\bar{u}}_j = \hat{\bar{x}}_j - \hat{\bar{x}}_{oj}$$

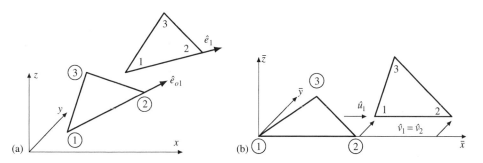

Figure 1.31: The corotational concept. (a) Global positions of an element. (b) Local positions of an element.

The local virtual displacements are related to the global variables by

$$\delta\{\hat{\bar{u}}\}_j = \delta\{[E^T]\{\hat{x}_j - \hat{x}_{o1}\} - \hat{\bar{x}}_{oj}\} = [E^T]\{\delta\hat{u}_j\} + \delta[E^T]\{\hat{x}_j - \hat{x}_{o1}\}$$

Note that, for some vector $\hat{v}$,

$$\delta[E^T]\{v\} = [E^T][S(\hat{v})]\{\delta\beta\}$$

where $\{\beta\}$ is the small rotation spin and the notation $[S(\hat{v})]$ means the skew-symmetric rotation matrix [71] obtained with the components of the corresponding vector; that is, put the components of the vector $\hat{v}$ (i.e., v_x, v_y, v_z) into the rotation array. Using this, we get

$$\{\delta\hat{\bar{u}}\}_j = [E^T]\{\delta u\}_j + [E^T][S(\hat{x}_j - \hat{x}_{o1})]\{\delta\beta\}$$

This shows how the local displacements $\{\delta\bar{u}\}_j$, global spin $\{\delta\beta\}$, and global displacements $\{\delta u\}_j$ are interrelated. However, the spin is not independent of the displacements because we require that the local spin be zero (since it is rotating with the element).

Establishing this constraint [71] then leads to

$$\{\delta\bar{u}\} = [T]\{\delta u\}, \qquad [T] = [P][E^T]$$

where we refer to $[P]$ as a *projector matrix*; it depends only on the local coordinates. Discussions of the projector matrix can be found in References [130, 134, 135].

The virtual work in global variables must equal the virtual work in local variables, hence

$$\{F_e\}^T\{\delta u\} = \{\overline{F}\}^T\{\delta\bar{u}\} = \{\overline{F}\}^T[T]\{\delta u\}$$

From this we conclude that

$$\{F_e\} = [T]^T\{\overline{F}\} = [E][P]^T\{\overline{F}_e\}$$

What this highlights is the possibility (at least when equilibrium is achieved as part of an iterative process) that local quantities (including the stiffness matrix) have a simple coordinate rotation relation to their global counterparts. We will keep that in mind when we develop the stiffness relations.

At the global level, the variation of the nodal forces leads to

$$\{\delta F_e\} = \left[\frac{\partial F_e}{\partial u}\right]\{\delta u\} = [K_T]\{\delta u\}$$

where $[K_T]$ is the element tangent stiffness matrix in global coordinates. Substituting for $\{F_e\}$ in terms of local variables, we get

$$\{\delta F_e\} = [T]^T\{\delta\overline{F}_e\} + \delta[T]^T\{\overline{F}_e\} = [T]^T[\bar{k}]\{\delta\bar{u}\} + \delta[T]^T\{\overline{F}_e\}$$

We see that the tangent behavior is comprised of two parts: one is related to the elastic stiffness properties of the element, the other is related to the rotation of the element. Replacing the local variables in terms of the global variables gives

$$\{\delta F_e\} = [\ T\ ]^{\mathrm{T}}[\ \bar{k}\ ][\ T\ ]\{\delta u\} + [\cdots]\{\delta u\}$$

Each term is postmultiplied by $\{\delta u\}$, and therefore we can associate each term with a stiffness matrix. The latter contribution (which is relatively complex) is the geometric stiffness matrix.

The first set of terms

$$[\ k_E\ ] \equiv [\ T\ ]^{\mathrm{T}}[\ \bar{k}\ ][\ T\ ] = [\ E\ ][\ P\ ]^{\mathrm{T}}[\ \bar{k}\ ][\ P\ ][\ E\ ]^{\mathrm{T}}$$

gives the elastic stiffness. We therefore recognize the global stiffness matrix as the components of the local stiffness matrices transformed to the current orientation of the element. However, it is not just the local stiffness itself but the local stiffness times the projector matrix. The remaining set of terms gives the geometric contribution to the stiffness matrix. The contributions to the geometric stiffness reduce to

$$[\ k_G\ ] = [\ E\ ][[\bar{k}_{G1}] + [\bar{k}_{G2}] + [\bar{k}_{G3}]][\ E^{\mathrm{T}}\ ]$$

The core terms are referred only to the local coordinates.

An important point learned from the earlier simple results is that the more accurate the tangent stiffness, the better the convergence rate, but a consequence of using Newton–Raphson equilibrium iterations is that it is not essential that the actual exact tangent stiffness be used. Consequently, if it is convenient to approximate the tangent stiffness, then the basic nonlinear formulation is not affected, only the convergence rate (and radius of convergence) of the algorithm is affected. With this in mind, the geometric stiffness for a frame with member loads $\bar{F}_o$, $\bar{V}_o$, $\bar{V}_o$, is

$$[\bar{k}_G]_{11} = \frac{\bar{F}_o}{L}\begin{bmatrix} 0 & 0 & 0 \\ 0 & 1 & 0 \\ 0 & 0 & 1 \end{bmatrix} + \frac{\bar{V}_o}{L}\begin{bmatrix} 0 & -1 & 0 \\ -1 & 0 & 0 \\ 0 & 0 & 0 \end{bmatrix} + \frac{\bar{W}_o}{L}\begin{bmatrix} 0 & 0 & -1 \\ 0 & 0 & 0 \\ -1 & 0 & 0 \end{bmatrix}$$

with the remainder of the $[6 \times 6]$ array given by

$$[\bar{k}_G]_{12} = [\bar{k}_G]_{21} = -[\bar{k}_G]_{22} = -[\bar{k}_G]_{11}$$

The contributions associated with the rotational DoF are zero. The $[3 \times 3]$ submatrix of the $[9 \times 9]$ geometric stiffness for a plate with in-plane loads $\bar{N}_{xx}$, $\bar{N}_{xy}$, $\bar{N}_{yy}$, is

$$[\bar{k}_G]_{ij} = \frac{\bar{N}_{xx}}{4A}\begin{bmatrix} 0 & 0 & 0 \\ 0 & 0 & 0 \\ 0 & 0 & b_i b_j \end{bmatrix} + \frac{\bar{N}_{xy}}{4A}\begin{bmatrix} 0 & 0 & 0 \\ 0 & 0 & 0 \\ 0 & 0 & b_i c_j + b_j c_i \end{bmatrix} + \frac{\bar{N}_{yy}}{4A}\begin{bmatrix} 0 & 0 & 0 \\ 0 & 0 & 0 \\ 0 & 0 & c_i c_j \end{bmatrix}$$

where the coefficients are those of the triangular interpolations Again, the contributions associated with the rotational DoF are zero. Thus, in both cases, we ignore flexural effects on the geometric stiffness and say it depends only on the axial/membrane behavior.

Nonlinear Dynamics

When inertia effects are significant in nonlinear problems, as in such problems as wave propagation, we have the choice of using either an explicit integration scheme or an implicit scheme. We will see that the explicit scheme is simpler to implement (primarily because it does not utilize the tangent stiffness matrix) and therefore is generally the more preferable of the two.

We assume the solution is known at time t and wish to find it at time $t + \Delta t$. The applied loads in this case include the inertia and damping forces. For the explicit scheme, we begin with the dynamic equilibrium equation at time t

$$[\,M\,]\{\ddot{u}\}_t + [\,C\,]\{\dot{u}\}_t = \{P\}_t - \{F\}_t$$

Using difference expressions for the nodal velocities and accelerations at the current time leads to

$$\left[\frac{1}{2\Delta t}C + \frac{1}{\Delta t^2}M\right]\{u\}_{t+\Delta t} = \{P\}_t - \{F\}_t \tag{1.41}$$

$$- \left[-\frac{2}{\Delta t^2}M\right]\{u\}_t - \left[\frac{1}{\Delta t^2}M - \frac{1}{2\Delta t}C\right]\{u\}_{t-\Delta t}$$

Generally, the mass and damping matrices are arranged to be diagonal and hence the solution for the displacements is a trivial computational effort (a simple division). Furthermore, there are no equilibrium iteration loops as will be needed in the implicit schemes.

A very important feature of this scheme is that the structural stiffness matrix need not be formed [22]; that is, we can assemble the element nodal forces directly for each element. This is significant in terms of saving memory and also has the advantage that the assemblage effort can be easily distributed on a multiprocessor computer. Furthermore, since the nodal forces $\{F\}$ are obtained directly for each element, there is flexibility in the types of forces that can be allowed. For example, Reference [175] uses a cohesive stress between element sides in order to model the dynamic fracture of brittle materials. In this context, we can also model deformation dependent loading effects (such as follower forces or sliding friction) without changing the basic algorithm.

As indicated in Figure 1.27, our meaning of structural dynamics is the frequency range that begins close to zero and extends a moderate way, perhaps 5 to 50 of the modal frequencies. Since this encompasses nearly static problems, we see that we must use some sort of implicit integration scheme that utilizes the tangent stiffness [71]. The dynamic equilibrium equation at time $t + \Delta t$ is

$$[\,M\,]\{\ddot{u}\}_{t+\Delta t} + [\,C\,]\{\dot{u}\}_{t+\Delta t} = \{P\}_{t+\Delta t} - \{F\}_{t+\Delta t} \tag{1.42}$$

The element nodal loads are not known but, as was done in the static case, can be approximated as

$$\{F\}_{t+\Delta t} \approx \{F\}_t + \{\Delta F\} = \{F\}_t + [\,K_{\mathrm{T}}\,]\{\Delta u\}, \qquad [\,K_{\mathrm{T}}\,] \equiv \left[\frac{\partial F}{\partial u}\right] \tag{1.43}$$

where $\{\Delta F\}$ is the increment in element nodal loads due to an increment in the displacements. Using the finite differences of the constant acceleration method allows the increment in displacements to be obtained by solving

$$\left[K_T + \frac{4}{\Delta t} C + \frac{4}{\Delta t^2} M \right] \{\Delta u\} = \{P\}_{t+\Delta t} - \{F\}_t \qquad (1.44)$$

$$+ [\, C \,] \left\{ \frac{2}{\Delta t} u + \dot{u} \right\}_t + [\, M \,] \left\{ \frac{4}{\Delta t^2} u + \frac{4}{\Delta t} \dot{u} + \ddot{u} \right\}_t$$

Just as in the static case, we need Newton–Raphson iterative improvements to the solution; these are obtained by

$$\left[K_T + \frac{4}{\Delta t} C + \frac{4}{\Delta t^2} M \right]^{i-1} \{\Delta u\}^i = \{P\}_{t+\Delta t} - \{F\}_{t+\Delta t}^{i-1} \qquad (1.45)$$

$$+ [\, C \,] \left\{ \frac{-2}{\Delta t} \left(u_{t+\Delta t}^{i-1} - u_t \right) + \dot{u} \right\}_t$$

$$+ [\, M \,] \left\{ \frac{-4}{\Delta t^2} \left(u_{t+\Delta t}^{i-1} - u_t \right) + \frac{4}{\Delta t} \dot{u} + \ddot{u} \right\}_t$$

The iteration process is started (at each increment) using the starter values

$$\{u\}_{t+\Delta t}^0 = \{u\}_t , \qquad [\, K_T \,]_{t+\Delta t}^0 = [\, K_T \,]_t , \qquad \{F\}_{t+\Delta t}^0 = \{F\}_t$$

This basic algorithm is essentially the same as for the static case.

The implicit integration scheme, when applied to nonlinear problems, has a step size restriction much more restrictive than for the corresponding linear problem [71]. Schemes for improving the algorithm are discussed in References [46, 84, 111], but all require additional computations at the element level and therefore are not readily adapted to general-purpose finite element programs. The best that can be done at present is to always test convergence with respect to step size and tolerance.

As done for linear problems, generally, we take the damping as proportional in the form

$$[\, C \,] = \alpha [\, M \,] + \beta [\, K \,]$$

The easiest way to see the effect of damping is to observe the attenuation behavior of a wave propagating in a structure. From this it becomes apparent that when $\alpha \neq 0$, $\beta = 0$ (mass proportional), the low frequencies are attenuated, whereas, when $\alpha = 0$, $\beta \neq 0$ (stiffness proportional), the high frequencies are attenuated.

These results have an important implication for our choice of time integration scheme. If we choose an explicit integration scheme, then the stiffness matrix is not readily available and we must use a mass proportional damping. Consequently, the high-frequency components will not be attenuated. If, in order to get some sort of high-frequency attenuation, [C] is approximated somehow in terms of an initial stiffness matrix, then we loose the advantage of dealing with diagonal matrices, and the computational cost increases significantly.

1.9 Nonlinear Materials

Figure 1.32 shows two examples of nonlinear behavior: one is nonlinear elastic, and the other is elastic/plastic. We elaborate on both of these in this section.

The key to the FEM formulation for nonlinear materials is to write the element stiffness relations for the increment of strain rather than the total strain. In this sense, it has a good deal in common with the incremental formulation of the geometrically nonlinear problem. For the explicit integration scheme where equilibrium is imposed at step n to get the displacement at $n + 1$, we need to know the element nodal forces $\{F\}_n$. These are obtained as part of the updating after the new displacements are computed. For example, for each element

$$\{F_e\}_n = \{F_e\}_{n-1} + \{\Delta F_e\} = \{F_e\}_{n-1} + [K_T]\{u_n - u_{n-1}\}$$

If the stresses change significantly over the step, then $\{\Delta F_e\}$ is obtained by integrating over the step. For the implicit integration scheme, $\{u\}_{n+1}$ and $\{F_e\}_{n+1}$ are obtained simultaneously as part of the Newton–Raphson iterations.

The constitutive description is simplest in the corotational context (because the rotations are automatically accounted for) and that is the one used here. We consider both nonlinear elastic and elastic-plastic materials and the task is to establish $[\,\bar{k}\,]$.

Hyperelastic Materials

Hyperelastic materials are elastic materials for which the current stresses are some (non-linear) function of the current total strains; a common example is the rubber materials. For these materials, the stresses are derivable from an elastic potential, which can be written (from Equation (1.18)) for small strain as

$$\sigma_{ij} = \frac{\partial \mathcal{U}}{\partial \epsilon_{ij}}$$

A simple 1-D example is

$$\mathcal{U} = E[\tfrac{1}{2}\epsilon^2 - \tfrac{1}{6}\epsilon^3/\epsilon_o], \qquad \sigma = f(\epsilon) = E[\epsilon - \tfrac{1}{2}\epsilon^2/\epsilon_o], \qquad E_T = E[1 - \epsilon/\epsilon_o]$$

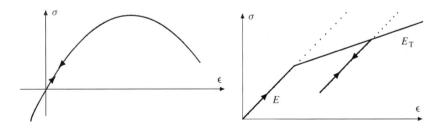

Figure 1.32: Examples of nonlinear material behavior. (a) Elastic. (b) Elastic-plastic.

and is depicted in Figure 1.32(a). Even this simple example shows a characteristic of nonlinear materials (and nonlinear problems in general), namely, the possibility of (static) instabilities. This would occur when $\epsilon = \epsilon_o$ giving a zero tangent modulus and hence a zero tangent stiffness; we will not be considering these cases.

In an incremental formulation over many steps n, given $\{\Delta u\}$, we compute the increment of strain as

$$\{\Delta \epsilon\} = \{\epsilon_{n+1} - \epsilon_n\}, \qquad \epsilon_{n+1} = \epsilon(u_{n+1}) \quad \epsilon_n = \epsilon(u_n)$$

It is necessary to store the previous strain $\{\epsilon\}_n$ at each time step. The tangent modulus relationship can be written, in general, as

$$\Delta \sigma_{ij} = \sum_{p,q} \frac{\partial^2 \mathcal{U}}{\partial \epsilon_{ij} \partial \epsilon_{pq}} \Delta \epsilon_{pq} = \sum_{p,q} D_{Tijpq} \Delta \epsilon_{pq} \qquad \text{or} \qquad \{\Delta \sigma\} = [D_T]\{\Delta \epsilon\}$$

Note that the fourth-order tensor $[D_T]$ is symmetric but generally not constant, being a function of the current strains. We need it to form the tangent stiffness matrix for an element, but the stresses (when needed) will be computed from the total strain relationship. The element stiffness in local coordinates is

$$[\bar{k}_T] = \int_V [\,B\,]^T [D_T][\,B\,] \, dV$$

and therefore has the same structure as the corresponding linear element—only the appropriate tangent modulus need be used. For example, the nonlinear rod has

$$[\bar{k}_T] = \frac{\hat{E}_T A}{L} \begin{bmatrix} 1 & -1 \\ -1 & 1 \end{bmatrix}, \qquad \hat{E}_T \equiv \int_{\epsilon}^{\epsilon + \Delta \epsilon} E_T(\epsilon) \, d\epsilon$$

If $\Delta \epsilon$ is large, then numerical methods may be used to compute the integral.

Elastic-Plastic Materials

Thin-walled structures, for example, have a plane state of stress where σ_{zz}, σ_{xz}, and σ_{yz} are all approximately zero; we will derive explicit elastic-plastic relations for this case.

Plasticity is a phenomenon that occurs beyond a certain stress level called the yield stress. The von Mises yield function, for example, is

$$f = [\sigma_{xx}^2 + \sigma_{yy}^2 - \sigma_{xx}\sigma_{yy} + 3\sigma_{xy}^2]^{1/2} - \sigma_Y \equiv \bar{\sigma} - \sigma_Y$$

where σ_Y is the yield stress for a uniaxial specimen. In this relation, $\bar{\sigma}$ is called the *effective stress*, which is a scalar measure of the level of stress. We would like to have a similar effective measure of the plastic strain.

The basic assumptions of the Prandtl–Reuss theory of plasticity [99, 101] is that plasticity is a distortion (no volume change) dominated phenomenon and that the plastic strain increments are proportional to the current stress deviation. Thus

$$\frac{\Delta \epsilon_{xx}^p}{\sigma_{xx} - \sigma_m} = \frac{\Delta \epsilon_{yy}^p}{\sigma_{yy} - \sigma_m} = \frac{\Delta \epsilon_{zz}^p}{\sigma_{zz} - \sigma_m} = \frac{\Delta \epsilon_{xy}^p}{\sigma_{xy}} = \frac{\Delta \epsilon_{yz}^p}{\sigma_{yz}} = \frac{\Delta \epsilon_{zx}^p}{\sigma_{zx}} = d\lambda \qquad (1.46)$$

where

$$\sigma_m = (\sigma_{xx} + \sigma_{yy} + \sigma_{zz})/3$$

is the mean hydrostatic stress. Specializing this for plane stress gives

$$\left\{ \begin{matrix} \Delta\epsilon^p_{xx} \\ \Delta\epsilon^p_{yy} \\ 2\Delta\epsilon^p_{xy} \end{matrix} \right\} = \frac{d\lambda}{3} \left\{ \begin{matrix} 2\sigma_{xx} - \sigma_{yy} \\ 2\sigma_{yy} - \sigma_{xx} \\ 6\sigma_{xy} \end{matrix} \right\} = \frac{d\lambda}{3} \{ s \}, \qquad \left\{ \begin{matrix} \sigma_{xx} \\ \sigma_{yy} \\ \sigma_{xy} \end{matrix} \right\} = \frac{1}{d\lambda} \left\{ \begin{matrix} 2\Delta\epsilon^p_{xx} + \Delta\epsilon_{yy} \\ 2\Delta\epsilon^p_{yy} + \Delta\epsilon_{xx} \\ \Delta\epsilon^p_{xy} \end{matrix} \right\}$$

The plastic work performed during an increment of plastic strain is

$$\Delta W^p = \sigma_{xx}\Delta\epsilon^p_{xx} + \sigma_{yy}\Delta\epsilon^p_{yy} + \sigma_{xy}2\Delta\epsilon^p_{xy} \equiv \overline{\sigma}\Delta\overline{\epsilon}^p$$

We can substitute for the strain increments in terms of the stresses to get

$$\Delta W^p = \tfrac{2}{3}d\lambda \left[\sigma^2_{xx} + \sigma^2_{yy} - \sigma_{xx}\sigma_{yy} + 3\sigma^2_{xy} \right] = \tfrac{2}{3}d\lambda\overline{\sigma}^2$$

Comparing the two work terms leads to the measure of effective plastic strain

$$\Delta\overline{\epsilon}^p = \tfrac{2}{3}d\lambda\overline{\sigma} \qquad \text{and} \qquad \Delta W^p = \frac{3}{2d\lambda}\Delta\overline{\epsilon}^{p2}$$

It remains now to determine $\Delta\overline{\epsilon}^p$ in terms of its components. Substitute for the stresses in terms of the strain increments in the plastic work expression to get

$$\Delta W^p = \frac{2}{d\lambda}[\Delta\epsilon^{p2}_{xx} + \Delta\epsilon^{p2}_{yy} + \Delta\epsilon^p_{xx}\Delta\epsilon^p_{yy} + \Delta\epsilon^{p2}_{xy}] = \frac{3}{2d\lambda}\Delta\overline{\epsilon}^{p2}$$

from which we conclude that

$$\Delta\overline{\epsilon}^p = \frac{2}{\sqrt{3}}[\Delta\epsilon^{p2}_{xx} + \Delta\epsilon^{p2}_{yy} + \Delta\epsilon^p_{xx}\Delta\epsilon^p_{yy} + \Delta\epsilon^{p2}_{xy}]^{1/2}$$

We also have the interpretation, analogous to Equation (1.46),

$$d\lambda = \frac{3}{2}\frac{\Delta\overline{\epsilon}^p}{\overline{\sigma}}$$

Thus, $d\lambda$ relates the increment of plastic stress to the current stress state. For uniaxial stress, these plasticity relations reduce to

$$\overline{\sigma} = \sigma_{xx}, \qquad \Delta\overline{\epsilon} = \Delta\epsilon_{xx}$$

and the plastic parameters become

$$d\lambda = \frac{\tfrac{3}{2}E\sigma_{xx}\Delta\epsilon_{xx}}{[E\sigma^2_{xx} + H']}, \qquad \Delta\epsilon_{xx} = \frac{3}{2}d\lambda\sigma_{xx}, \qquad \Delta\sigma_{xx} = E\left[1 - \frac{E}{E + H'}\right]\Delta\epsilon_{xx}$$

This elastic-plastic behavior is shown in Figure 1.32(b).

Most structural materials increase their yield stress under increased loading or straining; this is called *strain hardening* or *work hardening*. Let the yield stress change according to

$$f = \overline{\sigma}(\sigma) - \sigma_Y(\overline{\epsilon}^P), \qquad \overline{\epsilon}^P = \sum \Delta\overline{\epsilon}^P$$

Under uniaxial tension with σ_{xx} as the only nonzero stress, we have $\overline{\sigma} = \sigma_{xx} = \sigma_Y$. This leads to $\Delta\epsilon_{yy}^P = -\frac{1}{2}\Delta\epsilon_{xx}^P$ giving $\Delta\overline{\epsilon}^P = \Delta\epsilon_{xx}^P$; consequently, the relationship between $\overline{\epsilon}$ and $\overline{\sigma}$ can be taken directly off the uniaxial stress/strain curve. In particular,

$$\sigma_Y = \sigma_{Y0} + H'\overline{\epsilon}^P \qquad \text{or} \qquad H' = \frac{\partial\sigma_Y}{\partial\overline{\epsilon}^P} = \frac{\partial\sigma_{xx}}{\partial\epsilon_{xx}^P} = \frac{E_T}{1 - E_T/E}$$

This is shown in Figure 1.32(b).

With increasing stress, plastic flow occurs, and the yield surface expands so that the stresses always remain on the yield surface. This means that

$$\Delta f = \Delta\overline{\sigma}(\sigma) - \Delta\sigma_Y(\overline{\epsilon}^P) = \left\{\frac{\partial\overline{\sigma}}{\partial\sigma}\right\}^T \{\Delta\sigma\} + \frac{\partial\sigma_Y}{\partial\overline{\epsilon}^P}\Delta\overline{\epsilon}^P = \frac{1}{2\overline{\sigma}}\{s\}^T\{\Delta\sigma\} - H'\frac{2}{3}\overline{\sigma}d\lambda = 0$$

Solving for $d\lambda$, we then have an alternative expression for the increments of plastic strain

$$\{\Delta\epsilon^P\} = \left[\frac{\{s\}\{s\}^T}{\overline{\sigma}^2 H'}\right]\{\Delta\sigma\} = [C_p]\{\Delta\sigma\}$$

The [3 × 3] square bracket term is like an (inverse) modulus; in this case, however, it is a stress-dependent modulus.

At each stage of the loading, the increment of stress is related to the increment of strain. To establish this incremental stiffness relation, we need to replace the plastic strain increments with total strains. The elastic strain increment and the stresses are related by Hooke's law

$$\{\Delta\sigma\} = [D_e]\{\Delta\epsilon^e\}, \qquad [D_e] = \frac{E}{1-\nu^2}\begin{bmatrix} 1 & \nu & 0 \\ \nu & 1 & 0 \\ 0 & 0 & (1-\nu)/2 \end{bmatrix}$$

Replace the elastic strains with total strains

$$\{\Delta\sigma\} = [D_e]\{\Delta\epsilon - \Delta\epsilon^P\} = [D_e]\{\{\Delta\epsilon\} - \tfrac{1}{3}\lambda\{s\}\}$$

and substitute this into the strain-hardening relation to solve for $d\lambda$ as

$$d\lambda = \frac{3\{s\}^T[D_e]\{\Delta\epsilon\}}{\{s\}^T[D_e]\{s\} + 4H'\overline{\sigma}^2}$$

The incremental stress relation thus becomes

$$\{\Delta\sigma\} = [D_e]\left[\lceil 1 \rfloor - \frac{\{s\}\{s\}^T[D_e]}{\{s\}^T[D_e]\{s\} + 4H'\overline{\sigma}^2}\right]\{\Delta\epsilon\} = [D_T]\{\Delta\epsilon\}$$

The element tangent stiffness for the increments is then

$$[\,k_\mathrm{T}\,] = \int_V [\;B\;]^\mathrm{T}[\,D_\mathrm{T}\,][\;B\;]\,\mathrm{d}V = [\;B\;]^\mathrm{T}[\,D_\mathrm{T}\,][\;B\;]\,V$$

In the case of the CST element, it is thus just a matter of replacing the elastic material stiffness with the elastic/plastic stress-dependent relation. From a programming point of view, it is necessary to remember the state of stress and elastic strain from the previous step.

Discussion of Nonlinear Materials in FEM Codes

The purpose of this section, as is so with many other sections, is to give an intimation of the issues addressed in actual FEM codes. It is not intended to be a primer on how to program these codes—this is best left to specialist books. Indeed, as we will see in the ensuing chapters, we will develop inverse methods whereby we use FEM programs as stand-alone external processes such that FEM code is not part of the inverse methods. In this way, essentially, if an FEM program can handle a certain nonlinear material behavior, then the inverse method also can handle that material behavior and the particular implementation details are not especially relevant. This also applies to special material implementations such as those for composite materials.

2

Experimental Methods

A transducer is a device that transforms one type of energy into another. These are of immense importance in measurement because, generally, it happens that the original quantity (strain, say) to be measured is too awkward or too small to handle in its original form and instead an intermediate quantity is used. In modern measurements, the most common intermediate quantity is the electric voltage. The most significant change in recent years is the digital acquisition of electrical signals. While there are a great many sensors available to the experimental engineer, only two sensors are focused on in this chapter: electrical resistance strain gages and piezoelectric accelerometers.

In contrast to point transducers, the optical methods give information that are distributed continuously over an area; Figure 2.1 shows nicely how the main participants in a problem can be captured all at once. We cover Moiré, holography, and photoelasticity and since, nowadays, the data is recorded digitally and analyzed digitally, we also describe image-processing procedures.

References [43, 48, 107] give an in-depth treatment of the experimental methods considered here, and also cover a host of other methods.

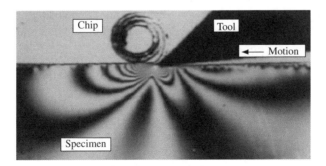

Figure 2.1: Photoelastic fringes formed during the machining of a polycarbonate specimen.

Modern Experimental Stress Analysis: completing the solution of partially specified problems.   James Doyle
© 2004 John Wiley & Sons, Ltd   ISBN 0-470-86156-8

2.1   Electrical Filter Circuits

All electrical measurement systems are inherently dynamic. This is due primarily to the feedback loops necessary to control the unwanted electrical noise. We therefore begin by looking at some of the basic circuits in use for measuring, conditioning, and filtering electrical circuits.

Measurement instruments can vary significantly in their dynamic responses. To get an understanding of this, we consider three electronic circuits that are widely used; these are operational amplifiers, RC circuits, and second-order circuits. As will be seen, these are characterized by the increasing order of differential relations.

Operational Amplifiers

An operational amplifier (op-amp) is a high-gain ($G > 10^5$), high-input impedance ($R_i > 10\,M\Omega$), low-output impedance ($R_o < 10\,\Omega$) amplifier circuit commonly used in signal-conditioning circuits. A typical schematic is shown in Figure 2.2(a). Schematic diagrams of op-amps usually leave off the power supply and null adjustment inputs, but, of course, they must always be actually provided.

Consider a simple use of an op-amp shown in Figure 2.2(b). As demonstrated in Reference [92], because the gain of the amplifier is so large, the output voltage E^* is related to the input voltages E by

$$E^* = -\frac{R_f}{R_1} E$$

which is a gain of R_f / R_1. Reference [92] also discusses many other op-amp configurations.

RC Circuits

In describing electrical circuits the two basic properties are resistance, R, and capacitance, C. The former refers to the resistance to current flow and is related to the voltage potential by $V = IR$, where I is the current. Capacitance is the ability of materials to hold a charge

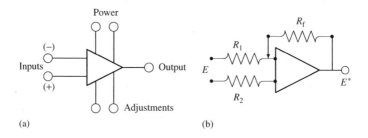

Figure 2.2: Operational amplifiers. (a) Basic circuit. (b) Application to provide gain.

and it is related to voltage by $V = Q/C$, where Q is the charge. All electrical components have these properties, so it is of interest to see how they affect dynamic circuits.

The RC differentiator circuit is shown in Figure 2.3(a). It is assumed that the input voltage source has zero resistance and the output voltage is measured with an instrument with infinite input impedance.

Let the signal input be E and output be E^*. The capacitor stores energy such that the voltage drop is

$$E_C = \frac{1}{C}Q \quad \text{or} \quad \frac{dE_C}{dt} = \frac{1}{C}\frac{dQ}{dt} = \frac{1}{C}I$$

The voltage drop across the resistor is $E_R = IR$, which is also the output voltage E^*. The voltage drop in the circuit is therefore

$$E = E_C + E_R \quad \text{or} \quad \frac{dE}{dt} = \frac{dE_C}{dt} + \frac{dE_R}{dt} = \frac{1}{C}I + R\frac{dI}{dt} = \frac{E^*}{RC} + \frac{dE^*}{dt}$$

Hence the output voltage is obtained from the input voltage by

$$RC\frac{dE^*}{dt} + E^* = RC\frac{dE}{dt}$$

Consider some limiting behaviors. When RC is very large, then $E^* = E$, but when RC is small, then $E^* \approx 0$ indicating a significant amount of distortion. Figure 2.4(a) shows the output due to a nearly rectangular input for a circuit with different values of RC. It is apparent that as RC is decreased, there can be a good deal of distortion. This is characteristic of dynamic circuits and as to understanding them is essential for accurate measurements.

The RC integrator circuit is shown in Figure 2.3(b). Following a procedure similar to the above, we can show that the equation describing the circuit behavior is

$$RC\frac{dE^*}{dt} + E^* = E$$

When RC is small, $E^* = E$, but when RC is very large, then $E^* = \int E\, dt$, again indicating a significant amount of distortion. Figure 2.4(b) shows its response to a rectangular input. This circuit can also cause a distortion in the output voltage, but the distortion is different than that for the differentiator circuit.

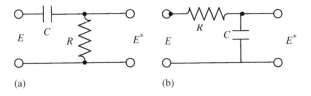

E C R E^* E R C E^*

(a)   (b)

Figure 2.3: Two RC circuits. (a) Differentiator. (b) Integrator.

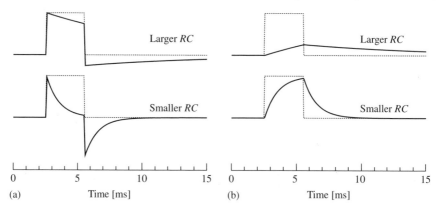

Figure 2.4: Response of RC circuits to a rectangular input voltage history. (a) Differentiator. (b) Integrator.

Forced Frequency Response

To get further insight into the functioning of the circuits, we will now consider response to sinusoidal inputs. Note that the basic behavior of a sinusoid is

$$E(t) = A e^{i\omega t}, \quad \frac{dE}{dt} = A i \omega e^{i\omega t} = i[A\omega] E(t), \quad \int E \, dt = A \frac{1}{i\omega} e^{i\omega t} = -i[A/\omega] E(t)$$

where ω is the angular frequency and $i = \sqrt{-1}$. Thus, differentiation corresponds to multiplication by ω, whereas integration corresponds to division by ω; the presence of i indicates a phase shift of $90°$. With this as background, consider a sinusoidal input in the form

$$E = E_o + \Delta E \sin \omega t$$

Substituting into the differentiator circuit gives

$$\frac{dE^*}{dt} + \frac{1}{RC} E^* = \Delta E \omega \cos \omega t$$

Notice that E_o does not appear; this means that the dc voltage will not appear in the output voltage. This is an inhomogeneous differential equation whose homogeneous solution is

$$E_\mathrm{h} = K e^{-t/RC}$$

where K is a constant. Assume a particular solution as $E_\mathrm{p} = A \cos \omega t + B \sin \omega t$ and substitute into the differential equation to determine A and B. The complete solution is then given by

$$E^* = K e^{-t/RC} + \Delta E \frac{\omega RC}{1 + (\omega RC)^2} [\cos \omega t + \omega RC \sin \omega t]$$

Defining a phase angle as

$$\tan \delta \equiv \frac{1}{\omega RC} \qquad \cos \delta \equiv \frac{\omega RC}{\sqrt{1 + (\omega RC)^2}} \qquad \sin \delta \equiv \frac{1}{\sqrt{1 + (\omega RC)^2}}$$

allows the output voltage to be written as

$$E^* = K e^{-t/RC} + \Delta E \frac{\omega RC}{\sqrt{1 + (\omega RC)^2}} \sin(\omega t + \delta)$$

Hence, apart from the initial transient, only the sinusoidal part (and not the dc part) is transferred to the output. The output signal is distorted in two ways:

- Amplitude modulation: $\dfrac{\omega RC}{\sqrt{1 + (\omega RC)^2}}$

- Phase delay: $\delta = \tan^{-1}\left(\dfrac{1}{\omega RC}\right)$

This means that at a given frequency, the relation between the driving signal and output signal is as shown in Figure 2.5(a).

It is evident from Figure 2.5(a) that for frequencies $\omega RC < 1.0$, the response magnitude increases linearly with ω so that the circuit essentially differentiates the input signal. This circuit is often called a *blocking circuit* since any dc signal is completely attenuated or blocked from passing through the system. For frequencies $\omega RC > 5.0$, there is essentially no distortion of the ac component of the signal and the output is equal to and in phase with the input.

A similar analysis for the integrator circuit gives

$$E^* = K e^{-t/RC} + E_0 - \Delta E \frac{1}{\sqrt{1 + (\omega RC)^2}} \sin(\omega t - \delta)$$

where the phase angle is defined as

$$\sin \delta \equiv \frac{\omega RC}{\sqrt{1 + (\omega RC)^2}} \qquad \cos \delta \equiv \frac{1}{\sqrt{1 + (\omega RC)^2}}$$

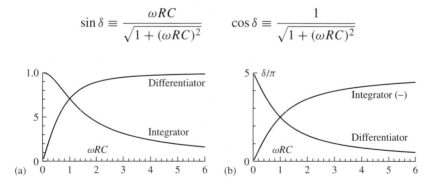

Figure 2.5: Frequency response of RC circuits. (a) Amplitude. (b) Phase.

As seen from Figure 2.5(a), when $\omega RC > 1.0$, the circuit begins to perform like an integrator since it depends on the reciprocal of ω. For practical purposes, this circuit performs like an integrator for values of $\omega RC > 5.0$ where it filters unwanted higher frequency signals and passes the lower frequency signals; that is, it is a useful circuit for signal conditioning.

This insight into the frequency behavior is important because we know from Fourier analysis that a general signal contains many frequency components that can be expressed analytically in a Fourier series; that is, the Fourier series can be written as

$$E(t) = a_1 \sin \omega_0 t + a_2 \sin 2\omega_0 t + \cdots + a_n \sin n\omega_0 t$$

where $\omega_0 = 2\pi/T$ is the fundamental frequency, and T is the fundamental period of the signal. The magnitude and phase of each of these frequency components are modified by the circuit to give an output signal that can be expressed as

$$E^*(t) = b_1 \sin(\omega_0 t - \delta_1) + b_2 \sin(2\omega_0 t - \delta_2) + \cdots + b_n \sin(n\omega_0 t - \delta_n)$$

Thus, each amplitude and each phase is modified according to frequency behavior displayed in Figure 2.5.

Generally, it is easier to characterize the performance of circuits (and hence devices) in the frequency domain. To that end, we will analyze the circuits by assuming the input and output voltages to be harmonic and of the form

$$E = \hat{E} e^{i\omega t}, \qquad E = \hat{E}^* e^{i\omega t}$$

Note that both $\hat{E}$ and $\hat{E}^*$ could be complex. Substituting into the RC differentiator circuit gives

$$i\omega \hat{E}^* e^{i\omega t} + \frac{1}{RC} \hat{E}^* e^{i\omega t} = i\omega \hat{E} e^{i\omega t}$$

Since the $e^{i\omega t}$ term is never zero, we can cancel it through allowing the output to be written in terms of the input as

$$\hat{E}^* = \frac{i\omega}{i\omega + \dfrac{1}{RC}} \hat{E} = \frac{i\omega RC}{1 + i\omega RC} \hat{E} = H(\omega)\hat{E}$$

where $H(\omega)$ is called the frequency response function (FRF). It is complex and can be written in alternative forms as

$$H(\omega) = \frac{i\omega RC}{1 + i\omega RC} \frac{1 - i\omega RC}{1 - i\omega RC} = \frac{\omega RC(\omega RC + i)}{1 + \omega^2 R^2 C^2} = \frac{\omega RC}{\sqrt{1 + \omega^2 R^2 C^2}} \frac{\omega RC + i}{\sqrt{1 + \omega^2 R^2 C^2}}$$

$$= \frac{\omega RC}{\sqrt{1 + \omega^2 R^2 C^2}} [\cos \delta + i \sin \delta] = \frac{\omega RC}{\sqrt{1 + \omega^2 R^2 C^2}} e^{i\delta}$$

where $\tan \delta \equiv 1/(\omega RC)$, and we recover what we already derived. This, however, is a more direct route.

A similar analysis for the integrator circuit gives the frequency response function as

$$\hat{E}^* = \frac{1}{1 + i\omega RC} \hat{E} = H(\omega)\hat{E}$$

and the phase is $\tan \delta \equiv \omega RC$.

Second-Order Filters

Second-order filter circuits can be easily constructed using op-amps. One such simple circuit is discussed here.

The single op-amp circuit for simulating a second-order filter is shown in Figure 2.6. The circuit parameters and equation variables are related by the following expressions:

$$R_1 C_1 R_2 C_2 \frac{d^2 E^*}{dt^2} - (R_1 + R_2) C_2 \frac{dE^*}{dt} + E^* = E$$

The corresponding FRF $H(\omega)$ is easily established to be

$$\hat{E}^* = \frac{1}{1 + i\omega[R_1 + R_2]C_2 - R_1 C_1 R_2 C_2 \omega^2} \hat{E} = H(\omega)\hat{E}$$

The FRF is often displayed as a Bode plot where the amplitude is given in decibels, dB, and the frequency term is plotted on a logarithmic scale. The decibel is defined as $10\log(P/P_r)$, where P is a powerlike quantity and P_r is a reference powerlike quantity.

For voltages, power is proportional to the voltage squared; therefore, for the integrator circuit

$$dB = 20\log\left(\frac{\hat{E}_o}{\hat{E}_i}\right) = -10\log(1 + (\omega RC)^2)$$

A Bode plot for the second-order filter is shown in Figure 2.7. The resonant-like behavior occurs at $\omega_0 = 1/\sqrt{R_1 C_1 R_2 C_2}$ and the peak is controlled by the $[R_1 + R_2]C_2$ term.

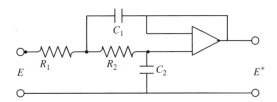

Figure 2.6: Second-order filter circuit.

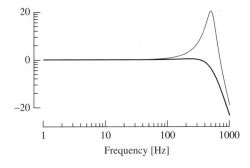

Figure 2.7: Bode plot for second-order filter.

This plot indicates that low-frequency signals ($\omega < \omega_0$) are not attenuated, while high-frequency signals ($\omega > \omega_0$) are attenuated at the rate of about 40 dB/decade.

Typically, the damping range is between 0.6 and 0.707. In this range, magnitude errors are small, and phase angle shift is nearly linear with respect to frequency. This latter property is needed so that complex signals maintain their wave shape.

2.2 Digital Recording and Manipulation of Signals

The analog-to-digital (A/D) converter takes the analog input signal, a voltage $E(t)$, and converts it at discrete times into an equivalent digital code. Once digitized, the signal can be stored in memory, transmitted at high speed over a data bus, processed in a computer, or displayed graphically on a monitor. Reproducing an accurate digital representation of the incoming analog waveform depends significantly on the sampling rate and the resolution. These two parameters, in turn, depend on the conversion technique; we therefore begin by considering some conversion techniques.

Binary Code Representation and Resolution

Each bit of a digital word can have two states: on/off, true/false, or 1/0. For a word n-bits long there will be 2^n combinations of on/off states. Consider a digital word consisting of a 3-bit array; these can be arranged to represent a total of eight numbers as follows:

$$
\begin{array}{cccc}
0: & 0 & 0 & 0 \\
1: & 0 & 0 & 1 \\
2: & 0 & 1 & 0 \\
3: & 0 & 1 & 1 \\
4: & 1 & 0 & 0 \\
5: & 1 & 0 & 1 \\
6: & 1 & 1 & 0 \\
7: & 1 & 1 & 1 \\
\end{array}
$$

For the number 6: [110], the 1 at the left is the most-significant bit (MSB) and the 0 at the right is the least-significant bit (LSB). The weights in a binary code give the least-significant bit a weight $2^0 = 1$, the next bit has a weight $2^1 = 2$, the next $2^2 = 4$, and the most-significant bit has a weight $2^{(n-1)}$. When all bits are 0, [000], the equivalent count is 0. When all bits are 1, [111], the equivalent count is $2^2 + 2^1 + 2^0 = 4 + 2 + 1 = 7$. An 8-bit word would have $2^7 + 2^6 + \cdots + 2^1 + 2^0 = 128 + 64 + \cdots + 2 + 1 = 255 = 2^8 - 1$

The result of coding a signal is a digital "word" that typically varies from 8 to 16 bits in length. The word length implies the number of parts that the voltage range of a conversion can be divided into. The length of this word is significant in the accuracy of the converter and the length of time it takes for the coding process to be completed. The longer the word length, the greater the resolution of the A/D, but the conversion process can be slower.

The least-significant bit represents the resolution R of a digital count containing n bits. Thus, the resolution of an 8-bit digital word (a byte) is 1/255 or 0.392% of full scale. If the input voltage to an 8-bit converter ranges from 0 to 10 V, then that voltage can be divided into 256 parts and the resolution of this combination will be 10/256 or 39 mV. For a converter that produces a 12-bit word, there are 4096 parts into which the input voltage can be divided. Again, for an input voltage range of 0 to 10 V, this converter would have a resolution of 10/4096 or 2.44 mV.

The uncertainty illustrated in Figure 2.8, when referenced to full scale, is

$$\Delta E = FS / 2^{n+1}$$

where n is the number of bits used in the approximation. This equation shows that analog-to-digital conversion results in a maximum uncertainty that is equivalent to $\frac{1}{2}$ LSB.

Description of a Typical Capture Board

Advances in developing high-performance and low-cost electronic components such as A/Ds, RAM chips, and multiplexers, have made digital data-acquisition systems readily available.

Our interest is in analog-to-digital conversion, but we start with the simpler digital-to-analog conversion to illustrate the design of the circuits. Consider the simple 4-bit D/A circuit shown in the Figure 2.9; commercial D/As are more complicated than indicated because they contain more bits (8, 12, and 16 are common), and have several regulated voltages, integrated circuits (IC) for switching, and on-chip registers. A reference voltage, E_R, is connected to a set of precision resistors (10, 20, 40, 80; i.e., doubling) arranged with binary values through a set of switches. The switches are operated by digital logic, with 0 representing open and 1 representing closed. Let the switches be numbered $n = 1$ to N, with $n = 1$ being the MSB, then,

$$R_n = 2^n R_f$$

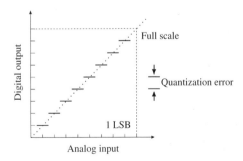

Figure 2.8: Discretization of an analog signal showing the uncertainty of a measured signal.

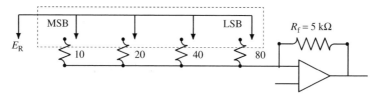

Figure 2.9: Simple 4-bit D/A circuit.

where R_n is the resistance of the nth bit and R_f is the feedback resistance on the operational amplifier.

A binary-weighted current, I_n, flows to the summing bus when the switches are closed, and is given by

$$I_n = \frac{E_R}{E_n} = \frac{E_R}{2^n R_f}$$

The op-amps convert these currents to voltages. The analog output voltage E_o is given by

$$E_o = -R_f \sum_n^N I_n$$

where I_n is summed only if the switch n is closed. Consider, as an example, the digital code [101] or 5 with the circuit shown in Figure 2.9. The currents are $I_1 = 1$, $I_2 = 1/2$, $I_3 = 1/4 \, \text{mA}$. Summing these currents and multiplying by $R_f = 5 \, \text{k}\Omega$ gives

$$E_o = -5[1 \times 1 + 0 \times \tfrac{1}{2} + 1 \times \tfrac{1}{4}] = -6.25 \, V$$

that is 5/8 of the full-scale (reference) voltage.

The simplest and fastest A/D converter is the parallel or flash converter. Flash-type A/Ds are employed on high-speed digital-acquisition systems such as waveform recorders and digital oscilloscopes. Conversion can be made at 10 to 100 MHz, which gives sampling times of 100 to 10 ns.

The flash converter, shown schematically in Figure 2.10(a), uses $2^n - 1$ analog comparators arranged in parallel. Each comparator is connected to the unknown voltage E_u. The reference voltage is applied to a resistance ladder so that when it is applied to a given comparator it is 1 LSB higher than the reference voltage applied to the lower adjacent comparator. When the analog signal is presented to the array of comparators, all those comparators for which $E_{rn}^* < E_u$ will go high and those for which $E_{rn}^* > E_u$ will stay low. As these are latching-type comparators, they hold high or low until they are downloaded to a system of digital logic gates that converts them into a binary coded word.

A 10-bit parallel A/D has $2^{10} - 1 = 1023$ latching comparators and resistances and about 5000 logic gates to convert the output to binary code. This is why it is the most expensive method.

The method of successive approximation is shown schematically in Figure 2.10(b). The bias voltage, E_b, is compared to a sequence of precise voltages generated by an

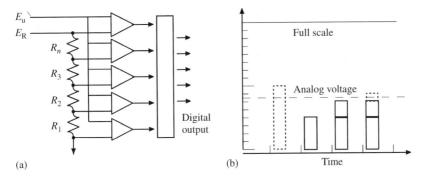

(a)   (b)   Time

Figure 2.10: Examples of A/D converters. (a) Flash. (b) Successive approximation.

input-controlled D/A, such as in Figure 2.9. The digital input to the D/A is successively adjusted until the known output, E_o, from the D/A compares closely with E_b.

We now describe the conversion process in more detail. At the start, the input to the D/A is set at [1000] (the MSB is on and the others are off). The analog voltage output $E_o = 8/16\,FS$, and a voltage comparison shows $E_o > E_b$. Since this overestimates E_b, the MSB is locked at 0. The second bit is now turned on to give a digital input of [0100] to the D/A. The output from this second approximation is $E_o = 4/16\,FS$. The voltage comparison shows $E_o < E_b$; hence, bit 2 is locked on. The third approximation involves turning bit 3 on to give a digital input of [0110] and $E_o = 6/16\,FS$. Since the voltage comparison shows $E_o > E_b$, bit 3 is also locked on. The final approximation turns bit 4 on to give the digital input of 0111 and $E_o = 7/16\,FS$. Since $E_o > E_b$, bit 4 is turned off to give a final result of [0110] at the input to the D/A after four approximations.

The speed of this conversion process depends primarily on the number of bits. Typically, 8- and 12-bit successive-approximation-type A/Ds require on the order of $1\,\mu s$ to convert an analog signal.

Analog-to-digital conversion by integration is the slowest of the three methods because it is based on counting clock pulses. The basic idea is that the unknown input voltage E_u is applied together with a reference voltage, E_R, to a summing amplifier which in turn is imposed on an integrator to obtain

$$E_i = \frac{1}{2}(E_u + E_R)\frac{\Delta t_1}{RC}$$

where Δt_1 is the fixed time of integration. A switch on the integrator now disconnects the summing amplifier and connects the reference voltage E_R to the integrator; the output voltage of the integrator then decreases with a slope of

$$\frac{\Delta E_i}{\Delta t_2} = -\frac{E_R}{RC}$$

The comparator monitors the output voltage E_i when it goes to zero. This zero crossing occurs when

$$\frac{1}{2}(E_u + E_R)\frac{\Delta t_1}{RC} = \frac{E_R \Delta t_2}{RC} \qquad \text{or} \qquad \frac{E_u}{E_R} = 2\frac{\Delta t_2}{\Delta t_1} - 1$$

We see that $\Delta t_2/\Delta t_1$ is a proportional count related to E_u/E_R. If a counter is initiated by the switch on the integrator, then the counter will give a binary number representing the unknown voltage E_u.

The integration method has several advantages. First, it is very accurate, as its output is independent of R, C, and clock frequency because these quantities affect the up and down ramps in the same way. Second, the influence of noise is significantly attenuated because of the integration process. Its main disadvantage is the speed of conversion, which is less than Δt_1 conversions per second. For example, to eliminate 60-Hz noise, $\Delta t_1 > 16.67\,\text{ms}$ and the rate of conversions must be less than 30 per second. This is too slow for general-purpose data-acquisition systems, and certainly for any dynamic recordings. However, it satisfactory for digital voltmeters and smaller data-logging systems.

Multifunction I/O boards can be composed of many components and capabilities. Some allow analog-to-digital conversion, digital-to-analog conversion, digital input, digital output, external and internal triggering, direct-memory access and internal and external clocking of the processes. The input end consists of several sections as shown in Figure 2.11. An analog multiplexer selects the analog input, an instrumentation amplifier amplifies the input signal and removes the unwanted common-mode signal, a sample-hold amplifier "freezes" the input signal, and an A/D converter converts the sampled analog voltage to a digital equivalent.

Spectral Analysis of Signals

As discussed earlier in this chapter, an arbitrary time signal can be thought of as the superposition of many sinusoidal components, that is, it has a distribution or *spectrum* of components. Working in terms of the spectrum is called *spectral analysis* or sometimes *frequency domain analysis*. Functions in the time domain range from $-\infty$ to $+\infty$ and

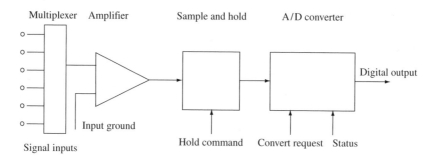

Figure 2.11: A schematic of a typical data-acquisition board.

are represented by a continuous distribution of components that is known as its continuous Fourier transform (CFT). However, the numerical manipulation of signals requires discretizing and truncating the distribution in some manner—in recent years, the discrete Fourier transform (DFT) and it numerical implementation in the form of the fast Fourier transform (FFT) has become the pervasive method of analyzing signals. We now give an introduction to some of their characteristics, especially as to how it applies to signal processing.

I: Fourier Transforms

The continuous transform is a convenient starting point for discussing spectral analysis because of its exact representation of functions. Only its definition and basic properties are given here; a more complete account can be found in References [38, 159, 160].

The continuous Fourier transform pair of a function $F(t)$, defined on the time domain from $-\infty$ to $+\infty$, is given as

$$2\pi F(t) = \int_{-\infty}^{\infty} \hat{C}(\omega)e^{+i\omega t}\, d\omega, \qquad \hat{C}(\omega) = \int_{-\infty}^{\infty} F(t)e^{-i\omega t}\, dt \qquad (2.1)$$

where $\hat{C}(\omega)$ is the continuous Fourier transform (CFT), ω is the angular frequency, and i is the complex $\sqrt{-1}$. The first form is the *inverse* transform while the second is the *forward* transform—this arbitrary convention arises because the signal to be transformed usually originates in the time domain.

The process of obtaining the Fourier transform of a signal separates the waveform into its constituent sinusoids (or spectrum) and thus a plot of $\hat{C}(\omega)$ against frequency represents a diagram displaying the amplitude of each of the constituent sinusoids. The spectrum $\hat{C}(\omega)$ usually has nonzero real and imaginary parts. We will consistently use the superscript "hat" to indicate the frequency domain spectrum of a function. The continuous transform has infinite limits; this makes it unsuitable for computer implementation. We now look at discretized forms.

Consider the representation of a function $F(t)$, extended periodically, in terms of its transform over a single period; that is,

$$F(t) = \frac{1}{T} \sum_{n=-\infty}^{n=+\infty} \hat{C}_\infty(\omega_n)e^{+i\omega_n t}, \qquad \hat{C}_\infty(\omega_n) = \int_0^T F(t)e^{-i\omega_n t}\, dt \qquad (2.2)$$

Except for a normalizing constant, this is the complex Fourier series representation of a periodic signal. The Fourier series gives the exact values of the continuous transform, but it does so only at discrete frequencies. The discretization of the frequencies is given by

$$\omega_n = 2\pi n / T$$

Therefore, for a given pulse, a larger time window gives a more dense distribution of coefficients, approaching a continuous distribution in the limit of an infinite period. This, of course, is the continuous transform limit. The finite time integral causes the transform to be discretized in the frequency domain.

The discrete coefficients in the Fourier series are obtained by performing continuous integrations over the time period. These integrations are replaced by summations as a further step in the numerical implementation of the continuous transform. An approximation for the Fourier series coefficients is

$$F_m = F(t_m) \approx \frac{1}{T} \sum_{n=0}^{N-1} \hat{D}_n e^{+i\omega_n t_m} = \frac{1}{T} \sum_{n=0}^{N-1} \hat{D}_n e^{+i2\pi nm/N}$$

$$\hat{D}_n = \hat{D}(\omega_n) \approx \Delta T \sum_{m=0}^{N-1} F_m e^{-i\omega_n t_m} = \Delta T \sum_{m=0}^{N-1} F_m e^{-i2\pi nm/N} \qquad (2.3)$$

where both m and n range from 0 to $N-1$. These are the definition of what is called the *discrete Fourier transform* (DFT). It is interesting to note that the exponentials do not contain dimensional quantities; only the integers n, m, N appear. In this transform, both the time and frequency domains are discretized, and as a consequence, the transform behaves periodically in both domains. The dimensional scale factors $\Delta T, 1/T$ have been retained so that the discrete transform gives the same numerical values as the continuous transform. There are other possibilities for these scales found in the literature.

The discrete transform enjoys all the properties of the continuous transform; the only significant difference is that both the time domain and frequency domain functions are now periodic. To put this point into perspective, consider the following: A discrete Fourier transform seeks to represent a signal (known over a finite time T) by a finite number of frequencies. Thus it is the continuous Fourier transform of a periodic signal. Alternatively, the continuous Fourier transform itself can be viewed as a discrete Fourier transform of a signal but with an infinite period. The lesson is that by choosing a large signal sample length, the effect due to the periodicity assumption can be minimized and the discrete Fourier transform approaches the continuous Fourier transform. The fast Fourier transform (FFT) is simply a very efficient numerical scheme for computing the discrete Fourier transform. This is not a different transform — the numbers obtained from the FFT are exactly the same in every respect as those obtained from the DFT. The fast Fourier transform algorithm is so efficient that it has revolutionized the whole area of spectral analysis.

There are many codes available for performing the FFT, as can be found in Reference [153], and well-documented FORTRAN routines are described in Reference [143]. The following examples serve to show the basic procedures used in applying the FFT to transient signals. The FFT is a transform for which no information is gained or lost relative to the original signal. However, the information is presented in a different format that often enriches the understanding of the information. In the examples to follow, we will try to present that duality of the given information.

II: Transforms of Time-Limited Functions

The complementarity of information between the time and frequency domains is illustrated in Figure 2.12. The top three plots are for triangles. On comparing the first two, we notice that the longer duration pulse has a shorter main frequency range; however, both exhibit side lobes that extend significantly along the frequency range. The third trace shows a smoothed version of the second triangle; in the frequency domain, it is almost identical to the second, the only difference being the reduction in the high-frequency side lobes. Thus time domain smoothing (using moving averages, say) acts as a high-band filter in the frequency domain.

The remaining two plots are for *modulated* signals — in this case, a sinusoid modulated (multiplied) by a triangle. The sinusoid function on its own would give a spike at 15 kHz, whereas the triangular modulations on their own are as shown for the top traces. The product of the two in the time domain is a convolution in the frequency domain. The convolution tends to distribute the effect of the pulse. A point of interest here is that if a very narrow-banded signal in the frequency domain is desired, then it is necessary to extend the time domain signal as shown in the bottom example. The modulated waves are called wave packets or *wave groups* because they contain a narrow collection of frequency components.

Digital Processing of Experimental Signals

The signals we collect will be the functions of time. Even nominally static problems will have time-varying signals due, at least, to the noise. Spectral analysis can give insight into how all are treated.

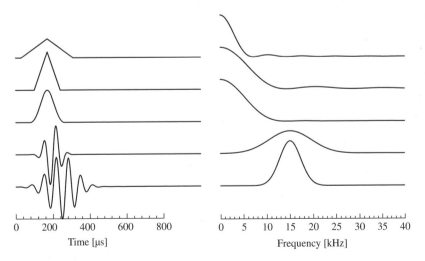

Figure 2.12: Comparison of some pulse-type signals in both the time and frequency domains.

I: Moderate Rate Testing

The specifications of an A/D board should be evaluated with regard to the converter range, converter resolution, sampling speed, the frequency of the event, and the number of input signals, which will be connected to the board. Consider an event with a characteristic frequency of 100 Hz and the need to monitor eight channels of analog data. What sampling speed should the A/D be capable of sustaining? Assume that the converter can handle eight channels of input and requires a clock pulse for each conversion. Further, assume that we want to have 20 samples during each period of the incoming signal. The A/D sampling rate of $\Delta T = 1/(8 \times 20)$ means the clock pulse rate will have to be $100 \times (20 \times 8) = 12,000$ Hz. This illustrates that the sampling rate specified for an I/O board must be substantial to monitor dynamic events even if they contain only low or moderate frequencies.

II: Measurement Techniques for Transients

Testing associated with transients arising from impact and stress wave propagation is different from that of vibration testing. Some of the characteristics are:

- High-frequency content (1 kHz–1 MHz).
- Only the initial portion ($\sim$5000 µs) of the signal is analyzed.
- Only a small number of "runs" are performed.
- The number of data recording channels are usually limited (2–8).

There is quite a range of measurement techniques available for studying wave propagation and high-frequency structural dynamics problems, but we restrict ourselves to strain gages and accelerometers, both of which are discussed in the next sections.

It is impossible to be categorical about the minimum setup necessary for waveform recording, but the following pieces of equipment definitely seem essential. Good preamplifiers are essential for dynamic work; they not only provide the gain for the low voltages from the bridge but also perform signal conditioning. The former gives the flexibility in choice of gage circuit as well as in reducing the burden on the recorder for providing amplification. A separate preamplifier is needed for each strain gage circuit and a typical one should have up to 100-K gain, dc to 1 MHz frequency response, and selectable band pass filtering. The analog filtering (which is essential) is performed at or below the Nyquist frequency associated with the digitizing. Tektronix 505A preamplifiers were used in all the dynamic measurements to be reported.

Historically, oscilloscopes were the primary means of recording (and displaying) the waveform but nowadays digital recorders are used. For high-speed recording, we use the Norland 3001 [181]; this has four channels, each storing 4 K data points, with up to 100 ns per sample. The DASH-18 [180] is an example of the newer breed of recorders; it is like a digital strip chart recorder and notebook computer rolled into one. The data is continuously recorded and stored to the computer hard drive. Its fastest rate of recording is 10 µs per sample. The DASH-18 can record up to 18 channels and has a very flexible interface for manipulating the recorded data. There are also relatively inexpensive recorders available that can be installed in desktop computers. The Omega Instruments DAS-58 data-acquisition card is an example; this is a 12-bit card capable of 1-MHz sampling rate and storing up to 1 M data points in the on-board memory. A total of eight multiplexed channels can be sampled giving the fastest possible sampling rate of 1M/8 = 125 K samples per second.

III: Smoothing

The measured signal can contain unwanted contributions from at least two sources. The first is the ever present noise. Analog filters can be used to remove much of this, although care must be taken not to remove some of the signal itself. The second may arise in the case of transient analyzes from unwanted reflections. This is especially prevalent for dispersive signals as the recording is usually extended to capture as much as possible of the initial passage of the wave. There are many ways of smoothing digital data but the simplest is to use moving averages of various amounts and points of application. Figure 2.13 shows an example of smoothing a signal that has an exaggerated amount of noise and reflections present. The filtering sequence was used on data values sampled at every 1 μs. The last averaging brings the signal smoothly to zero in the vicinity of 1000 μs. Also shown in the figure is that part of the signal removed; this can be monitored to detect if too much of the signal is being filtered.

IV: Windowing and Leakage

By necessity, actual signal records are finite in length, and this can add some difficulties to the scheme of estimating its spectral content. This section reviews the major effects that should be considered.

When we compare the truncated measured signal to the (imagined) infinite trace, we say the former was "windowed"—only a small portion of the infinite trace is visible

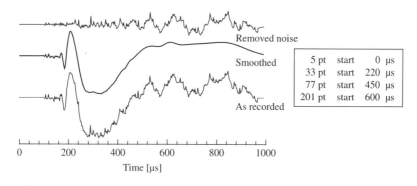

5 pt	start	0 μs
33 pt	start	220 μs
77 pt	start	450 μs
201 pt	start	600 μs

Figure 2.13: Effect of moving-average smoothing.

through the window. Examples of time windows commonly used are the rectangle and
the Hanning window (which is like a rectangle with smoothed edges); see Figure 2.14
and References [29, 38] for more examples of windows. None of the windows usually
used in vibration studies seem particularly appropriate for transient signals of the type
obtained from propagating waves.

The estimated spectrum is given only at discrete frequencies and this can cause a
problem known as *leakage*. For example, suppose there is a spectral peak at 17 kHz and
the sampling is such that the spectrum values are only at every 2 kHz. Then the energy
associated with the 17 kHz peak "leaks" into the neighboring frequencies and distorts
their spectral estimate.

Leakage is most significant in instances where the signal has sharp spectral peaks and
is least significant in cases where the signal is flat and broad-banded. In the analysis of
the impact of structures, the signals generated generally do not exhibit very sharp spectral
peaks. However, it can become a problem if there are very many reflections present in
the signal.

Since discrete Fourier analysis represents a finite sample of an infinite signal on a
finite period, then schemes for increasing the apparent period must be used. The simplest
means of doing this is by *padding*. There are various ways of padding the signal and
the one most commonly used is that of simply adding zeros. Parenthetically, it does not
matter if the zeros are added before the recorded signal or after. Thus, even if the signal
never decreases to zero, padding is justified because it represents the early quiescent part
of the signal.

Figure 2.14: Some common windows.

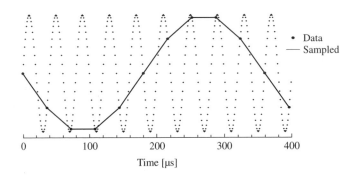

Figure 2.15: Aliasing caused by coarse sampling; the sampled high-frequency signal appears as a
lower frequency signal.

V: Sampled Waveforms and Aliasing

If the sampling rate is ΔT, then the highest detectable frequency using the DFT is $1/(2\Delta T)$. Thus, if ΔT is too large, the high frequency appears as a lower frequency (its alias) as shown in Figure 2.15. This phenomenon of *aliasing* can be avoided if the sampling rate is high enough. By "high enough", is implied a rate commensurate with the highest significant frequency in the signal. From a practical point of view, this is also avoided by recording with the analog filters set so as to remove significant frequencies above the Nyquist frequency.

2.3 Electrical Resistance Strain Gages

Probably the most ubiquitous and reliable of all the tools of experimental stress analysis is the electrical resistance strain gage. The principle of its action is that the electrical resistance of a conductor changes proportionally to any strain applied to it. Thus, if a short length of wire were bonded to the structure in such a way that it experiences the same deformation as the structure, then by measuring the change in resistance, the strain can be obtained.

While this principle was known by Lord Kelvin back in 1856, it was not until the late 1930s that Simmons and Ruge made practical use of it; in spite of its simple concept, there are many features of its behavior that must be understood in order to obtain satisfactory results.

Basic Theory

Consider a conducting wire of length L and area A as shown in Figure 2.16(a). Its resistance (neglecting temperature effects) is proportional to its length but inversely proportional to the area, that is,

$$R = \rho \frac{L}{A}$$

where ρ is a constant of proportionality, called the specific resistivity and is essentially inversely proportional to the number of mobile electrons per unit volume (N/V). Letting

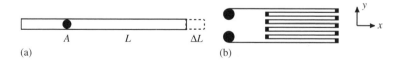

Figure 2.16: Concept of the electrical resistance strain gage. (a) A long piece of conducting wire changes resistance under strain. (b) A typical gage shape is equivalent to a long wire wound back and forth.

$\rho = K/(N/V)$ where K is a proportionality factor, the resistance is written as

$$R = K\frac{VL}{NA} = K\frac{L^2}{N}$$

It is interesting that the area does not figure in this equation.

Suppose this wire is stretched; there will be a small change in resistance given by

$$\Delta R = K\frac{1}{N}2L\Delta L - K\frac{L^2}{N^2}\Delta N \quad \text{or} \quad \frac{\Delta R}{R} = 2\frac{\Delta L}{L} - \frac{\Delta N}{N}$$

The number of mobile electrons is generally proportional to stress (and hence strain $\epsilon = \Delta L/L$), that is,

$$\frac{\Delta N}{N} = \Pi\epsilon$$

where Π is a proportionality constant. This gives

$$\frac{\Delta R}{R} = [2 - \Pi]\epsilon \quad \text{or} \quad \frac{\Delta R}{R} = S_A\epsilon$$

where S_A is called the *material sensitivity*. Thus, the unit change in resistance can be related to the imposed strain on the conductor.

There are two main contributions to the resistance change (or material sensitivity), namely, dimensional changes (2) and the electron mobility (Π). Correspondingly, two types of gages are on the market, differentiated by the contribution that is emphasized. The metal gages have negligible stress-induced electron mobility giving material sensitivities of order 2; whereas for the semiconductor gages that utilize the piezoresistive effect, Π dominates and sensitivities up to ± 150 can be obtained.

An actual gage, however, is not a straight length of wire but is in the form of a grid as shown in Figure 2.16(b). (The original wire gages were wound back and forth over a paper former so as to maintain a large resistance but small size.) The actual gage sensitivity may not, therefore, be the same as the material sensitivity.

Three sensitivities can be defined depending on the type of strain field:

Longitudinal: $\quad S_L = \left.\dfrac{\Delta R}{R}\right/\epsilon_{xx}, \quad \epsilon_{yy} = 0$; uniaxial strain

Transverse: $\quad S_T = \left.\dfrac{\Delta R}{R}\right/\epsilon_{yy}, \quad \epsilon_{xx} = 0$; uniaxial strain

Gage: $\quad S_g = \left.\dfrac{\Delta R}{R}\right/\epsilon_{xx}, \quad \epsilon_{yy} = -\nu\epsilon_{xx}$; uniaxial stress

The gage sensitivity or *gage factor* is measured in a uniaxial stress field and this will give rise to an error when the gage is used in other than the calibrated situation. This is known as the transverse sensitivity error. In general, this error is not very large and therefore the extension in the gage direction (even in multiaxial strain fields) is assumed to be given by

$$\epsilon = \left.\frac{\Delta R}{R}\right/S_g$$

It is of interest to note that a strain gage measures only extension, that is, the normal component of strain. It does not measure shear strain directly—this must be obtained by deduction from a number of extension measurements. This is considered later when we discuss strain gage rosettes.

Temperature Effects

The discussion of the last section made no reference to the temperature effects. This topic must be treated separately because it is potentially the source of most serious errors in the measurement of static strains. When the thermal environment changes, both the gage and specimen are affected and this must be taken into account in deriving the expression for the change of resistance because these changes could be misinterpreted as strain.

Suppose the load-induced strain is negligible, then there are three main contributions to the change of length $\Delta L/L$ when the temperature changes:

Specimen elongates causing extension of gage: $\quad (\Delta L/L) = \alpha \Delta T$
Gage grid elongates: $\quad (\Delta L/L) = \beta \Delta T$
Resistivity of gage material changes: $\quad (\Delta R/R) = \gamma \Delta T = S_g(\Delta L/L)$

where α and β are coefficients of thermal expansion of the specimen and gage, respectively, and γ is the temperature coefficient of resistivity of the gage. The total apparent strain is then given by

$$ \frac{\Delta L}{L} = \left[(\alpha - \beta) + \frac{\gamma}{S_g} \right] \Delta T $$

This relation is not necessarily linear in temperature because the coefficients themselves may be functions of temperature. Thus, the apparent strain depends not only on the nature of the gage but also on the material to which it is bonded. The values of the coefficients for some relevant gage materials are given in Table 2.1.

The manufacturers adjust β and γ for a particular α to produce what are called *self-compensated* gages, that is, gages that show little apparent strain when the temperature changes. However, these gages usually have a very narrow range of application with regard to both temperature and specimen material.

Temperature compensation can also be effected by special circuitry design, and this will be discussed later.

Table 2.1: Properties of some gage materials.

Material	S_A	β	γ	[$\mu\epsilon/°$F]
Constantan	2.1	8	50	
Isoelastic	3.6	2	125	
Nichrome	2.1	9	250	
P-Silicone	110	0.3	–	

Gage Construction and Mounting

In practice, the active sensing element of the gage is bonded to the specimen so as to form an integral part of it. In this way, the deformation in the specimen is experienced exactly by the gage. Figure 2.17 shows the basic components of a strain gage installation.

The core of the gage is the sensing element, which may be of three basic types. The original gages were made by wrapping a very fine wire around a form. These gages have been superseded by foil gages that are formed by photoetching thin sheets (usually less than 0.005 mm (0.0002 in.) thick) of heat-treated metallic alloys to obtain the desired grid patterns and dimensions. Foil gages give excellent flexibility in the design of grid shapes. They also can be obtained in sizes ranging from 0.8 to 152 mm (0.032 to 6 in.). The newer form of gage is of the piezoresistive type, which uses a doped crystal element. These give high signal outputs but are quite inflexible with regard to design and shape.

Two of the more popular materials used in strain gages are Constantan, a copper–nickel alloy used primarily in static strain analysis because of its low and controllable temperature coefficient, and Isoelastic, a nickel–iron alloy recommended for dynamic tests where its large temperature coefficient is inconsequential, and advantage can be made of its high gage factor. Silicone is used primarily in semiconductor gages where the specially processed crystals are cut into filaments. They exhibit large resistance changes, but are very temperature sensitive and cannot be self-compensated.

The backing or carrier material is primarily a handling aid; it allows proper alignment and supports the lead wires. It also serves the function of insulating the gage from the specimen. A typical backing material is nitro-cellulose-impregnated paper, which is easy to handle and good up to about 80 °C (180 °F).

The adhesive must possess sufficient shear strength after curing to accurately transmit the specimen strain to the gage. Adhesives capable of maintaining a shear strength of about 10 MPa (1500 psi) are generally acceptable for strain gage work. The major adhesives are acrylic or epoxy based.

The lead wire system should not introduce significant resistance or generate and transmit signals that are not related to the gage output. The junction between the lead-in wire and the gage is particularly vulnerable and for this reason a step-down approach is always used.

Protective coatings are often used to prevent mechanical or chemical damage to the gage installation. The main source of this damage is moisture that can cause deterioration of the bond/backing system.

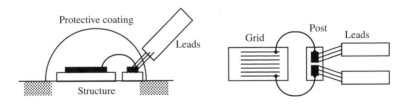

Figure 2.17: Strain gage construction and mounting.

The variety of gages available is enormous and all of them are designed for particular applications. The manufacturer generally lists them with designations such as the following from Micro-Measurements:

EA-06-250BG-120

where the above letters and numbers indicate, respectively: gage type, temperature compensation, sensing length, resistance, and options. There are also many special purpose gages that can be welded, or have high elongation or can be embedded in concrete. When ordering gages, the individual manufacturers, handbook should be consulted.

The semiconductor gage consists of a small thin filament of single crystal silicon. Depending on the type and amount of impurity diffused into the pure silicon, the sensitivity can be either positive or negative—Boron is used in producing P-type (positive) while Arsenic is used to produce N-type (negative) gages. They exhibit high strain sensitivity (on the order of 150) but have high cost, limited strain range (about $3000\,\mu\epsilon$), and large temperature effects. They also have problems when mounted on curved surfaces.

Strain Gage Rosettes

At a particular point on the surface of a deformed structure, there are, in general, three unknown strains; either $(\epsilon_{xx}, \epsilon_{yy}, \gamma_{xy})$ or $(\epsilon_1, \epsilon_2, \theta)$. Therefore, three independent pieces of experimental data must be recorded if the strain state is to be specified. Since one strain gage cannot measure shear directly, three gages must be arranged to give three independent extension measurements and the transformation equations used to determine the shear component. This combination of three (sometimes four) gages is called a *strain gage rosette*. In practice, rosettes are usually mounted with respect to a reference axis (not necessarily the principal strain axes).

It is possible to mount three separate gages on the specimen but this can be awkward if the surface area is small. Having the three separate gages oriented correctly on the same backing allows all three gages to be easily mounted simultaneously. Note, however, that each gage must have its own circuit and be wired separately.

To see how the data reduction is achieved, consider the three gages shown in Figure 2.18 measuring the three extensions: $(\epsilon_A, \epsilon_B, \epsilon_C)$. Using the transformation law,

$$\epsilon_{\theta\theta} = \epsilon_{xx} \cos^2\theta + \epsilon_{yy} \sin^2\theta + \gamma_{xy} \sin\theta \cos\theta \tag{2.4}$$

these normal components can be written in terms of the components referred to the x–y reference axes as

$$\begin{Bmatrix} \epsilon_A \\ \epsilon_B \\ \epsilon_C \end{Bmatrix} = \begin{bmatrix} \cos^2\theta_A & \sin^2\theta_A & \cos\theta_A \sin\theta_A \\ \cos^2\theta_B & \sin^2\theta_B & \cos\theta_B \sin\theta_B \\ \cos^2\theta_C & \sin^2\theta_C & \cos\theta_C \sin\theta_C \end{bmatrix} \begin{Bmatrix} \epsilon_{xx} \\ \epsilon_{yy} \\ \gamma_{xy} \end{Bmatrix}$$

or $\{\epsilon_{\text{data}}\} = [\,C\,]\{\epsilon_{\text{coord}}\}$. Since there are three equations and three unknowns, these equations can be solved, but rather than doing this in general, consider the following special cases.

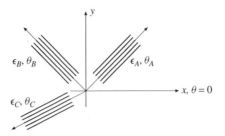

Figure 2.18: Strain gage orientations for a general rosette.

For the special case of the rectangular rosette, $\theta_A = 0°$, $\theta_B = 45°$, $\theta_C = 90°$, the coordinate strains are obtained as

$$\epsilon_{xx} = \epsilon_A, \qquad \epsilon_{yy} = \epsilon_C, \qquad \gamma_{xy} = 2\epsilon_B - \epsilon_A - \epsilon_C \qquad (2.5)$$

For the special case of the delta rosette: $\theta_A = 0°$, $\theta_B = 120°$, $\theta_C = 240°$, the coordinate strains are obtained as

$$\epsilon_{xx} = \epsilon_A, \qquad \epsilon_{yy} = \frac{2}{3}\left[\epsilon_B + \epsilon_C - \frac{1}{2}\epsilon_A\right], \qquad \gamma_{xy} = \frac{2}{\sqrt{3}}[\epsilon_C - \epsilon_B] \qquad (2.6)$$

The general equation can be specialized for other cases also.

2.4 Strain Gage Circuits

Consider a foil gage mounted on a bar of steel experiencing a stress of 10000 psi (E/300). The change in resistance is

$$\frac{\Delta R}{R} = S_g\epsilon = 2\frac{\sigma}{E} = 2\frac{10000}{30 \times 10^6} = 0.00066 \; \Omega/\Omega$$

This change in resistance is a very small quantity, especially when we realize that the resistance before extension is 120.0000 and after is 120.0007. The problem is that any ohmmeter would utilize all its accuracy on the 120.000 portion of the measurement. What is needed is to measure the difference directly. The main circuits for accomplishing this will be reviewed here.

Wheatstone Bridge

The Wheatstone bridge was the first circuit designed to measure ΔR and is very accurate because it is based on a null principle.

Referring to Figure 2.19, from the simple circuit equation $V = IR$, we get the two currents as

$$I_{abc} = \frac{V}{(R_1 + R_2)}, \qquad I_{adc} = \frac{V}{(R_3 + R_4)}$$

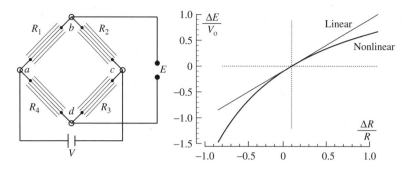

Figure 2.19: The Wheatstone bridge. (a) Circuit diagram. (b) Nonlinearity of Wheatstone bridge.

From these, the voltage drops to b and d can be obtained as

$$V_{ab} = I_{abc} R_1 = \frac{V R_1}{(R_1 + R_2)}, \qquad V_{ad} = I_{adc} R_4 = \frac{V R_4}{(R_3 + R_4)}$$

Therefore, the voltage drop across bd is

$$E = V_{ab} - V_{ad} = V \left[\frac{R_1}{(R_1 + R_2)} - \frac{R_4}{(R_3 + R_4)} \right] \tag{2.7}$$

Suppose the resistances in the bridge circuit change by amounts ΔR_1, ΔR_2, etc., then the output voltage is

$$E + \Delta E = V \left[\frac{1 + \delta_1}{(1 + \delta_1 + r_{12}\delta_2)} - \frac{1 + \delta_4}{(1 + \delta_3 + r_{34}\delta_4)} \right], \qquad \delta_i \equiv \frac{\Delta R_i}{R_i}, \qquad r_{ij} \equiv \frac{R_j}{R_i}$$

From the measurement point of view, we wish to infer strain (δ) from the measurement of E and $E + \Delta E$. This is a highly nonlinear relation and it requires computing differences. It is seen from Equation (2.7) that E can be made zero (called *Balancing*) by having

$$\frac{R_1}{(R_1 + R_2)} = \frac{R_4}{(R_3 + R_4)} \qquad \text{or} \qquad r_{12} = \frac{R_2}{R_1} = \frac{R_3}{R_4} = r_{34} = r$$

This feature will be used shortly to eliminate the need for differences. This gives

$$\Delta E = V \left[\frac{r[\delta_1 - \delta_2 + \delta_3 - \delta_4 + \delta_1\delta_2 - \delta_3\delta_4]}{[(1 + r) + \delta_1 + r\delta_2][(1 + r) + \delta_4 + r\delta_3]} \right] \tag{2.8}$$

This is highly nonlinear. Consider when only one resistor is active; then

$$\Delta E = V \frac{r}{(1 + r)^2} \left[\frac{\delta_1}{1 + \delta_1/(1 + r)} \right] = V \frac{r}{(1 + r)^2} \left[\frac{\Delta R_g/R_g}{1 + \Delta R_g/R_g/(1 + r)} \right]$$

Figure 2.19(b) shows the case when $r = 1$ and $V_o \equiv Vr/(1 + r)^2$. It is apparent that if the nonlinear effects were neglected, then $\delta = \Delta R/R$ must be less than 0.1 in order to have an error less than about 5% error.

Let the nonlinear terms in ΔR be neglected, then ΔE is obtained directly and is related to the resistance changes by

$$\Delta E = \frac{Vr}{(1+r)^2}\left[\frac{\Delta R_1}{R_1} - \frac{\Delta R_2}{R_2} + \frac{\Delta R_3}{R_3} - \frac{\Delta R_4}{R_4}\right], \qquad r = \frac{R_2}{R_1} = \frac{R_3}{R_4} \qquad (2.9)$$

This is the basic Wheatstone bridge equation. The remarkable thing is that ΔE is the sole output from the balanced circuit, that is, the change in voltage is measured directly. It is important to note that the voltage measured is $E + \Delta E$, in general, but when the bridge is initially balanced then E is zero and the change ΔE is measured directly. It is also noted that some of the resistance changes are positive while others are negative. It is this occurrence that will be used to a great advantage as another means of temperature compensation.

Suppose R_1 is the only active resistance, that is, it is the gage, then $\Delta R_2 = \Delta R_3 = \Delta R_4 = 0$ and

$$\Delta E = \frac{Vr}{(1+r)^2}\frac{\Delta R_1}{R_1}$$

Consequently, the strain can be obtained from

$$\epsilon = \frac{\Delta R_1}{R_1}/S_g = \frac{(1+r)^2}{rS_g}\frac{\Delta E}{V} \qquad (2.10)$$

Hence the strain is easily determined by measuring the voltage output from the circuit.

Circuit Parameters and Temperature Compensation

There are many possibilities in utilizing the Wheatstone bridge, so it is of advantage now to establish a number of parameters that will aid in the evaluation of the different circuits.

The circuit sensitivity is defined as the change in output voltage for a given strain; that is,

$$S_c = \frac{\Delta E}{\epsilon} = \frac{V}{\epsilon}\frac{r}{(1+r)^2}\left[\frac{\Delta R_1}{R_1} - etc\right] = \frac{I_g R_g}{\epsilon}\frac{r}{(1+r)}\left[\frac{\Delta R_1}{R_1} - etc\right]$$

Suppose only R_1 is active, then with $R_1 = R_g$, we get

$$S_c = \frac{1}{\epsilon}\frac{r}{(1+r)^2}I_g R_g S_g, \qquad V = I_g R_g(1+r)$$

The circuit sensitivity is seen to be made up of two terms:

- $r/(1+r)$: This is called the circuit efficiency; since for a given gage the rest of the circuit (R_2 or R_4) can be changed to make this ratio approach unity. In practice, however, there is an upper limit to r since the voltage required to drive the circuit becomes excessive. A value of $r = 1$ is usually chosen.

- $I_g R_g S_g$: This relates specifically to the gage specifications. Note that there is usually an upper limit put on the current since otherwise the gage may overheat, that is, the power to be dissipated by the gage is $P = I^2 R$.

The ability of the gage to carry a high current is a function of its ability to dissipate heat, which depends not only on the gage itself but also on the specimen structure. Currents may reasonably be restricted to 10 milliamps (mA) for long-term work on metallic specimens, but up to 50 mA are used for dynamic (transient) work. On poor heat sinks such as plastics, the currents are correspondingly decreased.

Gage resistances are grouped around values such as 60, 120, 350, 500, and 1000 ohms. The higher resistances are not necessarily preferable to the lower resistances unless they can also carry currents such that IR is high. Often, several low-resistance gages on a specimen can be placed in series and thus dissipate larger amounts of heat.

The above parameters can be used to evaluate the circuits shown in Figure 2.20. Case (a) of the figure has an active gage in position 1. From before

$$S_c = \frac{r}{(1+r)} I_g R_g S_g \qquad S_{cmax} = I_g R_g S_g \qquad S_{cnom} = \tfrac{1}{2} I_g R_g S_g$$

To achieve the maximum response, a large r (and hence V) is required. The nominal is when all resisters have the same values; we will use this as a reference. Case (b) has an active gage in position 1 and dummy gage in position 4. Since the analysis is the same as in case (a)

$$S_c = \frac{r}{(1+r)} I_g R_g S_g \qquad S_{cmax} = I_g R_g S_g \qquad S_{cnom} = \tfrac{1}{2} I_g R_g S_g$$

This circuit is as effective as case (a). The difference, however, can be utilized if Gage 4 is put in the same temperature field as Gage 1 but is not subject to strain, then the change in voltage is given by

$$\Delta E = \frac{Vr}{(1+r)^2} \left[\frac{\Delta R_g^S}{R_g} + \frac{\Delta R_g^T}{R_g} - \frac{\Delta R_g^T}{R_g} \right] = \frac{Vr}{(1+r)^2} \frac{\Delta R_g^S}{R_g}$$

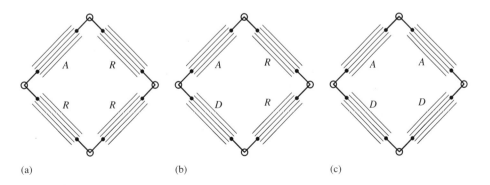

(a) (b) (c)

Figure 2.20: Sample circuits. A = active, D = dummy, R = completion resistor.

The superscripts S and T refer to strain and temperature effects, respectively. Thus, the temperature effects of the two gages cancel each other leaving a voltage change that is due only to the strain. This is an effective way of providing temperature compensation.

Consider a similar case with an active gage in position 1 and dummy gage in position 2. The analysis is the same as the above except that $r = 1$ giving $S_{cmax} = I_g R_g S_g/2$. While this circuit is also capable of temperature compensation, its maximum sensitivity is only half that of case (b) and therefore we could conclude that (b) is the better circuit design.

Case (c) has two active and two dummy gages. The sensitivity is given by

$$S_c = \frac{\Delta E}{\epsilon} = \frac{V}{\epsilon}\frac{r}{(1+r)^2}\left[\frac{\Delta R_1}{R_1} - \frac{\Delta R_2}{R_2}\right] = \frac{I_g R_g}{\epsilon}\frac{r}{(1+r)}\left[\frac{\Delta R_1}{R_1} - \frac{\Delta R_2}{R_2}\right]$$

and since the two active gages experience the same strain (and have the same gage factor),

$$S_c = \frac{r}{(1+r)}2I_g R_g S_g \qquad S_{cmax} = I_g R_g S_g \qquad S_{cmax} = I_g R_g S_g$$

Thus, this circuit is not only capable of temperature compensation but gives a doubling of circuit sensitivity.

Note that for some arrangements, the resistance changes cancel. This can be utilized in two ways as shown in Figure 2.21. Suppose a specimen is in simple bending with the top in tension and bottom in compression, then placing the gages at the top and bottom will double the output. Furthermore, if this is put in a transducer, then any unwanted axial components of strain would be eliminated. Note that making R_3 instead of R_2 active allows axial strains to be measured while canceling any unwanted bending strains. As a final case, consider when all gages are active. This could be used in an axial transducer using the Poisson's ratio effect to contribute the negative strains for arms 2 and 4.

Potentiometer and Constant-Current Circuits

The Wheatstone bridge circuit is adept at measuring small resistance changes; and further, it can easily be adapted for temperature compensation without adversely affecting circuit performance. In some cases, however, even simpler circuits can be used. This is especially true of dynamic problems and two useful circuits will now be reviewed.

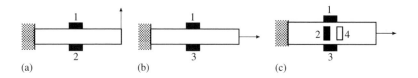

(a)   (b)   (c)

Figure 2.21: Some circuit arrangements. (a) Amplifies bending, cancels axial effects. (b) Amplifies axial, cancels bending effects. (c) Amplifies axial, cancels bending effects and has temperature compensation.

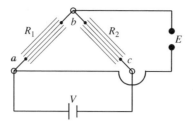

Figure 2.22: Potentiometer circuit.

If R_3 and R_4 are removed from the Wheatstone bridge, then with $V = (R_1 + R_2)I$, the voltage across R_1 is

$$E = R_1/(R_1 + R_2)$$

The potentiometer circuit of Figure 2.22 has the same nonlinearities as the full Wheatstone bridge. The change in voltage for small changes in resistance reduces to

$$\Delta E = \frac{Vr}{(1+r)^2}\left[\frac{\Delta R_1}{R_1} - \frac{\Delta R_2}{R_2}\right] = I_g R_g \frac{r}{(1+r)}\left[\frac{\Delta R_1}{R_1} - \frac{\Delta R_2}{R_2}\right]$$

Note that here too, $E + \Delta E$ is actually measured, but unlike the Wheatstone bridge E is not zero. However, if the event is dynamic and the output connected to the ac option on an oscilloscope or an RC differentiator circuit, the constant dc component, E, can be blocked. In this way, only the variable part (which is due to the change in resistance) will be recorded. The sensitivity of the circuit is given by

$$S_c = \frac{r}{(1+r)}I_g R_g S_g$$

which is the same as case (a) of the Wheatstone bridge. But if this circuit is used for temperature compensation ($r = 1$) then its performance is not as good as case (b), and so the Wheatstone bridge would be preferable.

So far, the power sources in the circuits have been constant voltage sources. However, from the analysis of the circuits it seems that the current (and not the voltage) is the relevant quantity, and so a constant current source seems more appropriate. To see this, reconsider the potentiometer circuit. The current through the gage is

$$I_g = \frac{V}{R_g + R_2}$$

If R_2 (the ballast resister) is large then for small changes in R_g (due to strain) the current will remain essentially constant since

$$\Delta I_g = -\frac{V}{(R_g + R_2)^2}\Delta R_g = -\frac{I_g}{R_g + R_2}\Delta R_g \approx 0$$

This is a simple way of achieving constant current, but a large voltage source is required to maintain $V/R_2 = $ constant as R_2 is increased. With the advent of technology that made operational amplifiers readily available, efficient and economical constant-current sources can be made. The function of the op-amp is to essentially maintain a zero potential across *ab* and thus allow all the current to go through R_g which will then be the same as in R_2. Basically, the constant-current power supply is a high-impedance (1 to 10 MΩ) device that changes the output voltage with changing resistive load to maintain a constant current; that is,

$$I_g = \frac{V}{R_2}$$

Irrespective of the value of R_g or how much it changes, the current will remain essentially constant.

The change in voltage across the gage is

$$\Delta E = I_g \Delta R_g = I_g R_g S_g \epsilon$$

The sensitivity is therefore

$$S_c = I_g R_g S_g$$

which, in comparison to the Wheatstone bridge parameters, has an efficiency of 100%. This circuit is also linear in ΔR even for large changes in resistances, which makes it quite suitable for use with semiconductor gages.

This circuit, however, cannot be used for temperature compensation but it was concluded that the potentiometer circuit was poor in this respect anyway. Also, temperature compensation is generally not as important in dynamic testing (especially if self-compensated gages can be used).

It is also possible to use a constant-current source in a regular Wheatstone bridge for static testing, but not much (except a slight decrease in circuit nonlinearity) is achieved. Constant-current sources are primarily used for dynamic work because of their simple circuit design, high efficiency, high linearity, and low noise.

Commercial Strain Indicators

Perhaps the most commonly used bridge arrangement for static strain measurements is that of the reference bridge. In this arrangement, the bridge on one side contains the strain gage or gages, and the bridge on the other side contains variable resistors. When gages are attached, the variable resistance is adjusted to obtain initial balance between the two bridges. Strains producing resistance changes cause an unbalance between the two bridges, which can then be read by a meter. In some cases, this meter reading is obtained by further nulling the adjustments of the variable resistors.

The advantage of the reference-bridge arrangement is that the gage bridge is left free for the strain gages and all adjustments for the initial and null balance are performed on the other bridge.

Recent advances in electronics have lead to the development of direct-reading Wheatstone bridge arrangements. The output of the gage bridge is the input of an instrument amplifier. Potentiometer adjustments are used to initially balance the amplifier and to null the bridge output. In contrast to the reference-bridge arrangement, the nulling is achieved by adjusting a voltage on the amplifier rather than adjusting a resistance in one arm of the bridge. The analog output from the amplifier is converted into digital format for display. The sensitivity of the bridge is adjusted through the amplifier gain—typical gain settings are ×4 and ×40 [48].

Often, it is required that many strain gages be installed and monitored several times during testing. In these circumstances, it would be very costly to use a separate recording instrument for each gage. An effective economic solution is to use a single recording instrument, and let the gages be switched in and out of this instrument. Two different methods of switching are commonly used in multiple-gage installations.

One method involves switching each active gage, in turn, into arm R_1 of the bridge as shown in Figure 2.23. Since the switch is located within arm R_1 of the bridge, a high-quality switch must be used. The advantage of this arrangement is that it is simple and inexpensive, the disadvantages are that the gages must be nominally the same, only a single compensating gage can be used, and the gage outputs cannot be individually nulled.

The second method switches the complete bridge. The advantage is that the switch is not located in the arms of the bridge, and hence switching resistance is not so important. Other advantages are that each gage can be different and can be individually nulled. However, the arrangement is clearly more expensive to implement.

Experiments using Strain Gages

Most types of electrical resistance strain gages are usually adequate for dynamic work and since the events are of short duration, temperature compensation is usually not a serious issue. The largest gage size allowable is dependent on the highest significant frequency (or shortest wavelength) of the signal, but generally, gages equal to or less than 3 mm (0.125 in.) are good for most situations. A rule of thumb for estimating the maximum length is

$$L = \frac{1}{10f}\sqrt{\frac{E}{\rho}} \qquad (2.11)$$

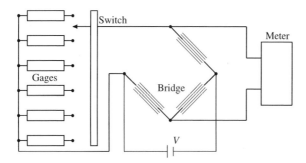

Figure 2.23: Schematic of the switching unit.

where L is the gage length, f is the highest significant frequency, and E and ρ are the Young's modulus and density, respectively, of the structural material. A typical gage of length 3 mm (0.125 in.) is good to about 160 kHz. Other aspects of the use of strain gages in dynamic situations are covered in References [27, 131]. Either constant-current, potentiometer or Wheatstone bridge gage circuits can be used.

Experiment I: Nonlinearity of a Wheatstone Bridge

The Wheatstone bridge is an inherently nonlinear device and this must be taken into account when large resistance changes (as with semiconductor gages) are expected. We compare $\Delta R/R$ to 1.0. If the strain is $\epsilon = 1000\,\mu\epsilon$ and the gage factor is $S_g = 2$, then

$$\Delta R/R = 2 \times 1000 \times 10^{-6} = 0.002$$

which can be reasonably neglected in comparison to unity. If, however, the gage is of the semiconductor type with $S_g = 150$ then

$$\Delta R/R = 150 \times 1000 \times 10^{-6} = 0.15$$

which is significant. Hence, semiconductor gages are usually restricted to small-strain situations.

Figure 2.24(a) shows the recorded voltage taken from a rod impact experiment. The set up is essentially that of References [65, 149]. As the wave propagates, the pulse alternates in sign with a gradually decreasing amplitude due to various dissipation mechanisms. The magnitude of $\Delta R/R$ (which is proportional to $\Delta E/V_o$) is not very large. Figure 2.24(b) shows the square of the recorded voltage. Clearly, there is a consistent difference between the compressive and tensile responses with the former overestimating and the latter underestimating the average voltage.

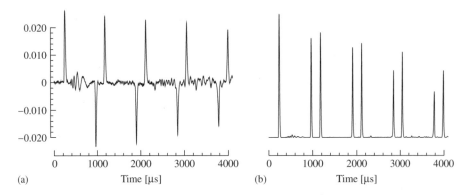

Figure 2.24: Bridge nonlinearity observed in a wave propagation problem. (a) Recorded voltage $\Delta E/V_o$. (b) Recorded voltage squared.

The nonlinear behavior is approximated as (with $r \approx 1$)

$$\frac{\Delta E}{V_{\text{o}}} \approx \delta_1[1 - \delta_1/2], \qquad \left(\frac{\Delta E}{V_{\text{o}}\delta_1}\right)^2 \approx [1 - \delta_1]$$

Since $\delta_1 \approx 0.02$, we see that this is indeed the order of magnitude of the differences observed.

Experiment II: Force Transducer for Static Measurements

Figure 2.25 shows two possible force transducers and their gage arrangements for static measurements.

The use of a ring design converts the axial load into a bending load and therefore increases the strain. Additionally, since there are multiple tension and compression points on the inside of the ring, a complete (four-arm) bridge arrangement can be used. This further increases the sensitivity of the load cell, provides temperature compensation, and avoids the need for completion resistors.

The C-shaped design is a little simpler. It also induces both tension and compression strains due to bending and therefore has attributes similar to the ring. However, (at least in the case shown in the figure), only two gages are used. Note that if a commercial strain indicator is used, then completion resistors are supplied automatically.

Figure 2.25(c) shows the calibration results for the C-shaped transducer; the calibration was done under dead weight loading. For this simple geometry it is possible to calculate the transducer sensitivity, but nonetheless, a calibration is usually in order. Typically, a (commercial) strain indicator is used; therefore a strain gage factor must be set. The gage factor need not be the actual one, however, each time the transducer is connected up, the same one must be used.

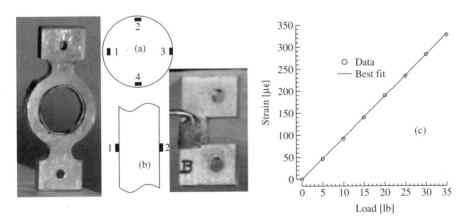

Figure 2.25: Two force transducers and their gage layouts. (a) Ring with four-arm bridge. (b) C-shape with two-arm bridge. (c) Calibration curve for the C-shaped transducer.

2.5   Motion and Force Transducers

Seismic transducers (accelerometers, velocimeters) make motion measurements by measuring the relative motion between the object and a seismic mass. Pressure and force transducers have the same mechanical characteristics as seismic transducers and hence we will treat them together here. Additional details on the material covered here can be found in References [52, 120].

Basic Theory

The mechanical behavior of force, pressure, and motion-measuring instruments can be modeled using a single degree of freedom mechanical model with viscous damping as shown in Figure 2.26(a).

The equation of motion for the seismic mass (M) is obtained from the free-body diagram and Newton's second law of motion to give

$$M\frac{d^2 u_m}{dt^2} + C\frac{d}{dt}(u_m - u) + K(u_m - u) = P(t) \tag{2.12}$$

where K is the linear spring constant, C is the viscous damping constant, $u(t)$ is the base motion, $u_m(t)$ is the seismic mass motion, and $P(t)$ is the time-dependent external force or pressure applied to the transducer.

The sensing element measures the relative motion between the base and the seismic mass. These motions can be expressed as $w = u_m - u$ so that Equation (2.12) can then be written as

$$M\frac{d^2 w}{dt^2} + C\frac{dw}{dt} + Kw = P(t) - M\frac{d^2 u}{dt^2} = R(t) \tag{2.13}$$

The right-hand side is a general external forcing function that is dependent on the base motion, and the external forces (including pressure).

Before proceeding with a discussion of the effects of damping, we would first like to get a measure of what is meant by small amounts of damping. To this end, consider the free vibration of the system with viscous damping and look for particular solutions of

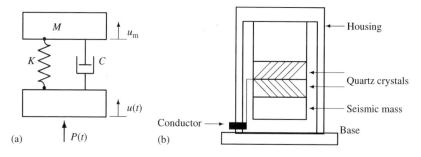

Figure 2.26: Schematic of a seismic transducers. (a) Notation. (b) Construction of an accelerometer.

the form $u(t) = Ae^{i\alpha t}$. Substitute into the differential equation and get the characteristic equation

$$A[K + iC\alpha - M\alpha^2] = 0 \qquad \text{or} \qquad \alpha = \frac{iC}{2M} \pm \frac{1}{2M}\sqrt{4MK - C^2}$$

The time response of the solution is affected by the sign of the radical term for α as

$$\begin{aligned} C^2 > 4MK; & \qquad \text{overdamped} \\ C^2 = 4MK; & \qquad \text{critical damping} \\ C^2 < 4MK; & \qquad \text{underdamped} \end{aligned}$$

Let the critical damping be given by $C_c \equiv \sqrt{4MK} = 2M\omega_o$, then the characteristic values of α are given by

$$\alpha = \omega_o[i\zeta \pm \sqrt{1 - \zeta^2}]$$

where $\zeta \equiv C/C_c$ is the ratio of the damping to critical damping. The free vibration solutions are

$$u(t) = e^{-\zeta\omega_o t}[Ae^{-i\omega_[\text{ D }]t} + Be^{+i\omega_[\text{ D }]t}]$$

where $\omega_d \equiv \omega_o\sqrt{1 - \zeta^2}$ is called the *damped natural frequency*. The critical point occurs when $\zeta = 1$, thus we say that the structure is lightly damped when $\zeta \ll 1$. This is the situation of most interest to us in structural analysis, however, in instrument design, we set it much higher.

Consider the motion of the mass after it is displaced from its initial position and released. Using initial conditions that at $t = 0$

$$u(0) = u_o, \qquad \dot{u}(0) = 0$$

gives the solution

$$u(t) = \frac{1}{2}u_o e^{-\omega_o \zeta t}\left[\left(1 + \frac{i\omega_o\zeta}{\omega_d}\right)e^{-i\omega_d t} + \left(1 - \frac{i\omega_o\zeta}{\omega_d}\right)e^{+i\omega_d t}\right]$$

which is shown plotted in Figure 2.27. Note that the response eventually decreases to zero, but oscillates as it does so. The frequency of oscillation is $\omega_d = \omega_o\sqrt{1 - \zeta^2} \approx \omega_o(1 - \frac{1}{2}\zeta^2)$. Hence, for small amounts of damping this is essentially the undamped natural frequency. The rate of decay is dictated by the term $e^{-\omega_o\zeta t} = e^{-Ct/2M}$.

Motion and Force Transducers

Seismic displacement transducers measure the displacement of the base relative to the seismic mass. This type of transducer behavior can be illustrated by assuming that the excitation force $P(t)$ is zero. Let the steady state response to a sinusoidal excitation be of the form

$$R(t) = \hat{R}_o e^{i\omega t}, \qquad w(t) = \hat{w}_o e^{i\omega t}$$

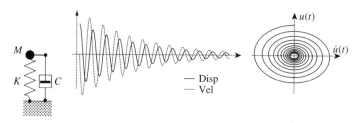

Figure 2.27: Damped response due to initial displacement.

then the corresponding solution from Equation 2.12 becomes

$$\hat{w} = \frac{M\omega^2\hat{u}}{K + i\omega C - M\omega^2} = \frac{(\omega/\omega_o)^2\hat{u}}{1 + i\omega2\xi - (\omega/\omega_o)^2} \qquad (2.14)$$

This is the instrument FRF. Figure 2.28(a) shows the magnitude ratio $|\hat{w}/\hat{u}|$ as a function of the frequency ratio for different damping ratios. The magnitude ratio is seen to approach a value of unity regardless of the amount of damping. These results show that the natural frequency should be as low as possible so that the operating frequency range is as large as possible under most measurement situations.

For high frequencies, Equation (2.14) shows that $\hat{w} = -\hat{u}$. This means that the base vibrates relative to the stationary seismic mass. Thus, a large rattle space with soft springs is required to accommodate the large relative motion that takes place within the instrument. The type of sensing elements in these transducers are Linear Voltage Differential Transformers (LVDTs) and are often strain gages on flexible elastic support members.

Seismic velocity transducers are obtained by employing velocity-dependent magnetic sensing elements in displacement transducers to measure the relative velocity between the base and the seismic mass. Differentiation of Equation (2.14) with respect to time

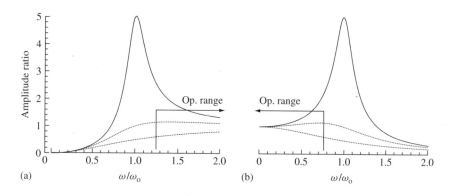

Figure 2.28: Frequency response function. (a) Displacement transducer. (b) Accelerometer.

(multiply by iω) gives the FRF for a seismic velocity transducer to be

$$\hat{v} = \frac{i\omega M\omega^2 \hat{u}}{K + i\omega C - M\omega^2} = \frac{(\omega/\omega_o)^2 \hat{v}_b}{1 + i\omega 2\xi - (\omega/\omega_o)^2} \qquad (2.15)$$

which reduces to $\hat{v} = -\hat{v}_b$ and $\delta = \pi$ for high frequency. The corresponding FRF is identical to that shown for the displacement transducer in Figure 2.28(a). The major disadvantages of this type of instrument are size, weight, and sensitivity to stray magnetic fields.

Seismic accelerometers are constructed by using displacement sensors with a stiff spring to give a high natural frequency. For frequencies far below the natural frequency of the transducer, Equation (2.13) shows that the relative motion is proportional to the base acceleration ($\hat{a} = -\omega^2 \hat{u}$). Thus,

$$\text{force:} \quad K\hat{w} = \frac{K M\omega^2 \hat{u}}{K + i\omega C - M\omega^2} = \frac{-\hat{a}}{1 + i\omega 2\xi - (\omega/\omega_o)^2} \qquad (2.16)$$

Figure 2.28(b) shows the FRF magnitude of Equation (2.16) where it is seen that the output is near-unity when the frequency ratio is near-zero regardless of the amount of damping. The peak response occurs around $\omega/\omega_o = 1.0$ for small values of damping, and decreases for increasing values of damping in the range between 0 and 0.707.

There are no magnitude peaks for damping ratios greater than 0.707. The usable frequency range for this type of instrument is limited to frequencies below 20% of the natural frequency for magnitude and errors less than 5% for a range of lightly damped instruments where $\xi < 0.01$ to 0.06. The phase shift is also small under these conditions.

For ω/ω_o near zero, Equation (2.16) reduces to

$$\text{force:} \quad K\hat{w} = -M\hat{a} \quad \text{or} \quad Kw(t) = -Ma(t)$$

which shows that the basic sensing mechanism is that of an inertia force Ma_o being resisted by the spring force Kw. This also shows that the high natural-frequency requirement is detrimental to obtaining large sensitivities since the relative motion is inversely proportional to the square of the natural frequency. Fortunately, an accelerometer with a piezoelectric sensing element can simultaneously provide the required stiffness and sensitivity.

Force and pressure transducers require stiff structures and high natural frequencies. Consequently, they are seismic instruments that respond to base acceleration as well as force or pressure. The FRF for these transducers is given by Equation (2.13), so that they have the same magnitude and phase response curves as those of the accelerometer. Thus, force and pressure transducers have characteristics that are very similar to accelerometers.

Construction of Accelerometers

Figure 2.26(b) shows a schematic for the construction of a piezoelectric accelerometer. Piezoelectric accelerometers function to transfer shock and vibratory motion into high-level, low-impedance (100 Ω) voltage signals compatible with readout, recording or analyzing instruments. They are quite sensitive (typically 10 mV/g) sensors and operate reliably over wide amplitude and frequency ranges.

They are structured with permanently polarized compression-mode quartz elements and a microelectronic amplifier housed in a lightweight metal case. The built-in electronics operate over a coaxial or two- conductor cable; one lead conducts both signal and power. They typically have a resonance on the order of 50 kHz giving a useful frequency range (for 5% error) of 0–10 kHz.

During operation, a change in acceleration of the base exerts a force on the seismic mass through the crystals, resulting in an electrostatic charge (ΔQ) proportional to the change in input acceleration. This charge is collected in the crystal shunt capacitance (C), resulting in a voltage (ΔV), in accordance with the electrostatic equation

$$\Delta V = \frac{1}{C} \Delta Q$$

This voltage is impressed across the gate of the FET (Field Effect Transistor) source follower, resulting in the same voltage change at the source terminal, but at a much lower impedance level. This signal at the source terminal is added to a dc bias voltage of approximately +11 volts. The signal is separated from the bias voltage in the power units.

Good high-frequency transmissibility into the base of the accelerometer is dependent upon intimate contact between the base and the mounting surface. The mating surface, in contact with the accelerometer base must be machined as smooth and flat as possible. Surface grinding, while not always possible, is ideal, however, spot facing with a sharp tool, ground flat is also acceptable. A light film of silicone grease between mating surfaces will enhance transmissibility of high frequencies. For more permanent installations, a thin layer of epoxy between mating surfaces will ensure good high-frequency performance without fear of the accelerometer becoming loose during the use.

Petro wax is a convenient temporary mounting method for small accelerometers used in place of studs and permanent adhesives. Test results show that frequency responses comparable to stud mounting is achievable with a thin coating.

Accelerometers for Structural Testing

A wide variety of small accelerometers are available. There is a problem with accelerometers, however, in that if the frequency is high enough, ringing will be observed in the signal. This can often occur for frequencies as low as 20 kHz. Generally speaking, accelerometers do not have the necessary frequency response of strain gages, but they are movable, thus making them more suitable for larger, complex structures.

The accelerometers used are PCB 303A and 305 [182]. These are light weight (2 and 4 g, respectively) and have good sensitivity (1.0 mV/g).

Experiment: Force Transducer for Impact Measurements

Figure 2.29(a) shows a force transducer used for impact studies. The commercial PCB force transducer (which has flat faces) is sandwiched between two masses. The back mass is necessary in order to register the force.

The front and back masses are machined from 5/8 in. steel rod and the two pieces are bonded to the transducer with epoxy cement. The front mass was machined with a near-spherical cap, so that the impacts would be point impacts. The transducer was made part of a light weight (but stiff) pendulum system of about 500 mm (20 in.) radius.

Because of the attached masses, it is necessary to recalibrate the transducer. To this end, a mass with an attached accelerometer was freely suspended by strings and used as the impact target. The accelerometer (Kistler, Piezotron model 818) has a sensitivity of 9.94 mV/g and a resonance of 35 kHz. The weight of the combined mass and accelerometer was 0.55 lb. Metal-on-metal contact can generate forces with a high-frequency content; however, both the force transducer and accelerometer have an upper resonance limit that prevents them from accurately registering these signals. It was therefore necessary to place a masking tape on the mass so as to soften the impact. The recorded accelerometer and force voltage output are shown in Figure 2.30(a).

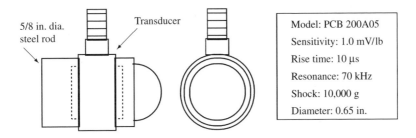

Model: PCB 200A05	
Sensitivity: 1.0 mV/lb	
Rise time: 10 μs	
Resonance: 70 kHz	
Shock: 10,000 g	
Diameter: 0.65 in.	

Figure 2.29: Schematic of the construction of a force transducer for impact studies.

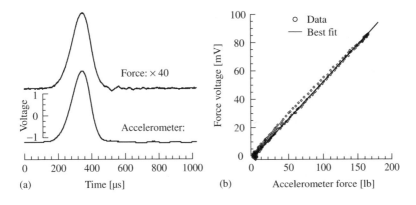

Figure 2.30: Calibration experiment for force transducer. (a) Voltage histories. (b) Calibration plot.

When wave propagation effects are not significant, the relation between the force and acceleration of the mass is simply Newton's second law

$$P = M\ddot{u} = \frac{W}{g}a$$

where W is the weight of the impacted mass. The accelerometer output was converted to force as above and used for the horizontal axis of Figure 2.30(b); that is, a plot of $P(t)$ against force transducer voltage should give a straight line if propagation (resonance) effects are negligible. Figure 2.30(b) shows that there is a slight difference between the loading and unloading paths but there is a reasonable linearity.

A best-fit of all the data gave the transducer sensitivity as 0.546 mV/lb. This compares to the manufacturer-reported value of 1.0 mV/lb. The difference is that the quasistatic calibration involved forces at both ends of the transducer, whereas the dynamic calibration involves a force only at the front end. The location of the active part of the transducer (near the mid-length) is such that it experiences only about half of the load applied at the tip.

2.6   Digital Recording and Analysis of Images

Digital cameras have become commonplace because they are inexpensive, and because they allow the manipulation of image data in ways that are impossible with traditional cameras and films. This has had a profound effect on the optical methods of stress analysis, and therefore we begin with a discussion of the digital analysis of images.

Concept of Light and Color

As shown in Figure 2.31, light can be considered as tiny bundles of energy traveling in space at a speed of about 300×10^6 m/s. Associated with these bundles are electromagnetic fields that vary in all directions (spherical) with a 1/R singularity. In addition, the field strength is strongest along the radius vector. Hence, a string of these bundles (a light ray) will give rise to fields that vary perpendicularly to the direction of propagation; that is, a sensor placed in the vicinity of the stream would detect a periodic variation (about 10^{15} Hz) in intensity. It is this periodic variation that allows light to also be considered as a wave phenomenon. Reference [90] is a very readable introduction to all aspects of optics.

While the bundles always have a constant speed (c) in a vacuum, their spacing (wavelength, λ) and hence their frequency (f), can be different; that is, different light sources can emit bundles at different rates and consequently at different wavelengths. It is only necessary that they satisfy the relation

$$c = f_1\lambda_1 = f_2\lambda_2 = \cdots = \text{constant}$$

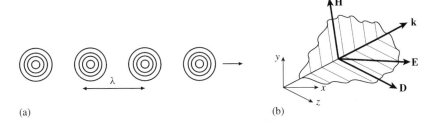

Figure 2.31: Dual aspect of light. (a) Light as a stream of particles with varying electromagnetic fields. (b) Electromagnetic triad for propagating light wave.

Table 2.2: Some light sources and the wavelength of light produced.

Arc Source	Color	λ [nm]	Laser Source	Color	λ [nm]
Sodium	Yellow	589	Ruby	Red	694
Mercury	Yellow	577	He–Ne	Red	633
Mercury	Green	546	Argon	Green	515
Mercury	Blue	435	Argon	Blue	488

There is quite a range of electromagnetic waves varying from the radio waves of 10^3 m at one end to X rays of 10^{-12} m at the other. The occurrence of such a wide range can be explained by the need to carry different amounts of energy.

The visible spectrum is a much narrower range of the electromagnetic spectrum and is centered around frequencies of 500×10^{12} Hz and wavelengths of 600×10^{-9} m. An idea of the range can be seen in the following table:

f:	380		480		500		520		610		660		770
		red		orange		yellow		green		blue		violet	
λ:	780		620		600		580		490		455		390

where f is in terahertz (10^{12} Hz) and λ in nanometers (10^{-9} m). The wavelengths of some light sources is given in Table 2.2.

Any color can be matched by mixing suitable proportions of any three independent colors. In practice, the widest range of colors can be matched by using primaries consisting of highly saturated red, green, and blue. These are called the additive primaries where color addition results when two or more colors are superimposed. Examples of color addition are three filtered lanterns projected onto a screen, a spinning color disk, and colors caused by the limited resolution of the eye. When all of the additive primaries are mixed, the result is white; Figure 2.32 illustrates this.

The more common form of color mixing is color subtraction, which occurs when light passes first through one color and then another. The mixing of pigments is the most

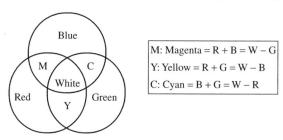

Figure 2.32: Color addition and subtraction.

common example of color subtraction. Yellow, cyan, and magenta are the subtractive primaries as illustrated in Figure 2.32. When all of the subtractive primaries are used, they transmit no color, and the result is in black.

A regular light source is excited at many frequencies and therefore emits white light (many colors). Filters can be used to absorb all but a narrow band of frequencies thus giving *monochromatic* light. Their action can be thought of as a resonance effect; most of the light particles impinging on the filter are absorbed and their energy converted to heat. For some frequencies, however, the electrons are caused to resonate, become unstable, and reemit their gained energy. The reemitted energy contains frequencies around the excitation frequency.

Lenses and Cameras

As a light wave travels through different media its velocity changes. The index of refraction, n, is a measure of the "optical density" of a medium and is given as

$$n = \frac{\text{velocity } in \text{ } vacuo}{\text{velocity in medium}} = \frac{c}{v_m}$$

Thus, n is always greater than unity. Typical values are given in Figure 2.33.

Light rays will bend (refract) when traveling through media with a changing index of refraction. At an interface between the two media (air and glass, say, as shown in

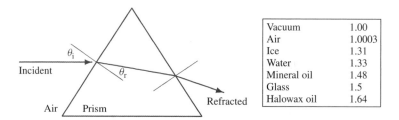

Vacuum	1.00
Air	1.0003
Ice	1.31
Water	1.33
Mineral oil	1.48
Glass	1.5
Halowax oil	1.64

Figure 2.33: Refraction of light through a prism.

Figure 2.33), the incident and transmitted (refracted) directions are related to the indices of refraction by Snell's law: $n_i \sin\theta_i = n_t \sin\theta_t$. It is this refraction that allows the lenses to focus by bending all rays in a predetermined way so that they come to a single point (the focus).

Consider the lens of Figure 2.34(a). The position of the image can be determined by noting that all parallel rays entering the lens must pass through the focal point on the far side, while all rays passing through the center continue straight through. Then, by similar triangles

$$\frac{H}{u} = \frac{h}{v}, \qquad \frac{H}{f} = \frac{h}{v-f} \qquad \text{or} \qquad \frac{1}{f} = \frac{1}{u} + \frac{1}{v}$$

The diverging lens, as well as curved mirrors, can be described by these relations.

A camera is an instrument that focuses the image of an object. The simplest camera contains a lens to form the image, a shutter to control the length of the time that light enters the camera, and a light-sensitive material (film).

In scientific work, there are basically two types of cameras in use. One is the view camera, which is so called because it allows the photographer to view the image on a ground glass screen before exposing the film. After the image is suitably focused, the shutter is closed and the ground glass replaced with a film holder that is made, so that the film lies in the same plane previously occupied by the ground glass. The other type is the reflex camera, which incorporates a combination mirror and screen in such a way that when the shutter is tripped, the mirror moves out of the way and the image is formed on the film. The mirror is used to project the image onto a screen or into a prism where it can be focused.

View cameras are generally big and use sheet film. The reflex cameras on the other hand are small and use 35 mm roll film.

Cameras generally have a compound lens system, but an idea of the essential parameters can be obtained by considering the single lens camera of Figure 2.34(b); this analysis is taken from Reference [115]. Assume the lighted object source is a flat body of area A_s and distance R from the lens. Further, let the object and lens dimensions be small compared to R. If the luminance (emitted lumens per unit area) of the source is E_s, then

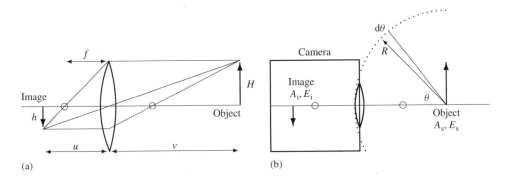

(a)   (b)

Figure 2.34: Lens and camera. (a) Lens relations. (b) Camera intensity.

the total flux from the source is $E_s A_s$. Let the intensity (or flux density) at the middle of the lens be I; then the intensity at an angle θ is $I \cos \theta$ (since a smaller projected area of the source is seen). The area of an annulus at this angle is $R \, d\theta \, 2\pi R \sin \theta$, hence the flux through it is (flux density x area)

$$I \cos \theta \, 2\pi R^2 \sin \theta \, d\theta$$

Since all the flux from the source must go through the hemisphere of radius R and $\theta = \pm \pi/2$ surrounding it

$$E_s A_s = \int_0^{\pi/2} I \cos \theta 2\pi R^2 \sin \theta \, d\theta = I \pi R^2 \quad \text{or} \quad I = \frac{E_s A_s}{\pi R^2}$$

This relation is true when the object and lens are small compared to R.

The flux through the lens is

$$\text{flux} = I \times \text{area} = I \frac{\pi D^2}{4} = \frac{E_s A_s}{\pi R^2} \frac{\pi D^2}{4}$$

where D is the diameter of the lens. All this flux goes to form the image, so the flux density or intensity of the image is flux/image area, that is,

$$I_i = \frac{E_s A_s \, \pi D^2}{\pi R^2} \frac{1}{4} \frac{1}{A_i} = E_s \left(\frac{D_s}{D_i}\right)^2 \left(\frac{D}{2R}\right)^2$$

where D_i is the diameter of the image. The ratio of the image dimension to the object dimension is called the magnification and is related to the focal length of the system by

$$\frac{D_i}{D_s} = m = \frac{f}{u - f} \approx \frac{f}{u} = \frac{f}{R}$$

when $u \gg f$. Hence

$$I_i = \tfrac{1}{4} E_s \left(\frac{D}{f}\right)^2$$

Thus, for a given object luminance, E_s, the image intensity is proportional to $(D/f)^2$. Note that this contains neither the object size nor the object distance. This relation forms the basis of giving the *speed* of a lens in terms of its focal ratio or F-number given by $F \equiv f/D$; that is, the image intensity is inversely proportional to the F-number. It is possible that two different lenses can have the same F-number as, for example,

$$[D, \, f, \, F] = [1, \, 5, \, 5] \quad \text{or} \quad [D, \, f, \, F,] = [.4, \, 2, \, 5]$$

In each case, while the image will have the same intensity, its size will be smaller in the latter case.

Most lens systems are equipped with a light stop or iris diaphragm, which controls the amount of light transmitted to the film. Stops are generally calibrated in terms of the F-number, and have indicated progressions of approximately $1 : \sqrt{2}$ so that for each increase in F the image intensity is cut in half. A common F-stop progression is F/2, 2.8, 4, 5.6, 8, 11, 16, 22, 32.

Sampling and Quantizing in Digital Cameras

The analog data representing the intensity distribution in an image is transformed into digital images by using an analog-to-digital (A/D) converter analogous to time signals discussed earlier in this chapter. The brightness is sampled at intersections called sampling points. The digitized value is called the "grey level." In practice, the digitized grey value is not linearly proportional to the brightness because of the nonlinearity of the image sensor and the A/D converter.

An image is represented as a two-dimensional array of pixels—a *pixel* (short for "picture element") is the smallest entity in an image. Generally, the *grey level* is one byte of memory for each pixel representing the brightness of that point in the image. The number of pixel rows and columns in a camera determines the resolution for the field of view. The number of bits of memory storage associated with each pixel is called the pixel depth. For example, an 8-bit digitizer is capable of displaying 256 grey levels. The product of pixel depth times the number of rows and columns in the image memory gives the total image-storage capacity in bits.

While a monochromatic image is sufficient for almost all cases discussed in this and subsequent chapters, color image processing is becoming more popular. Color image processing uses grey (brightness) levels for each of the three primary colors, red, green, and blue (RGB). The pixel arrangement is that alternating RGB sensors are used; generally, there is not an equal number of each sensor type, and the camera software does some interpolation to provide the final output image.

The stored information of an image is obtained by multiplying the pixel number by the bit number of the grey level and by the number of primary colors. For example, when an image has a size of [512 × 480] pixels and there are 256 (8 bits) grey levels, a monochromatic output will require a total of [512 × 480 × 8] = 1,966,080 bits = 245,760 bytes. The corresponding color image has 737,280 bytes.

Figure 2.35 shows the consequence of using an insufficient number of pixels. When the pixel size is observable, the image is referred to as being *pixelated*. This is different from *aliasing*. When an object with stripes is sampled by a camera, the image sometimes shows a *moire* pattern; this is the aliasing in signal-processing theory and as discussed earlier. When the sampling frequency (the reciprocal of the pixel size) is smaller than two times of the highest spatial frequency of the object, aliasing appears. This is the Shannon theorem. The sampling pitch should therefore be smaller than a half of the smallest pitch of the object.

A typical digital image-processing system is composed of a camera for image input, a frame grabber or controller box, and a computer for control of the system and for the image processing. A schematic is shown in Figure 2.36.

The basic hardware components include *Camera*—solid-state CCD with minimum [384*H* × 488*V* × 8] bits, and RS-170 video interface; *Monitor*—high-resolution RGB video monitor; *Frame grabber*—single-board digitizer for image capture and frame store; *Computer*—should have large disk, and enhanced graphics adapter. Typical digitizing systems have a continuous cycle of 30 frames per second. These are therefore not suitable for recording dynamic event—specially designed (and therefore expensive) cameras must be used.

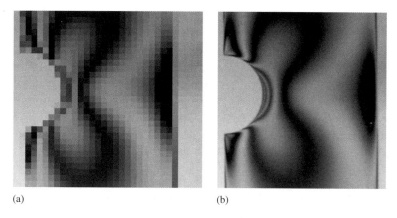

(a)   (b)

Figure 2.35: Effect of pixel size on image resolution. (a) Large pixels. (b) Pixel size is 1/8 of (a).

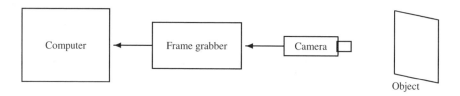

Figure 2.36: Digital imaging system.

Image capture and display boards which provide real-time video digitization frame store and pseudocolor image display are designed to be PC compatible. Resolution of single-board digitizers match the resolution of popular CCD cameras and usually display 256 shades of grey or 256 pseudohue colors. Also, real-time RS-170 display of the digitized image may be observed on standard video monitors.

Solid-state sensors such as the charge coupled device (CCD) type are most often used. Although the interference between the pixels of the sensor and the pixels of the image processor sometimes creates a moire pattern, dimensional accuracy is very good, sensor size is small, and variations due to temperature and time are very small. Such sensors are rugged, are operated with low power, and have high sensitivity at low light levels.

Basic Techniques of Digital Image Processing

With traditional cameras, experimentalists spent a good deal of time in the darkroom manipulating and enhancing images. Not surprising, a similar thing occurs with digital cameras; the difference is that the manipulations are done in the computer.

An image has a large amount of information; however, a good deal of it is redundant. Thus, there is a need for software to process the images as well as extract the required information. This section concentrates on processing the images. Many of the features are implemented in the program `Image`.

This section explains some popular general-purpose algorithms and presents sample applications. In general, the techniques for image processing are much more interactive than for time series data.

The numbers used for image data representation are often confined to 0 to 255 on grey values and to 0 to 511 on position because these are some typical hardware values. Further, when integer numbers are used instead of real numbers, the speed of calculation is improved, and the memory requirements are reduced. However, from a programming point of view it is better to abstract the image representation, and only at time of rendering are specific aspects of the hardware considered. Thus, the intensities range from 0.0 to 1.0 and the image size 0.0 to 1.0 square. This is the type of model used in the PostScript language.

We now look at some common operations. Rather than use an everyday image, we use the fringe pattern of Figure 2.35(b); fringe-specific operations are left until the end of the chapter.

I: Masking

Masking is the operation of producing a subset image. This can be as simple as drawing a box around the area or as complex as constructing a mask using some of the following operations. It is an advantage if the area size can be selected because processing the delimited area is faster than processing the whole area. On the other hand, by standardizing on an area ([512 × 512] pixels, or [640 × 640] pixels, say), the programming and manipulations will be easier.

II: Intensity Distributions

The fundamental information in a digitized image is the intensity. Intensity distributions show how neighboring pixels are related. Figure 2.37 shows the intensity plot on the minimum cross section. Clearly the image is very noisy; digitized images are generally much noisier than their time-signal counterparts.

III: Histograms

A grey-value histogram is obtained by counting the number of times a given grey value occurs in an image. Figure 2.38(a) shows an example.

For a general image, if the grey values are almost equal in frequency, then the digitization is appropriate. When the frequency of occurrence of grey values are skewed then the illumination or the aperture of the camera may need to be changed. In Figure 2.38, there is a good distribution of grey levels but the range is restricted.

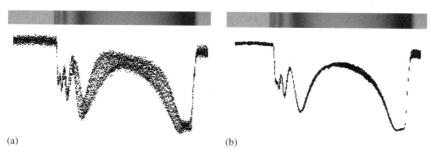

Figure 2.37: Intensity distribution showing noisy data. (a) Original image. (b) After smoothing by averaging.

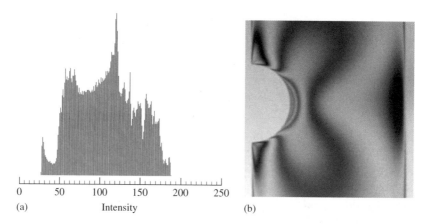

Figure 2.38: Effect of contrast. (a) Histogram of grey levels. (b) Image after rescaling.

IV: Grey-Value Transformations

When the original grey value $I(x, y)$ is processed in some way to produce a new grey image, we express it as

$$g(x, y) = G[I(x, y)]$$

For example, when the histogram is clustered around some values (and therefore showing low contrast), a high-contrast image is obtained by changing each grey value so that the new histogram has a distribution in the entire region of the grey values. A possible operation is

$$g = \frac{I - I_{min}}{I_{max} - I_{min}}$$

where $I_{min} \to I_{max}$ is the range of pixel greys. Figure 2.38(b) was obtained by changing the grey values according to the range of Figure 2.38(a).

While all points get improved contrast, this operation does not result in a uniform contrast. Sometimes it may be necessary to work on a subimage.

V: Thresholding

If we have *a priori* information about the expected grey level distribution in an image, then thresholding, combined with the histogram, can be used to bring out certain features. For example, the histogram of Figure 2.38(a) suggests thresholding at 110 to give a binary image. Figure 2.39(a) shows a binary image of the fringe pattern.

This transformation is sometimes used to extract the geometric characteristics of an image. For example, a white object is put on the black background and its image is recorded. By thresholding the image at the value of the histogram, the image is separated into two parts, white (object) and black (background). The geometric features are obtained by taking various moments of the white pixels. For example, the area is calculated simply by counting (zero moment) the number of white pixels.

VI: Color Separation and Pseudocolor

In a monochromatic image, the grey value is proportional to the brightness of the point. In a color image, the grey value of each primary color is proportional to the intensity (brightness) of the color. The color separations associated with the white light isochromatics of Figure 2.35 are shown in Figure 2.40.

There is a definite difference in the image patterns due to the isochromatics being sensitive to the particular wavelength of light. The noise level for each color is also different.

For a monochromatic image, a grey value can be associated with a color, thus giving an artificial coloring called a pseudocolor image. Images using pseudocolor are sometimes more effective in communicating variations than simple grey values. For example, white is made red indicating hot or intense, while black is made blue indicating cold or benign.

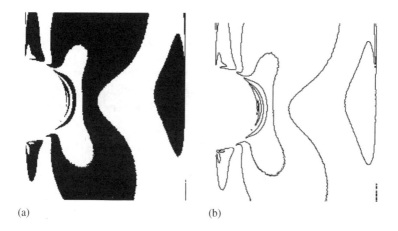

(a)   (b)

Figure 2.39: Some operations. (a) Binary image. (b) Edge detection.

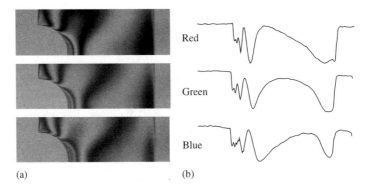

Figure 2.40: Color separations. (a) Grey scale images. (b) Intensity distributions along the line of symmetry.

VII: Montage

A montage creates a new image $g(x, y)$ by operating on two original images, that is,

$$g(x, y) = G[I_1(x, y), I_2(x, y)]$$

A simple operation is subtraction—this can be used to detect motion. The motion could be displacements of dots on a specimen or it could be the motion of a projectile. Another operation is multiplication—this, as we will see, is what happens in the superposition of Moiré grids.

Note that it is possible to get negative intensity values and hence some rule must be used to make them positive. Possible rules are to use only positive values, or to use wraparound for example, a value of -10 is the same as 245.

VIII: Filtering

For noise reduction and enhancement processing (e.g., filtering, averaging, edge detection), the relationship between the grey values at several neighboring pixels is considered. In two-dimensional analysis, square masks of $[3 \times 3]$ pixels, $[5 \times 5]$ pixels, and so on are used.

The general form for a $[3 \times 3]$ mask is

$$g_{ij} = \sum_{i,j} f_{ij} I_{ij}, \qquad f_{ij} = \begin{bmatrix} f_{11} & f_{12} & f_{13} \\ f_{21} & f_{22} & f_{23} \\ f_{31} & f_{32} & f_{33} \end{bmatrix}$$

This creates a new image intensity g_{ij} by sequentially moving across the whole image. The numbers f_{ij} are called the weights. We give some examples.

Averaging is expressed as

$$f_{ij} = \frac{1}{9}\begin{bmatrix} 1 & 1 & 1 \\ 1 & 1 & 1 \\ 1 & 1 & 1 \end{bmatrix}$$

This replaces the intensity at the center pixel with the average of itself and all the neighbors. Other averaging masks are also used. A related filter detects *outliers* (a pixel value significantly different from its neighbors); if a pixel is different from its neighbors by a certain percentage, it is replaced with the average of its neighbors.

The x and y differentials are expressed as

$$\frac{\partial}{\partial x} = I_x = \frac{1}{8}\begin{bmatrix} -1 & 0 & 1 \\ -2 & 0 & 2 \\ -1 & 0 & 1 \end{bmatrix}, \qquad \frac{\partial}{\partial y} = I_y = \frac{1}{8}\begin{bmatrix} -1 & -2 & -1 \\ 0 & 0 & 0 \\ 1 & 2 & 1 \end{bmatrix}$$

A less directional-dependent differential is usually expressed as the sum of the absolutes of the x- and y-directional differentials.

$$I = |I_x| + |I_y|$$

This differential is called the Sobel filter, which has double weights for the center line and row. This filter is useful to detect the edge lines of an object. An example is shown in Figure 2.38(b).

The Laplacian operator is

$$\nabla^2 g(x, y) = \frac{\partial^2 g(x, y)}{\partial x^2} + \frac{\partial^2 g(x, y)}{\partial y^2}$$

and has the filter coefficients

$$f_{ij} = \begin{bmatrix} 0 & 1 & 0 \\ 1 & -4 & 1 \\ 0 & 1 & 0 \end{bmatrix}$$

where the center line and row have four times the weight and the neighbor lines and rows have one time the weight. The Laplacian is also useful for detecting the edge lines of an object.

IX: Geometric Transformations

Enlargement, shrinkage, rotation, and translation are expressed as mappings. When an original point (x_0, y_0) with grey value $I_0(x_0, y_0)$ moves to another point (x_1, y_1) after geometric transformation, the grey values are not changed. Then

$$g(x_1, y_1) = I(x_0, y_0)$$

where

$$\begin{Bmatrix} x_1 \\ y_1 \end{Bmatrix} = \begin{bmatrix} a_{11} & a_{12} \\ a_{21} & a_{22} \end{bmatrix} \begin{Bmatrix} x_0 \\ y_0 \end{Bmatrix} + \begin{Bmatrix} a_{13} \\ a_{23} \end{Bmatrix}$$

The factors a_{ij} are constants whose values depend on the particular transform. For example, enlargement or shrinkage of b_x and b_y in the x- and y-directions respectively, yields

$$\left\{ \begin{array}{c} x_1 \\ y_1 \end{array} \right\} = \left[\begin{array}{cc} b_x & 0 \\ 0 & b_y \end{array} \right] \left\{ \begin{array}{c} x_0 \\ y_0 \end{array} \right\} + \left\{ \begin{array}{c} 0 \\ 0 \end{array} \right\}$$

For rotation by an angle θ,

$$\left\{ \begin{array}{c} x_1 \\ y_1 \end{array} \right\} = \left[\begin{array}{cc} \cos\theta & +\sin\theta \\ -\sin\theta & \cos\theta \end{array} \right] \left\{ \begin{array}{c} x_0 \\ y_0 \end{array} \right\} + \left\{ \begin{array}{c} 0 \\ 0 \end{array} \right\}$$

Sometimes when doing geometric transformations, it is best to transform to a hypothetical space that is [10000 × 10000], say, and only as a final step (e.g., painting on a screen or sending to a printer) is the image rendered.

X: Other Operations

There are a host of other operations that are becoming standard parts of image-manipulation packages. See References [128, 138] for additional methods of edge detection; Reference [125] has a good collection of citations of digital imaging as applied to various problems in experimental mechanics. References [28, 166] give good discussions of image restoration algorithms for removal of blurs due to motion or misfocusing.

Fringe Analysis

The procedure for fringe analysis is basically as follows:

1. Recording images.

2. Filtering to reduce noise.

3. Identifying geometry coordinates to set image scaling.

4. Thresholding to enhance image contrast.

5. Fringe identification and counting.

The histogram of grey values is useful to check whether the lighting conditions are good. For many fringe patterns, the best histogram should have all the grey values from the maximum to the minimum value with the distribution being somewhat cosinusoidal. These conditions are usually adjusted at the time of image recording.

Noise filtering is always necessary. This ranges from the removal of outliers (a single white point, say, surrounded by black points), to using moving averages to smooth the distributions. Fixed noise sources such as stains or markings sometimes may be erased by working with the difference images of before and after deformation. Contrast enhancement can also be done at this stage.

It is possible to obtain the specimen geometry and scaling information from the outline of the model in the image. It is preferable, however, to obtain the geometric information separately and use the image solely for the scaling. Well placed, and accurate, fiducial marks on the model are essential for this.

Figure 2.41 shows examples of typical fringe patterns. Identifying the fringe order in these patterns is sometimes very difficult, and additional information must be brought to bear on the problem. In photoelasticity, fringes form either closed loops or start at a boundary and end at a boundary. Fringe counting can begin by identifying areas of low stress (reentrant corners, for example); this is easier to do with white light because the zero fringe is black while others are colored but with increasing "wash out" as the fringe order increases. Sometimes counting the fringes as the specimen is loaded is a simple effective method.

In Moiré, the sign of the fringe is arbitrary, but the fringe derivative does have a specific sign. If the sign is not known from external evidence, it can be determined by shifting the reference grating with respect to the specimen grating: if the reference grating is shifted perpendicular to its grating lines, the derivative is positive if the fringes move in the direction of the shift. In more complicated patterns it is often necessary to consider the continuity, uniqueness, and direction of the displacements.

From the first identified fringe value, it is sometimes possible to assign values to the entire pattern by a simple count. Considerations of the symmetry and the load can also aid in doing this. However, it may well be that the fringe order identification process becomes iterative and part of the solution process itself.

Fringe Data Recording

Once the fringe orders have been identified, they must be digitized for use in the computer postprocessing. How this is done depends on the purpose of the data, the type of fringe data, and the amount of automation required. There is no one best method, and we outline three of them.

The crux of the problem relates to the issues as to what constitutes a fringe. To us as observers, a fringe is obvious because our eyes automatically associate the various

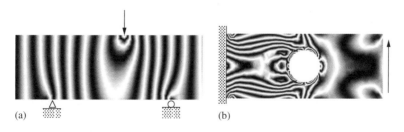

(a)   (b)

Figure 2.41: Sample fringe patterns. (a) In-plane $v(x, y)$ Moiré fringes for a beam in three-point bending. (b) Photoelastic isochromatic fringes for a cantilevered bar with a hole.

grey levels into spatially related groups; to the digital camera, each pixel grey level is unrelated to every other pixel. We review some of the methods.

I: Discrete Point Digitizing

Perhaps the simplest method of fringe data recording is to use the equivalent of a digitizing pad and "click" on the center of each fringe at a density that adequately represents the fringe distribution. Nowadays, a mouse and a screen image is adequate. A refinement is to associate a different code with each series of clicks. Essentially, this method accesses the $[i, j]$ coordinate of the clicked pixel and therefore requires that geometry information must also be input so as to transform the screen pixel coordinates to physical coordinates.

This method has the significant advantage of allowing the user maximum control of the number and location of data.

II: Discrete Fringe Grouping and Thinning

Here, we identify one complete fringe at a time. In other words, we wish to form groupings of collections of pixels.

The first step is to form a bilevel image; this creates islands with distinct boundaries. Next, a pixel is chosen inside an island and, in a recursive fashion, all surrounding similar pixels are identified. The identified pixels can then act as a mask on the original image to identify the subset.

Thinning is an operation for obtaining the centers of thick lines such as fringes. This can work on grey level or binary images. An effective scheme is illustrated in Figure 2.42. The process has two operations to it. First, we identify the boundary pixels; an interior pixel has eight connections and a boundary is any pixel that has less than eight connections. Figure 2.42(a) illustrates boundary pixels with five, four, and three connections.

In an iterative fashion, preferably random, we remove boundary pixels of a certain level and below. Level 3 will remove only protrusions and leave a smooth edge; Level 4 will also remove corners, and Level 5 erodes even straight edges. Repeated use of this

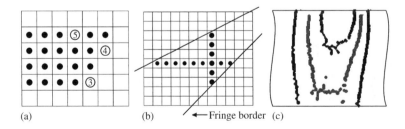

(a)   (b)   ←—Fringe border (c)

Figure 2.42: Thinning algorithm. (a) Level of boundary pixels. (b) Boundary associated with interior pixel. (c) Some identified and thinned fringes.

process will uniformly erode the boundary pixels. A minimum width can be set as illustrated in Figure 2.42(b); the criterion is that if either set (horizontal or vertical) of pixels is at the minimum, the boundary is not eroded.

Sometimes, a single fringe can range from being very thin to being very fat, and operating only on boundary pixels can then be very slow. Figure 2.42(b) shows how an identified interior pixel can be used to suggest candidate boundary pixels for removal; it is the boundary of the smaller range of pixels that is removed. When the interior pixels are chosen randomly, this scheme is biased toward the fat portions of the fringe.

Two black and one white fringe of Figure 2.41(a) are shown identified and thinned in Figure 2.42(c) (these were done in separate stages). Note that it is possible to have isolated clusters form, but this is not a concern since the data will be utilized in the form of triads x, y, N.

III: Whole-Field Phase Unwrapping

The most sophisticated of the fringe identification schemes involves working with phase maps directly. The basic idea is to record four (or five) separate images with phase shifts of $\pi/2$ between them; then the phase information is given as

$$\phi(x, y) = \tan^{-1} \left[\frac{I_4(x, y) - I_2(x, y)}{I_1(x, y) - I_3(x, y)} \right]$$

This determines $\phi \pm N\pi$. A plot of $|\phi + \pi|$ constitutes the phase map and shows sharp intensity changes at the $N\pi$ boundaries.

A second stage involves the unwrapping of the phase, that is, determining N. What makes this possible is that there is a directionality to the phase maps; the jumps show the direction of the phase increase or decrease and hence it is a simple operation to know if to add or subtract π along a contour. This is done for all pixel lines. A periodic check is done by unwrapping along a closed loop—the net phase should be zero. This works best with fringe patterns that have a simple topology.

This procedure is not pursued here for two reasons. First, it produces data at every pixel location that is good for plotting but generally unsuitable for use in our inverse methods. For example, an FEM solution node point may be surrounded by a [20 × 20 = 400] cluster of pixel data and only one data point can be used. Secondly, following one of the themes of this book, when performing inverse operations all available information should be used simultaneously. For example, as seen from the formula above, the phase is determined at each point independent of what neighboring points are doing, but clearly, closely located points must be similar. In the next chapter, we introduce regularization methods that allow us to utilize this supplementary information.

2.7   Moiré Analysis of Displacement

When two grids of regularly spaced lines are overlayed, an interference fringe pattern, called Moiré fringes, is observed. An example is shown in Figure 2.43. A characteristic feature of these fringes are that they are of low frequency caused by two (or more) sets of (relatively) high-frequency grids. There are many ways of forming a Moiré; indeed the word itself is French and used to describe a fabric with a wavy pattern. The grids may be superposed by reflection, by shadowing, by double-exposure photography, or by direct contact. A comprehensive review of Moiré with a substantial bibliography is given in Reference [154]. The Moiré method is attractive for experimental stress analysis because it is objective in that it does not require calibration, as for strain gages, or a model material, as for photoelasticity—the analysis is purely geometric.

Fringe Analysis

The following fringe analysis is applicable irrespective of the particular mode of Moiré fringe generation, and it is applicable to large displacements and rotations [117].

Consider a specimen grid of spacing s, orientation ϕ, and numbering $j = \ldots, -2, -1,$ $0, 1, 2, \ldots$. Overlay on this a master (or interrogating) grid of spacing m, orientation θ and numbering $i = \ldots, -2, -1, 0, 1, 2, \ldots$. At those points where the grid lines overlay, a maximum brightness is observed, and in between a maximum darkness is observed. The eye averages over the area and bright and dark fringes are detected. These fringes have spacing $f = \Delta x_\gamma / \Delta N$, orientation γ and numbering $N = \ldots, -2, -1, 0, 1, 2, \ldots$. These relationships are depicted in Figure 2.43.

Along any line connecting the intersections of grids (bright fringes), we have that

$$N = i - j$$

where N is the (bright) fringe order and can be either positive or negative. This relation is easily verified for Figure 2.43. Let N, i and j be considered as continuously distributed, then

$$N(x, y) = i(x, y) - j(x, y)$$

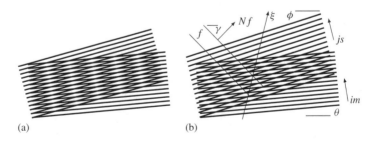

(a)   (b)

Figure 2.43: Moiré fringes from the superposition of two grids. (a) Rotation only. (b) Rotation plus stretching.

and this relation is valid at any point (not just the grid intersections). Along a straight line ξ at an angle ρ, we can alternatively write the distance as

$$\xi = \frac{im}{\cos(\rho - \theta)}, \qquad \xi = \frac{js}{\cos(\rho - \phi)}, \qquad \xi = \frac{Nf}{\cos(\rho - \gamma)} = N\frac{\Delta x_\rho}{\Delta N}$$

The density of fringes along this line is

$$\frac{\Delta N}{\Delta \xi} = \frac{i - j}{\xi} = \frac{1}{m}\cos(\rho - \theta) - \frac{1}{s}\cos(\rho - \phi) \qquad \text{and} \qquad \frac{\Delta N}{\Delta \xi} = \frac{\Delta N}{\Delta x_\gamma}\cos(\rho - \gamma)$$

These are the basic equations for fringe interpretation.

Choose ξ to be along the coordinate directions, then

$$\rho = 0: \qquad \frac{\Delta N}{\Delta \xi} = \frac{\Delta N}{\Delta x} = \frac{1}{m}\cos\theta - \frac{1}{s}\cos\phi = \frac{\Delta N}{\Delta x_\gamma}\cos\gamma$$

$$\rho = 0 + \pi/2: \qquad \frac{\Delta N}{\Delta \xi} = \frac{\Delta N}{\Delta y} = \frac{-1}{m}\sin\theta + \frac{1}{s}\sin\phi = -\frac{\Delta N}{\Delta x_\gamma}\sin\gamma$$

This can be rewritten as

$$\left\{ \begin{array}{c} \Delta N/\Delta y \\ \Delta N/\Delta x \end{array} \right\} = \frac{\Delta N}{\Delta x_\gamma} \left\{ \begin{array}{c} \sin\gamma \\ \cos\gamma \end{array} \right\} = \frac{1}{m} \left\{ \begin{array}{c} \sin\theta \\ \cos\theta \end{array} \right\} - \frac{1}{s} \left\{ \begin{array}{c} \sin\phi \\ \cos\phi \end{array} \right\}$$

This form is sometimes more convenient to use. Alternative forms for the fringe information are

$$\tan\gamma = \frac{m\sin\phi - s\sin\theta}{m\cos\phi - s\cos\theta}, \qquad \left[\frac{\Delta N}{\Delta x_\gamma}\right]^2 = \frac{1}{m^2}\left[1 + \left(\frac{m}{s}\right)^2 - 2\left(\frac{m}{s}\right)\cos(\phi - \theta)\right]$$

These two equations give the relationship between the fringe orientation and spacing, and the grid parameters.

Consider the special case when $\phi = \theta = 0$, that is, there is no rotation, only stretching. Then

$$\tan\gamma = 0 \qquad \text{or} \qquad \gamma = 0$$

and

$$\left[\frac{\Delta N}{\Delta x_\gamma}\right]^2 = \frac{1}{m^2}\left[1 + \left(\frac{m}{s}\right)^2 - 2\left(\frac{m}{s}\right)\right] \qquad \text{or} \qquad \frac{\Delta N}{\Delta x_\gamma} = \frac{1}{m} - \frac{1}{s}$$

The fringes are formed parallel to the grids and are related to the change of spacing of the specimen grid. Now consider the special case where $m = s$ and $\theta = 0$, that is, there is only a rigid-body rotation. Then

$$\tan\gamma = \frac{\sin\phi}{\cos\phi - 1} = \frac{-1}{\tan\phi/2} \qquad \text{or} \qquad \gamma = \frac{1}{2}\phi \pm \frac{1}{2}\pi$$

and

$$\frac{\Delta N}{\Delta x_\gamma} = \frac{1}{m}[2 - 2\cos(\phi)]^{1/2} = \frac{2}{m}\sin(\phi/2)$$

The fringes are predominantly at right angles to the grids. In the case of small rotations, the number of observed fringes is directly proportional to the rotation.

This analysis shows how the fringes are formed for a given deformation; in stress analysis and especially in inverse problems we are interested in obtaining information about the displacements from the measured fringe data. We take this up next.

In-plane Displacements

At each point on the fringe pattern, we assume the deformation is locally homogeneous, then we have the relation

$$\frac{1}{s}\left\{ \begin{array}{c} \sin\phi \\ \cos\phi \end{array} \right\} = \frac{1}{m}\left\{ \begin{array}{c} \sin\theta \\ \cos\theta \end{array} \right\} - \frac{\Delta N}{\Delta x_\gamma}\left\{ \begin{array}{c} \sin\gamma \\ \cos\gamma \end{array} \right\}$$

Letting $1/f = \Delta N/\Delta x_\gamma$ the alternative forms are

$$\tan\phi = \frac{f\sin\theta - m\sin\gamma}{f\cos\theta - m\cos\gamma}, \qquad s = \frac{mf}{\sqrt{m^2 + f^2 - 2mf\cos(\gamma - \theta)}} \qquad (2.17)$$

Each fringe pattern gives two pieces of information: a stretch and a rotation. However, a general 2-D deformation requires four pieces of information: two stretches and two rotations. Therefore, two sets of fringes are required.

With reference to Figure 2.44, let the initial grids have orientation $\theta = 90°$ and $\theta = 0°$, respectively; then the information obtained is

$$\tan\phi_2 = \frac{-m\sin\gamma_1}{f_1 - m\cos\gamma_1}, \qquad s_1 = \frac{mf_1}{\sqrt{m^2 + f_1^2 - 2mf_1\cos\gamma_1}}$$

$$\tan\phi_1 = \frac{-m\sin\gamma_2}{f_2 - m\cos\gamma_2}, \qquad s_2 = \frac{mf_2}{\sqrt{m^2 + f_2^2 - 2mf_2\cos\gamma_2}}$$

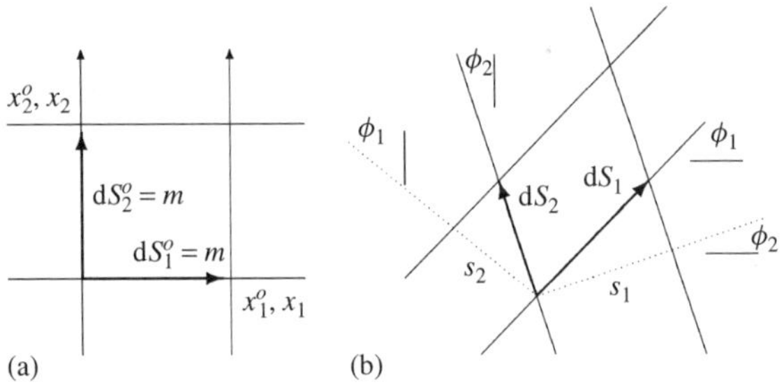

Figure 2.44: Generalized plane homogeneous deformation. (a) Grids before deformation. (b) Grids after deformation.

Note that fringe pattern 1 gives information about rotation ϕ_2 and pattern 2 gives information about rotation ϕ_1. Also, the reference orientation for pattern 1 is vertical. In processing the fringe patterns, it may be convenient to get the orientation and spacing information from the coordinate direction information as

$$\tan \gamma = \frac{f_x}{f_y} = \frac{dx/dN}{dy/dN}, \qquad f = \sqrt{f_x^2 + f_y^2} = \sqrt{\frac{dx}{dN}^2 + \frac{dy}{dN}^2}$$

These forms highlight the fact that the fringe patterns must be space differentiated to get the required geometric information.

Assuming $f(x, y)$ and $\gamma(x, y)$ can be obtained at a sufficient number of points from both patterns, we have $s_1(x, y)$, $\phi_1(x, y)$, $s_2(x, y)$, and $\phi_2(x, y)$. Once this geometric information is obtained, we need to relate it to the deformation. Consider the locally homogeneous deformation shown in Figure 2.44. For convenience, the original orientation of the grids is taken along the coordinate directions and both master gratings are assumed similar. Initially, perpendicular line segments dS_1^o and dS_2^o deform into line segments dS_1 and dS_2, respectively. From the geometry, we have that

$$dS_1 = \frac{s_1}{\cos(\phi_1 - \phi_2)}, \qquad dS_2 = \frac{s_2}{\cos(\phi_1 - \phi_2)}$$

The four quantities s_1, s_2, ϕ_1, ϕ_2 are obtained from the Moiré measurements using two grids; the deformation analysis requires the simultaneous use of two patterns.

Introducing extensions defined as

$$E_1 \equiv \frac{dS_1 - dS_1^o}{dS_1^o}, \qquad E_2 \equiv \frac{dS_2 - dS_2^o}{dS_2^o}$$

we can write the locally homogeneous deformation as

$$u_1 = (1 + E_1) \cos \phi_1 \, x_1^o - (1 + E_2) \sin \phi_2 \, x_2^o - x_1^o$$
$$u_2 = (1 + E_1) \sin \phi_1 \, x_1^o + (1 + E_2) \cos \phi_2 \, x_2^o - x_2^o$$

The deformation gradient is

$$\left[\frac{\partial x_i}{\partial x_j^o} \right] = \begin{bmatrix} (1 + E_1) \cos \phi_1 & -(1 + E_2) \sin \phi_2 \\ (1 + E_1) \sin \phi_1 & (1 + E_2) \cos \phi_2 \end{bmatrix}$$

and the Lagrangian strain components [56] are obtained as

$$[2E_{ij}] = \begin{bmatrix} 2E_1 + E_1^2 & (1 + E_1)(1 + E_2) \sin(\phi_1 - \phi_2) \\ (1 + E_1)(1 + E_2) \sin(\phi_1 - \phi_2) & 2E_2 + E_2^2 \end{bmatrix}$$

In this way, the four measurements ϕ_1, s_1, ϕ_2, s_2, are sufficient to determine the strain state.

When the strains and rotations are small the above analysis can be simplified considerably. For a given master grid, the fringe spacing f becomes very large, that is, $f \gg m$.

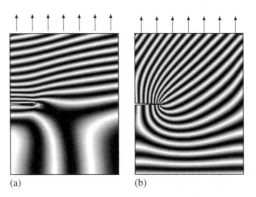

(a)   (b)

Figure 2.45: In-plane Moiré fringes for a cracked plate. For small deformations, the fringes can be interpreted as displacement contours. (a) Horizontal u displacements. (b) Vertical v displacements.

Substituting this into the fringe relations gives

$$\phi_2 = \frac{-m \sin \gamma_1}{f_1}, \qquad \frac{s_1 - m}{m} = \frac{m \cos \gamma_1}{f_1}$$

$$\phi_1 = \frac{-m \sin \gamma_2}{f_2}, \qquad \frac{s_2 - m}{m} = \frac{m \cos \gamma_2}{f_2}$$

The deformation terms are

$$dS_1 \approx s_1, \qquad dS_2 \approx s_2, \qquad E_1 = \frac{s_1 - m}{m}, \qquad E_2 = \frac{s_1 - m}{m}$$

Substituting these into the displacement relations then leads to (replacing $u_1(x_1^o, x_2^o) \rightarrow u(x, y)$, $u_2(x_1^o, x_2^o) \rightarrow v(x, y)$)

$$u(x, y) \approx E_1 x + \phi_2 y = \frac{m \cos \gamma_1}{f_1} x + \frac{m \sin \gamma_1}{f_1} y = m \frac{dN_1}{dx} x + m \frac{dN_1}{dy} y = N_1(x, y)$$

$$v(x, y) \approx E_1 x + \phi_2 y = -\frac{m \sin \gamma_1}{f_2} x + \frac{m \cos \gamma_1}{f_2} y = m \frac{dN_2}{dx} x + m \frac{dN_2}{dy} y = N_2(x, y)$$

where the fact that the terms $f / \cos \gamma$ and $f / \sin \gamma$ are the fringe spacings in the coordinate directions was used. The final form has the simple interpretation that the observed fringes are contours of the respective displacements.

The small-strain approximation gives

$$\epsilon_{xx} = \frac{\partial u}{\partial x} = m \frac{\partial N_1}{\partial x}, \qquad \epsilon_{yy} = \frac{\partial v}{\partial y} = m \frac{\partial N_2}{\partial y}, \qquad 2\epsilon_{xy} = \frac{\partial u}{\partial y} + \frac{\partial v}{\partial x} = m \left(\frac{\partial N_1}{\partial y} + \frac{\partial N_2}{\partial x} \right)$$

In comparison to the large strain case, the fringe patterns can be processed directly to give the strain. Figure 2.45 shows an example of a fringe pattern.

The analysis can be generalized by having an initial mismatch between the specimen and master grids. At zero displacement, an initial fringe pattern is observed and in terms of this, the displacements simply become

$$u = m(N_1 - N_{10}), \qquad v = m(N_2 - N_{20})$$

With this approach, the displacement at a point is just related to the difference of the fringe orders observed before and after deformation.

In the discussions above, the master grid was placed in contact with the specimen grid in order to produce the Moiré effect. But this is not essential, and there are a number of alternatives. For example, an image of the specimen grid itself can be recorded by a digital camera and the Moiré formed in the computer [125]. This has the advantage of variable sensitivity (mismatch); however, for high-density grids a very high-resolution camera would be required. As a rule of thumb, eight pixels are required per grid spacing m.

The Grid and Its Application

The Moiré method has many positive attributes for measuring in-plane strains. Its greatest limitation, however, is its sensitivity. For example, suppose the average strain in a specimen is $\epsilon = 1000\,\mu\epsilon$, and the specimen is about 250 mm (10 in.) in extent then the maximum displacement is about 0.25 mm (0.01 in.). This is quite small. Now suppose we would like to observe about ten fringes, then the grid spacing must be $u/N = 0.025\,\text{mm}$ (0.001 in.) or 40 lines per mm (1000 lines per inch). Most grids are less than this. Furthermore, applying such a grid to our extended specimen would be very difficult. This restricts the method to structural materials near or beyond their yield point, to plastics that have relatively large displacements before failure, and to displacement dominated structural problems.

The single most difficult part of the Moiré analysis is the application of the grid to the specimen surface. To produce meaningful fringes, the grids must be applied free of initial distortion; be of uniform pitch, contrast and line-to-space ratio; and have the necessary line density.

Gratings can be scribed, etched, ruled, printed, stamped, photographed, or cemented onto the specimen. Coarse gratings (about 1 to 5 lines per mm) can be obtained from graphic arts suppliers in sheets up to 500 mm square. Finer gratings (up to 40 lines/mm) are available from suppliers to lithographers and photoengravers in sizes up to 200 mm square.

The three most common methods of applying grids to the surface are photoprinting, transfer bonding, and the hybrid printing-bonding method. Photoprinting involves applying a photographic emulsion to the surface, making a contact print of the grid in the emulsion, developing as for a regular photographic negative, and finally dyeing the pattern. The disadvantage of this method is that it is difficult to get good grids greater than 10 lines per millimeter.

For the transfer bonding method, a grid in a carrier material is bonded to the test area. After the cement has cured, the carrier is stripped away, leaving a distortion-free grid applied to the specimen. Bondable grids are available in densities up to 40 lines per millimeter and sizes up to 100 mm square. The cost, however, is generally too expensive for routine work.

The hybrid method uses a photographic negative (available from photographic suppliers), which instead of being on the thick cellulose backing, has a very thin backing supported by paper. A grid is contact printed onto the negative and developed in the usual manner. This is then bonded to the specimen surface and the paper removed. The major disadvantage of the method is that distortion can arise during the developing stage.

Out-of-Plane Displacements

There are numerous Moiré methods for analyzing out-of-plane displacements, but the one, that is of particular interest here is called *shadow moire*. This is a convenient method because it uses only one grid and it need not be attached to the specimen.

The point light source in Figure 2.46 casts a shadow of the point x_{gs} of the master grid (with lines parallel to the y axis) onto the point x on the specimen; the camera sees a Moiré interference of the shadow grid (labelled j) with the master grid (labelled i) at point x_{gc} on the grid. Thus, the shadow of the master grid becomes the specimen grid and will be distorted depending on the contour of the specimen.

Consider the plane containing the light source at x_s and camera at x_c. The Moiré interference of the deflection at x is observed at x_{gc} on the grating; by similar triangles, we have that

$$\frac{x - x_c}{L_c + w} = \frac{x - x_{gc}}{w} \quad \text{or} \quad x = x_{gc} + (x_{gc} - x_c)\frac{w}{L_c}$$

There is a similar relation for y given as

$$y = y_{gc} + (y_{gc} - y_c)\frac{w}{L_c}$$

For the light source ray

$$\frac{x_s - x_{gs}}{L_s} = \frac{x_s - x}{L_s + w} \quad \text{or} \quad x = x_{gs} + (x_{gs} + x_s)\frac{w}{L_s}$$

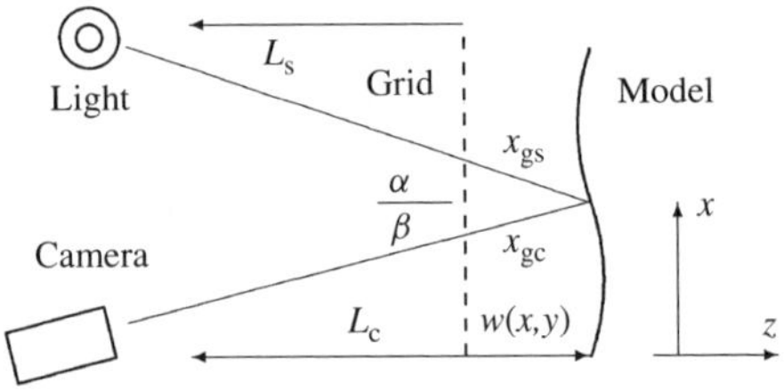

Figure 2.46: Shadow Moiré method.

The fringes observed on the grid plane are given by $N = i - j$ where $x_{gc} = im$ and $x_{gs} = jm$. Substituting for i and j and eliminating x and x_{gs} gives

$$N = \frac{1}{m}\left[(x_s - x_{gc}) + \frac{L_s}{L_c}(x_{gc} - x_c)\right]\frac{w}{L_s + w}$$

as the fringe order observed at point (x_g, y_g) on the grid. Note that at those points where $w = 0$, $N = 0$; this means that points in contact with the grid are white.

The displacement at (x, y) is given by

$$w = \frac{mN}{\left[\dfrac{x_s - x_{gc}}{L_s} + \dfrac{x_{gc} - x_c}{L_c} - \dfrac{sN}{L_s}\right]}$$

These equations are the basic equations of the shadow Moiré method and are applicable even for large deflections. Note that the absolute value of N must be known in order to determine w. Furthermore, the fringes do not have interpretation in terms of contours of equal displacement. While only the $x - z$ section was considered, nothing in the analysis (except the perspective of y) is altered by considering the dependence on y, since the grids are arranged in the y-direction.

The above equations are quite general but they can be simplified considerably by the addition of a couple of lenses. Let the source go to infinity in such a way that

$$\frac{x_s - x_g}{L_s} = \text{constant} = \tan\alpha, \qquad \frac{s}{L_s} \to 0$$

Also, let the camera go to infinity in such a way that

$$\frac{x_s - x_g}{L_c} = \text{constant} = \tan\beta$$

Then

$$w = \frac{mN}{\tan\alpha + \tan\beta}, \qquad x = x_g + w\tan\beta, \qquad y = y_g \qquad (2.18)$$

Here, the fringes have interpretation as contours of equal deflection. In practice, of course, neither the camera nor the source can be located at infinity; however, a good approximation is to place them at the focal points of field lenses.

A special case that is popular is to have the camera normal to the grid. This gives

$$w = \frac{mN}{\tan\alpha}, \qquad x = x_g, \qquad y = y_g$$

In this case, focusing of the fringes across the model becomes much easier and no adjustment for perspective need be made to the image. It is noted, however, that some sensitivity is lost. Two examples are shown in Figure 2.47.

In the small deflection of thin plates, the strains are obtained as

$$\epsilon_{xx} = \frac{1}{2}h\frac{\partial^2 w}{\partial x^2} = c\frac{\partial^2 N}{\partial x^2}, \qquad \epsilon_{yy} = \frac{1}{2}h\frac{\partial^2 w}{\partial y^2} = c\frac{\partial^2 N}{\partial y^2}, \qquad 2\epsilon_{xx} = \frac{1}{2}h\frac{\partial^2 w}{\partial x\partial y} = c\frac{\partial^2 N}{\partial x\partial y}$$

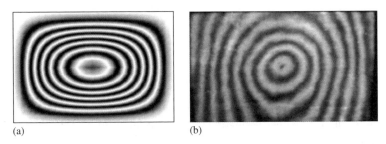

(a)   (b)

Figure 2.47: Examples of shadow Moiré fringes. (a) Synthetic data for a clamped plate with uniform pressure. (b) Experimental data for a cc-ss-cc-ss composite plate with a central point load.

with $c = m/\tan \alpha$ being the sensitivity of the set up. Taking double derivatives of experimental data is highly prone to errors; consequently, we revisit the question in later chapters after the inverse methods have been developed.

Sensitivity of the shadow Moiré method depends on illumination angle and grid density. Neither of these two, however, can be increased indefinitely because fringe contrast is lost. Suppose 10 fringes are observed using a grid of 1 line per millimeter and an illumination angle of $45°$, then

$$w = \frac{\frac{1}{1} \times 10}{1} = 10 \, \text{mm}$$

This is a relatively large deflection.

To obtain a shadow Moiré pattern, a matte rather than a polished surface is required. The surface is often painted with white tempera or flat paint. A grating is placed directly in front of the matte surface.

A variant of the shadow Moiré method is projection Moiré. Here, a grating is placed at the light source and its shadow projected with the use of lenses. A companion grating in front of the camera is used as a reference grating. References [43, 141] describe a number of shadow and projection Moiré variants, as well as many static and dynamic applications.

2.8 Holographic Interferometry

A hologram is capable of recording the complete information (intensity and phase) of light reflecting from an object. By double-exposure holography, it is possible to simultaneously record the before and after deformation positions of an object and since the two images contain phase information, they can interfere and the interference fringe pattern is related to the relative displacements. This phenomenon is based on the coherent properties of light, so we begin with a brief discussion of light waves and coherent optics.

Wave Nature of Light

One of the great events of the 19th century was the discovery that both electric and magnetic effects are intimately related. This was not obvious from statics, but the investigation of magnets and charges in motion lead to the connection. The change of magnetic field in the course of time induces an electric field strength even in the absence of any electric charges that might be considered the source of this field strength. This is Faraday's law. Similarly, a moving stream of charges or a changing electric field gives rise to a magnetic field; this is Ampere's law. These two dynamical equations form the basis of classical electromagnetic theory and are referred to collectively as Maxwell's equations

$$\hat{\nabla} \times \hat{E} = -\frac{1}{c}\frac{\partial \hat{H}}{\partial t}, \qquad \nabla(\hat{D}) = 0$$

$$\hat{\nabla} \times \hat{H} = \frac{1}{c}\frac{\partial \hat{D}}{\partial t}, \qquad \nabla(\hat{H}) = 0$$

where $\hat{E}$ is the electric field, $\hat{D}$ is the electric displacement field, and $\hat{H}$ is the magnetic field. The two divergence relations are derived from Gauss's law. To complete the Maxwell's equations, it is necessary to establish the constitutive relation between the electric displacement and electric field.

If the medium is static, it is usual to assume that the polarization is directly proportional to the applied field but that it may not necessarily be isotropic, that is,

$$\begin{Bmatrix} D_x \\ D_y \\ D_z \end{Bmatrix} = \begin{bmatrix} 1 + e_{xx} & e_{xy} & e_{xz} \\ e_{xy} & 1 + e_{yy} & e_{yz} \\ e_{xz} & e_{yz} & 1 + e_{zz} \end{bmatrix} \begin{Bmatrix} E_x \\ E_y \\ E_z \end{Bmatrix}$$

According to this, the restoring forces per unit displacement may not be the same in all directions. The tensor [e], called the *electric susceptibility* tensor, is symmetric since the restoring forces are conservative. In photomechanics, the field strength of the normal light levels used is ordinarily very weak, and consequently it contributes very little to the polarization. However, the materials used are such that under deformations polarization is mechanically induced, which then affects the transmitted light; that is, [e] is some function of the deformation and stress histories. It is this link between deformation and polarization that gives rise to the classical photoelastic effect, which we will discuss in the next section. The tensor $[1+e]$ is called the *dielectric* tensor and specific constitutive relations for it will be considered later.

In general, a wave propagating in the z-direction is given by

$$\Gamma = Ag(z - c_m t)$$

where A is the amplitude, c_m is the velocity of the wave, and $(z - c_m t)$ is the phase; Γ represents any property that behaves like a wave. Suppose the wave is a sinusoidal (harmonic) wave train of wavelength λ, then

$$\Gamma = A \cos\left[\frac{2\pi}{\lambda_m}(z - v_m t)\right] = A \cos\left[\frac{2\pi}{\lambda_m}z - \omega t\right] = A \cos[kz - \omega t]$$

where $\omega = 2\pi f = 2\pi c_m / \lambda_m$ is the angular frequency, $k = 2\pi / \lambda_m$ is called the wave number, and λ_m is the wavelength in the medium. The sinusoidal waves are of importance because arbitrary wave functions can always be expanded as a series of harmonic functions. We will find it more convenient to use the complex exponential description, that is,

$$\Gamma = A e^{-i[\frac{2\pi}{\lambda_m} z - \omega t]} = A e^{-i[kz - \omega t]}$$

When the wave is propagating in an arbitrary direction, the distance traveled is the vector dot product of the position vector with the direction of propagation $\hat{e} \cdot \hat{r}$ where $\hat{e}$ is the unit vector in the direction of propagation. It may happen that at $r = 0$ the phase is not zero and then an initial phase must be included. Also, in the later work it will be convenient to use the index of refraction. Hence, the general wave relation will be written as

$$\Gamma = A e^{-i[\frac{2\pi}{\lambda} n \hat{e} \cdot \hat{r} - \omega t + \delta_o]} = A e^{-i[k \hat{e} \cdot \hat{r} - \omega t + \delta_o]}$$

where λ is the wavelength in vacuum. It is noted that A may be a function of time and position.

A *plane wave* is one for which all disturbances are dependent on only one coordinate. All light waves are plane polarized, and what is commonly referred to as unpolarized light is actually the randomly polarized light. The planes are random because different parts of the light source emit the waves differently and because the emitter behaves differently in time. In the analysis of the behavior of light, it is thus sufficient to consider a generally oriented plane wave, and when necessary, the results can be averaged over all orientations. In practice, polarizing screens are used; these are made of tiny crystals (such as quinine iodosulfate) brought into alignment by chemical and mechanical methods so that the polarizing planes are parallel.

Suppose such a wave is time harmonic (i.e. $e^{i\omega t}$) and is propagating in the z-direction. Then, in the Maxwell's equations we have

$$\frac{\partial}{\partial t} \rightarrow i\omega \qquad \nabla \rightarrow -ik \frac{\partial}{\partial z} \rightarrow -ik \frac{d}{dz}$$

Further, to investigate the properties of the fields, it is necessary to consider only one of them. Choose $\hat{E}$ and eliminate $\hat{H}$ and substitute in the constitutive relation (with $\omega / c = 2\pi / \lambda$) to get

$$\frac{d^2}{dz^2} \left\{ \begin{array}{c} E_x \\ E_y \end{array} \right\} + \left(\frac{2\pi}{\lambda} \right)^2 \left[\begin{array}{cc} 1 + e_{xx} & e_{xy} \\ e_{xy} & 1 + e_{yy} \end{array} \right] \left\{ \begin{array}{c} E_x \\ E_y \end{array} \right\} \qquad (2.19)$$

since $E_z \approx 0$. This equation shows that the optical effect is dependent only on dielectric properties in the plane of the wave front. This comes about essentially because the anisotropy of the field is assumed small. Generally, $[1+e]$ is a function of z and therefore Equation (2.19) is a second-order differential equation with variable coefficients, which, except for trivial cases, does not have closed form solutions. These equations show that both $\hat{H}$ and $\hat{D}$ lie in the plane perpendicular to k (i.e. they are transverse to the direction of propagation). Also, $\hat{H}$, $\hat{D}$, and k form an orthogonal triad, and $\hat{E}$ has a component in the k direction, but lies in the plane formed by $\hat{D}$ and k. This is shown in Figure 2.31(b). A fuller discussion of this equation is given in Reference [1].

For an optically isotropic medium $e_{xx} = e_{yy} = e$, $e_{xy} = 0$, and assuming the medium is homogeneous, gives

$$\hat{E} = \hat{E}_0 e^{-i[\frac{2\pi}{\lambda} z\sqrt{1+e}-\omega t]}$$

Compare this with the general wave relation, and we conclude that $\sqrt{1+e} = n$, that is, the dielectric is the square of the index of refraction.

Coherent Optics: Interference and Diffraction

The possibility of applying coherent optics techniques became more widely available with the advent of lasers. This section is a review of some of the basic concepts and equations used in coherent optics as applied to photomechanics.

The general wave can be assumed to be of the form

$$\hat{E} = \hat{E}_0(\hat{r}) e^{-i\left[\frac{2\pi}{\lambda}\hat{e}\cdot\hat{r}-\omega t+\delta(r,t)\right]}$$

where $\hat{r}$ is the position vector, $\hat{E}_0(r)$ the vector amplitude, $\hat{e}$ the unit vector in the direction of propagation, and $\delta(r, t)$ is the time-dependent initial phase. The latter three terms are, in general, a function of the position vector.

Interference refers to the resulting effects of superimposing two or more wave trains. If the phase difference is $180°(\lambda/2)$ then the resultant amplitude is zero (destructive interference) and no light is observed. This is the fundamentally important aspect of interference.

Consider two wave trains overlapping in same region of space. The resultant wave is given by

$$\hat{E} = \hat{E}_1 + \hat{E}_2$$

and the intensity at any point by $\hat{E} \cdot \hat{E}^*$ where $\hat{E}^*$ is the complex conjugate of $\hat{E}$. On substituting for $\hat{E}$ we get

$$I = \hat{E} \cdot \hat{E}^* = (\hat{E}_1 + \hat{E}_2) \cdot (\hat{E}_1^* + \hat{E}_2^*)$$

$$= E_{10}^2 + E_{20}^2 + 2[\hat{E}_{10} \cdot \hat{E}_{20}]\cos\left[\frac{2\pi}{\lambda_1}\left(\hat{e}_1 - \frac{\lambda_1}{\lambda_2}\hat{e}_2\right) \cdot \hat{r} + \delta_1 - \delta_2\right]$$

This is the fundamental equation for the interference of two waves and it shows that there are three contributions; two are due to the irradiances of the separate waves, and one is due to the interference effect. The first two are independent of the wave orientations and cause a common background intensity; whereas the third, which is capable of a positive or negative contribution, causes an intensity variation depending on two factors:

Relative vector amplitude : $\quad [\hat{E}_{10} \cdot \hat{E}_{20}]$

Relative phases : $\quad \dfrac{2\pi}{\lambda_1}\left[\hat{e}_1 - \dfrac{\lambda_1}{\lambda_2}\hat{e}_2\right] \cdot \hat{r} + \Delta\delta$

The first term is due to the relative vector amplitudes of the two waves and can vary from zero to $E_{10}E_{20}$. Note that if $\hat{E}_{10}$ is from a reference (known) source then the first term contains information about $\hat{E}_{20}$. The second term is due to the relative phases which in turn is a function of the wave directions, the initial phase difference, the position in space, and the relative wavelengths.

The interference effect is thus a spatial distribution of a variation in light intensity. If a screen is placed in the region, a series of fringes will be observed on it as shown in Figure 2.48. The contrast, spacing, and distribution of these fringes depend on the above two factors.

If the initial phase difference is random, then the cosine term oscillates about zero very rapidly and no interference effects are observed. When the light waves have random initial phases, they are called incoherent and for these the observed intensity is just the sum of the separate irradiances. Laser light is predominantly coherent and therefore is an excellent light source for use in interferometry. For this light source, the initial phase difference can be considered as a constant dependent only on the position, that is, we take $\Delta\delta$ as constant.

By way of example, consider a plane wave interfering with a spherical wave on a screen in the $z = 0$ plane. For convenience, let the wave be polarized in the y direction. We have for the plane ($\hat{E}_1$) and spherical ($\hat{E}_2$) waves, respectively

$$\hat{E}_1 = E_0\hat{j}, \qquad \hat{e}_1 = \cos\alpha\,\hat{i} + \sin\alpha\,\hat{j}, \qquad \hat{E}_2 = \frac{\epsilon}{r}\hat{j}, \qquad \hat{e}_2 = \cos\beta\,\hat{i} + \sin\beta\,\hat{j}$$

The phase factor is

$$\delta = \frac{2\pi}{\lambda_1}\left[\left[\cos\alpha - \frac{\lambda_1}{\lambda_2}\cos\beta\right]x + \left[\sin\alpha - \frac{\lambda_1}{\lambda_2}\sin\beta\right]y\right]$$

The intensity distribution is given by

$$I = \left[E_0^2 + \frac{\epsilon^2}{r^2}\right] + 2\left[E_0\frac{\epsilon}{r}\right]\cos(\delta + \Delta\delta)$$

The interference term has its minimum contribution whenever

$$\delta + \Delta\delta = \pi, 3\pi, \ldots, (2N - 1)\pi$$

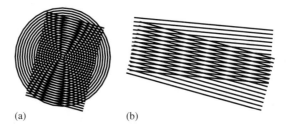

(a)   (b)

Figure 2.48: Interference of waves. (a) Plane and cylindrical waves. (b) Two plane waves.

or

$$\frac{1}{\lambda_1}\left[\left[\cos\alpha - \frac{\lambda_1}{\lambda_2}\cos\beta\right]x + \left[\sin\alpha - \frac{\lambda_1}{\lambda_2}\sin\beta\right]y\right] = N - \frac{1}{2} + \frac{\Delta\delta}{2\pi}$$

This gives an inhomogeneous distribution of fringes since β depend on x and y, respectively. In fact, they are curves of the form $\overline{y} = \pm\sqrt{(\overline{x}\lambda_2 + c_o)^2/\lambda_1^2 - \overline{x}^2}$ which are conic sections (oriented along $(\overline{x}, \overline{y})$) as shown in Figure 2.48(a). Note that the effect of $\Delta\delta$ is to give a horizontal shift to the fringes.

As another example, consider the special case as the point source is located at infinity in such a way that β is constant and $\epsilon/r = E_0$, that is, a plane wave is obtained. Then along $x = $ constant

$$I = 2E_0^2[1 + \cos\delta], \qquad \delta = -\frac{2\pi}{\lambda_1}\left[\left(\sin\alpha - \frac{\lambda_1}{\lambda_2}\sin\beta\right)y\right] + \Delta\delta$$

Zero-intensity fringes are observed whenever

$$\left[\sin\alpha + \frac{\lambda_1}{\lambda_2}\sin\beta\right]\frac{y}{\lambda_1} = N - \frac{1}{2} + \frac{\Delta\delta}{2\pi} \qquad \text{or} \qquad \Delta y = \lambda_1 \bigg/ \left[\sin\alpha + \frac{\lambda_1}{\lambda_2}\sin\beta\right]$$

This is a homogeneous set of fringes oriented nearly parallel to x with spacing Δy. This simple equation can be used to predict the fringe spacing observed when two beams at an angle are combined to interfere on a screen as shown in Figure 2.48(b). Suppose the two wavelengths are the same and the fringes are observed to be 1 mm apart, then the angle separation is

$$\sin\alpha + \sin\beta = \frac{\lambda}{\Delta y} = \frac{600 \times 10^{-9}}{10^{-3}} = 0.0006$$

The relative angle between α and β is very small.

Diffraction is the ability of rays of light to bend near sharp edges. This is caused by its wave nature, and can be demonstrated graphically using Huygen's wavelet construction as shown in Figure 2.49; that is, each wave front can be considered as an array of point sources at very short distance apart; the figure shows how an array of such sources can cause light to travel obliquely, but because of interference, will have an angle dependent intensity.

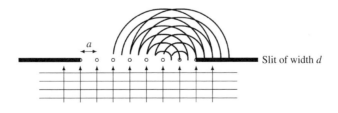

Figure 2.49: Diffraction by a single slit.

The following is a more mathematical description of Huygen's principle. Consider an array of identical point light sources, emitting waves of the form

$$\hat{E} = \hat{E}_o e^{-i[\frac{2\pi}{\lambda}\hat{e}\cdot\hat{r}-\omega t+\delta(r,t)]}$$

The wave arriving at a typical point P beyond the gap is given by

$$\hat{E}_p = \hat{E}_o \left[e^{-i\frac{2\pi}{\lambda}\hat{e}_1\cdot\hat{r}_1} + e^{-i\frac{2\pi}{\lambda}\hat{e}_2\cdot\hat{r}_2} + \cdots + e^{-i\frac{2\pi}{\lambda}\hat{e}_m\cdot\hat{r}_m} \right] e^{i[\omega t-\delta(r,t)]}$$

Suppose P is far away compared to the spacing a, then $\hat{e}_1 \approx \hat{e}_2 \approx \cdots \approx \hat{e}_n$ and

$$r_1 = r, \quad r_2 = r + a\sin\phi, \quad r_3 = r + 2a\sin\phi, \quad r_m = r + (m-1)a\sin\phi$$

Noting that

$$\sum_m^M e^{-i\frac{2\pi}{\lambda}r_m\hat{e}_m} = \sum_m^M e^{-i\frac{2\pi}{\lambda}[r+(m-1)a\sin\phi]} = e^{-i\frac{2\pi}{\lambda}r}\sum_m^M e^{-i\frac{2\pi}{\lambda}(m-1)a\sin\phi}$$

and

$$\sum_m^M e^{-ixm} = e^{-i(M-1)x/2}\left[\frac{\sin(Mx/2)}{\sin(x/2)}\right]$$

Therefore, the wave at P and its intensity are given by

$$\hat{E}_P = \hat{E}_o e^{-i[\frac{2\pi}{\lambda}r-\omega t+\delta]}\left[\frac{\sin[(M\pi a/\lambda)\sin\phi]}{\sin[(\pi a/\lambda)\sin\phi]}\right]$$

$$I = \hat{E}_p\cdot\hat{E}_p^* = E_o^2\left[\frac{\sin[(M\pi a/\lambda)\sin\phi]}{\sin[(\pi a/\lambda)\sin\phi]}\right]^2$$

The intensity is a trigonometric function of the orientation angle, as well as the number of sources and their spacing. Therefore, a distribution of maxima and minima intensities can be expected on a screen placed through P.

Light arriving in any region can be considered as an array of an infinite number of light sources very close together; that is, letting $d = Ma$ be the width of the slit so that as a becomes very small M becomes very large and the slit remains fixed. Let us simultaneously decrease the source strength of each point such that the total light emitted is constant, then in this limit, the intensity is approximated by

$$I \approx I_0 \left[\frac{\sin[(d\pi/\lambda)\sin\phi]}{(d\pi/\lambda)\sin\phi}\right]^2$$

This gives zero-intensity fringes wherever

$$\sin\left(\pi\frac{d}{\lambda}\sin\phi\right) = 0 \quad \text{or} \quad \sin\phi = N\frac{\lambda}{d}, \quad N = 1, 2, 3, \ldots$$

The maximum intensities occur wherever $\partial I/\partial\phi = 0$ giving

$$\tan\left(\pi\frac{d}{\lambda}\sin\phi\right) = \pi\frac{d}{\lambda}\sin\phi \qquad \text{or} \qquad \sin\phi = 1.43\frac{\lambda}{d},\ 2.46\frac{\lambda}{d},\ 3.47\frac{\lambda}{d},\ \dots$$

The intensities at these values are

$$I = I_0\,[1,\ 0.047,\ 0.016,\ 0.008,\ \dots]$$

This is called Fraunhofer diffraction and describes what would happen when light enters a narrow slit of width d: the bulk of the light passes through, but some of it is diffracted off the axis. There are several diffraction orders but each is at a lower intensity.

Holographic Setup and Recording

Consider an image-forming process for a single exposure hologram shown schematically in Figure 2.50(a). The exposed photographic plate records an interference pattern between light reflected from the object and the reference beam. For reconstruction, the developed photographic is replaced in the holder and the object removed; when the reference beam now strikes the plate and the plate is viewed as if looking through a window, then a virtual image of the object will be seen. A camera placed in the same location as the eye can take photographs of the image in exactly the same way as if the object was in its original position.

Because of the relatively long exposure times (order of seconds) in static holography, a stable support system must be provided in order to isolate the optical elements from the surroundings. Any random movement during exposure would result in a degrading of the holographic image. The laser may be either continuous or pulsed, depending on the problem. Surface preparation of models is not difficult for holography, and generally any diffuse-reflecting surface is sufficient. The hologram is essentially a very fine diffraction grating, and consequently, a high-resolution high-contrast film is needed. Typical types used with HeNe lasers are AGFA 10E75, KODAK 649-F, and KODAK 120-02, which

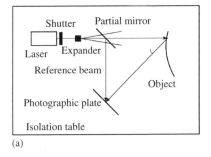

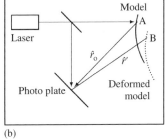

Figure 2.50: Recording holograms. (a) Typical setup on isolation table. (b) Recording displacements by double-exposure holography.

have the emulsion on glass plates and cost about \$2 per plate. It is also possible to get holographic film as 35 and 70 mm rolls, which is good for preliminary work but not for final results because of its flexibility. Developing a hologram is the same as that used in black and white photography; developer (D19), stop bath, and fixer. The laboratory room is usually used as the camera, therefore it must be light-tight and relatively free from building or air disturbances.

Most of the applications of holography (in experimental stress analysis) have been to measure surface displacement components of deforming bodies. A brief literature survey is given in Reference [145], a readable discussion of the use of holography in experimental mechanics is given in Reference [163], and a variety of other aspects are covered in Reference [133].

Unlike classical interferometry, which would require two models (one loaded, one unloaded), holography interferometry uses one model by the double-exposure technique. Consider the image-forming process for a double-exposure hologram shown schematically in Figure 2.50(b). An exposure of the model in state A is taken. The model is deformed to state B and a second exposure taken in the same emulsion with the same reference beam. When the processed photographic plate is illuminated with the reference beam, waves from both A and B are reconstructed simultaneously and interfere with one another (since they are coherent). The result is that fringes appear on the model, each fringe corresponding to a change in optical path length of nominally half a wavelength produced by surface displacements.

If the holographic plate is developed after the first exposure and replaced in exactly the same position, then the waves emanating from the illuminated object interfere with the diffracted waves (due to the reference beam) and fringes will be observed depending on the current position of the object relative to its original position. Thus the body can be loaded in real time, and the fringes observed as they are formed.

Fringe Formation under Motion

To make the analysis general, we will consider forming a hologram of a moving object. This is equivalent to a succession of double exposures and will allow the analysis of vibration problems.

Let two waves $\hat{E}_1$ and $\hat{E}_2$ interfere at the holographic plate, that is,

$$\hat{E} = \hat{E}_{10} e^{i\phi_1} + \hat{E}_{20} e^{i\phi_2}$$

The processed hologram is reconstructed by operating on it with the reference beam $\hat{E}_R$. The transmitted wave corresponds to three separate waves given by

$$\text{undiffracted}: \ [E_{10}^2 + E_{20}^2] \hat{E}_R$$

$$\text{real image}: \ [\hat{E}_{10} \cdot \hat{E}_{20}] \hat{E}_R e^{i(\phi_1 + \phi_R)} \int_0^T e^{-i\phi_2} \, dt$$

$$\text{virtual image}: \ [\hat{E}_{10} \cdot \hat{E}_{20}] \hat{E}_R e^{-i(\phi_1 - \phi_R)} \int_0^T e^{i\phi_2}$$

where the plate was exposed for a time T. The image of interest is that of the virtual image because it can be focused by a camera or an observer's eye. The intensity of this image is

$$I_{\text{virt}} \propto [\hat{E}_{10} \cdot \hat{E}_{20}]^2 E_R^2 \left| \int_0^T e^{i\phi_2} \, dt \right|^2$$

The intensity is seen to be a function of the time variation of the phase. This variation can come about for various reasons, for example: change in position, change in density, change in wavelength, and holography can be used to measure these phase changes. In the following, three special cases of time variation will be considered.

Consider a double exposure where ϕ_2 changes by a discontinuity

$$\phi_2 = \phi_0 \quad t < 0, \qquad \phi_2 = \phi_0 + \Delta\phi \quad t > 0$$

Therefore,

$$\int_0^T e^{-i\phi_2} \, dt = \frac{1}{2} T e^{-i\phi_0} [1 + e^{-i\Delta\phi}]$$

The transmitted intensity of the virtual image is then a function of the phase change

$$I_{\text{virt}} \propto [\hat{E}_{10} \cdot \hat{E}_{20}]^2 E_R^2 \cos^2(\Delta\phi/2)$$

Because of the presence of the cosine function, the intensity will show periodic maximums and minimums, the minimums occurring at the roots of the cosine function; that is, where

$$\Delta\phi = (2N - 1)\pi$$

fringes will appear on the image depending on the spatial variation of ϕ.

Now, consider a time-elapsed exposure where ϕ varies linearly with time over the time of exposure

$$\phi_2 = \phi_0 \quad t < 0, \qquad \phi_2 = \phi_0 + \Delta\phi \frac{t}{T} \quad t > 0; \qquad \int_0^T e^{-i\phi_2} \, dt = \frac{-iT}{\Delta\phi} e^{-i\phi_0} [1 - e^{-i\Delta\phi}]$$

The virtual image intensity is

$$I_{\text{virt}} \propto [\hat{E}_{10} \cdot \hat{E}_{20}]^2 E_R^2 \left| \frac{2}{\Delta\phi} \sin(\Delta\phi_0/2) \right|^2$$

This also shows a periodic variation in intensity with minimums whenever

$$\Delta\phi = 0, 2\pi, 4\pi, \ldots, 2N\pi$$

The fringes in this case occur at the zeros of the sine function, but will be of decreasing contrast because of the division by $\Delta\phi$ in the expression for I_{virt}.

As a final case, consider a time-elapsed exposure when ϕ_2 varies sinusoidally with time; that is,

$$\phi_2 = \phi_0 + \frac{1}{2} \Delta\phi [1 - \cos \omega t]$$

The intensity becomes

$$\int_0^T e^{-i\phi_2}\, dt = e^{-i[\phi_0 + \frac{1}{2}\Delta\phi]} \int_0^T \left[\cos\left[\frac{1}{2}\Delta\phi \cos(\omega t)\right] + i\sin\left[\frac{1}{2}\Delta\phi \cos(\omega t)\right] \right] dt$$

Since the exposure will be over many cycles, it is sufficient to consider only one complete cycle. We can perform the integration by first expanding the trigonometric functions in terms of Bessel functions; that is, using

$$\int_0^\pi \cos(a\cos\theta)\, d\theta = J_0(a), \qquad \int_0^\pi \sin(a\cos\theta)\, d\theta = 0$$

gives the intensity of the virtual image as

$$I_{\text{virt}} \propto [\hat{E}_{10} \cdot \hat{E}_{20}]^2 E_{\text{R}} J_0^2(\tfrac{1}{2}\Delta\phi)$$

The minimum intensity fringes occur at the zeros of the Bessel function; for N greater than 1, these are given as a simple function of π

$$\Delta\phi = 1.534\pi,\ 3.514\pi,\ 5.506\pi,\ 7.506\pi,\ \ldots,\ (N + \tfrac{1}{2})\pi$$

Again, there is a loss in fringe contrast. This is shown in Figure 2.51.

This last situation lies approximately halfway between the step change and the linear change. This is explained by noting that during the oscillation, the phase tends to dwell at the extremes and hence approximates a series of step functions.

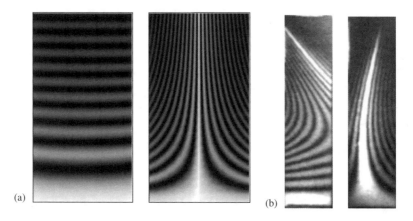

(a)   (b)

Figure 2.51: First bending and first torsion time average holographic fringes of a vibrating cantilevered plate. Contrast of the higher-order fringes deteriorates. (a) Synthetic data for an isotropic plate. (b) Experimental data for a fiber-reinforced composite plate.

Displacement Measurement

In general, the phase of the wave from the object is given by

$$\phi_2 = \frac{2\pi}{\lambda}\hat{e}_2 \cdot \hat{r}_2 - \delta_2 = \frac{2\pi}{\lambda}r_0 - \delta_s$$

where δ_s is the phase change in arriving from the source to the point. Small changes in phase can come about from

$$\Delta\phi = \frac{2\pi}{\lambda}\Delta r_0 - \Delta\delta_s - \frac{2\pi}{\lambda}r_0\frac{\Delta\lambda}{\lambda}$$

that is, through changes in position, direction, initial phase, and wavelength. This can be expressed in terms of the displacement by noting that $\hat{r}'_0 = \hat{r}_0 + \hat{u}$, or

$$\Delta r_0 = r'_0 - r_0 = |\hat{r}_0 + \hat{u}| - r_0 = \sqrt{r_0^2 + 2\hat{r}_0 \cdot \hat{u} + u^2} - r_0 \approx \frac{\hat{r}_0 \cdot \hat{u}}{r_0} = \hat{e}_0 \cdot \hat{u}$$

This relation was simplified by taking advantage of the fact that $\hat{u}$ is small. For the same source (during exposure), the initial phase is also changed because of the displacement, that is,

$$\delta_s = \frac{2\pi}{\lambda}r_s \qquad \text{or} \qquad \Delta\delta_s = \frac{2\pi}{\lambda}\hat{e}_s \cdot \hat{u} - \frac{2\pi}{\lambda}r_s\frac{\Delta\lambda}{\lambda}$$

A way of changing the wavelength is through changing the effective index of refraction, that is,

$$\lambda = \frac{c}{nf} \qquad \text{or} \qquad \frac{\Delta\lambda}{\lambda} = -\frac{\Delta n}{n}$$

Consequently, the total change in phase is given as

$$\Delta\phi = \frac{2\pi}{\lambda}\left[(\hat{e}_s + \hat{e}_0) \cdot \hat{u} + (r_s + r_0)\frac{\Delta n}{n}\right] \tag{2.20}$$

This is the basic equation for holographic interferometry. It shows that displacements on the order of a wavelength of light can be measured and changes in index of refraction of the order of λ/r. Both of these show the sensitivity of the holographic method.

The most convenient way of measuring displacements is by the double exposure technique; that is, by making an exposure before and after the displacement. This gives

$$(2N - 1)\pi = \Delta\phi = \frac{2\pi}{\lambda}(\hat{e}_s + \hat{e}_0) \cdot \hat{u} \qquad \text{or} \qquad (\hat{e}_s + \hat{e}_0) \cdot \hat{u} = (N - \tfrac{1}{2})\lambda$$

When the displacement is zero $N = 1/2$ and $I = \text{max}$, and consequently, zero displacements correspond to a bright fringe.

In general, the displacement $\hat{u}$ has three components and therefore three independent measurements must be made. This usually involves making three holograms from three different directions. While this procedure can work, it is very cumbersome and so only displacements for simpler situations are considered here. One such case of importance is the flexure of plates, where the displacement is dominated by the out-of-plane component.

The limitations on the holographic interferometry are quite numerous. It requires stringent mechanical and thermal stability of the system, it is too sensitive for most practical problems and while the data is recorded in a whole-field manner, it must be analyzed point by point in the general case. Further, the strains (which are the quantities of importance in stress analyses) are obtained by differentiation and this can be a source of large error. A problem not evident from the above analysis, but one of great practical importance, is that the fringes may localize anywhere in space depending on the deformation. This makes it very difficult to correlate the fringes with points on the surface of the object.

Other Interferometric Methods

The history of specklelike phenomena predates the laser, although the laser was primarily responsible for the development of engineering applications of speckle interferometry. This technique has been utilized as an effective measure of surface deformation measurements in experimental mechanics. Reference [145] gives an introduction to the method and References [42, 43] give an extensive analysis. In this method, an object is illuminated by two beams that optically interfere. Surface displacement information is obtained by comparing the speckle patterns photographed before and after the deformation process.

In recent years, direct processing of specklelike data has emerged as a technique in optical stress analysis. The principles of this method parallel the developments of speckle photography; small subimages of a discrete recording of a speckle type of image are correlated for undeformed and deformed configurations of a body. Neither method, however, will replace the other, because experiments to date indicate that speckle photography has greater sensitivity than image correlation. Therefore, problems that require a sensitive measurement technique will utilize the principles of holography and speckle photography.

2.9 Photoelasticity

Photoelasticity is an experimental method of stress analysis, which utilizes the birefringent phenomenon exhibited by some polymers under load, and has played a significant role in the history of experimental stress analysis. Long before computers were available, it was used to determine stress concentration factors for a variety of structural shapes, both two- and three-dimensional. Indeed, it is particularly useful in analyzing components of complex geometry and loading.

Its main limitation is that it is a model method; that is, a plastic model (and not the prototype) must be used. Since this limits the variety of constitutive behaviors, then by and large, photoelasticity is restricted to linear elastic problems. This section reviews the essentials of the method and concludes with a brief discussion of some of its extensions to make it more widely applicable.

Despite its drawbacks, photoelasticity remains a very popular method because with simple equipment difficult problems can be tackled, yielding good engineering solutions.

Plane Waves in Optically Anisotropic Media

Returning to Equation (2.19) which describes light propagating in a general medium, consider the case when the medium is optically anisotropic. Assume that a plane wave can propagate in this medium, that is,

$$\hat{E} = \hat{E}_0 e^{ikz}$$

where k is as yet undetermined. Substitute into the differential equation and get

$$\begin{bmatrix} k^2 + \left(\dfrac{2\pi}{\lambda}\right)^2 (1 + e_{xx}) & \left(\dfrac{2\pi}{\lambda}\right)^2 e_{xy} \\ \left(\dfrac{2\pi}{\lambda}\right)^2 e_{xy} & k^2 + \left(\dfrac{2\pi}{\lambda}\right)^2 (1 + e_{yy}) \end{bmatrix} \begin{Bmatrix} E_x \\ E_y \end{Bmatrix} = 0$$

This can have a nontrivial solution only if the determinant is zero, that is, only if

$$k^2 = \left(\frac{2\pi}{\lambda}\right)^2 \left[1 + \frac{1}{2}(e_{xx} + e_{yy}) \pm \frac{1}{2}\sqrt{(e_{xx} - e_{yy})^2 + 4e_{xy}^2}\right]$$

From the transformation of tensor components discussed in Chapter 1, this is recognized as the principal values of e_{ij} given by

$$k = \frac{2\pi}{\lambda}\sqrt{1 + e_1} = \frac{2\pi}{\lambda}n_1 \quad \text{or} \quad k = \frac{2\pi}{\lambda}\sqrt{1 + e_2} = \frac{2\pi}{\lambda}n_2$$

Thus, the medium behaves as if it has two indices of refraction and two waves will propagate in it. It is this phenomenon of two indices of refraction (or double refraction) that is called *birefringence*. Some materials exhibit it naturally, for example, the following [90]:

material:	Ice	Quartz	Mica	Calcite	Rutile	Sapphire
n_1:	1.309	1.544	1.598	1.658	2.616	1.768
n_2:	1.313	1.553	1.593	1.486	2.903	1.760

Some materials exhibit it artificially due to deformation or stress, and this will be used as the basis for the photoelastic method of stress analysis.

The orientation of each wave is obtained by substituting for k and rearranging to get

$$\tan \theta' = \frac{E_y}{E_x} = \frac{-e_{xy}}{1 + e_{yy} + \left(\dfrac{\lambda}{2\pi}\right)^2 k^2} \quad \text{or} \quad \tan 2\theta = \frac{2e_{xy}}{e_{xx} - e_{yy}}$$

that is, each wave has an orientation along of one of the principal directions. It is important to note that this orientation is independent of the wavelength.

In summary, a plane-polarized wave entering an optically anisotropic medium will resolve into two plane-polarized waves oriented along the principal directions. These will propagate with speeds proportional to the principal indices of refraction. On exiting the medium, there will be a phase difference between the two waves given by

$$\delta = \left(\frac{2\pi}{\lambda}n_1 h + \omega t\right) - \left(\frac{2\pi}{\lambda}n_2 h + \omega t\right) = \frac{2\pi}{\lambda}(n_1 - n_2)h$$

where h is the optical path length through the medium. The instrument that measures this phase difference is called a *polariscope*.

Polariscope Construction

The essential elements of the polariscope are a light source and two sheets of polaroid—one before and one after the model as shown in Figure 2.52. This arrangement is called a plane polariscope. There are two basic polariscope types: the diffuse and lens polariscope.

The diffused-light polariscope is one of the simplest and least expensive polariscopes to construct: a bank of fluorescent lamps plus two sheets of polaroid are the essentials. The light background can be made uniform by placing a translucent sheet in front of the lamps. Its field can be made very large since it depends only on the size of the available linear polarizers and quarter-wave plates. Diffused-light polariscopes with field diameters up to 450 mm can readily be constructed.

Lens polariscopes are used where parallel light over the whole field is required. A situation in which this is important is where the precise definition of the entire boundary is critical.

Several variations of the lens systems are possible. The arrangement shown in Figure 2.52 is one of the simpler types. The polarizer, quarter-wave plates, and analyzer are placed in the parallel beam between the two field lenses to avoid problems associated with internal stresses in the field lenses. A point source of light is required, and this is often obtained from a high-intensity mercury lamp with a very short arc. The photoelastic model causes a small amount of scattering of the light that disturbs the parallelism of the system. This effect can be minimized by placing a small diaphragm stop at the focal point of the second field lens.

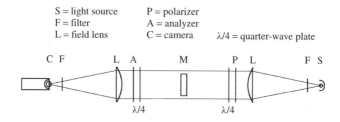

Figure 2.52: Elements of a lens polariscope.

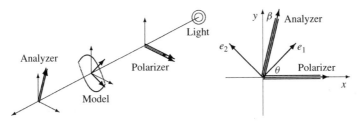

Figure 2.53: Orientation of the elements of a plane polariscope.

Consider the plane polariscope shown in Figure 2.53. To simplify the analysis, assume that the distances and medium between the elements do not affect the waveform. The polarizer is the element that converts a regular light source into one that is plane-polarized. Hence, after the polarization,

$$\hat{E} = E_0 \hat{i} e^{i\omega t}$$

The model behaves as an anisotropic dielectric and will propagate two waves. The initial conditions for these waves are

$$\hat{E}^1 = +E_0 \cos\theta [\cos\theta \hat{i} + \sin\theta \hat{j}] e^{i\omega t} = E_0 \cos\theta \, \hat{e}_\theta e^{i\omega t}$$

$$\hat{E}^2 = -E_0 \sin\theta [-\sin\theta \hat{i} + \cos\theta \hat{j}] e^{i\omega t} = -E_0 \sin\theta \, \hat{e}_{(\theta+\pi/2)} e^{i\omega t}$$

where $\hat{e}$ is a unit vector in the indicated direction and θ is the orientation of the principal axes. After passage through the model of thickness h, each wave receives an extra phase retardation; hence on exiting, the waves are given by

$$\hat{E}^1 = +E_0 \cos\theta \hat{e}_\theta e^{-i\delta_1}$$

$$\hat{E}^2 = -E_0 \sin\theta \hat{e}_{(\theta+\pi/2)} e^{-i\delta_2}$$

where $\delta = (\frac{2\pi}{\lambda} n_1 h + \omega t)$. The analyzer is of exactly the same construction as the polarizer, but it comes after the model. It will allow only those components parallel to its orientation to pass. Therefore, the light vector after an analyzer oriented in the $\hat{e}_\beta$ direction is given by

$$\hat{E}_A = E_0 [\cos\theta \hat{e}_\theta \cdot \hat{e}_\beta e^{-i\delta_1} - \sin\theta \hat{e}_{(\theta+\pi/2)} \cdot \hat{e}_\beta e^{-i\delta_2}] \hat{e}_\beta$$

The intensity of light recorded after the analyzer is (noting that $\hat{e}_\theta \cdot \hat{e}_\beta = \cos(\beta - \theta)$, $\hat{e}_{(\theta+\pi/2)} \cdot \hat{e}_\beta = \sin(\beta - \theta)$)

$$I = \hat{E}_A \hat{E}_A^* = \frac{1}{2} E_0^2 \left[1 + \cos 2\theta \cos 2(\beta - \theta) - \sin 2\theta \sin 2(\beta - \theta) \cos \frac{2\pi}{\lambda} (n_1 - n_2) h \right]$$

This is the fundamental equation for analyzing the plane polariscope. As is, the intensity is a rather complicated function of the angles and phase difference.

If the analyzer is perpendicular to the polarizer ($\beta = 0$), then the observed light intensity is given by

$$I = \left[E_0 \sin 2\theta \sin \frac{2\pi}{\lambda} (n_1 - n_2) h / 2 \right]^2$$

This is often called a crossed polariscope. Complete extinction can occur in two particular ways. If one of the principal directions of the dielectric coincide with the orientation of either the polarizer or analyzer, then

$$\theta = 0 \quad \text{or} \quad \pi/2$$

and hence $I = 0$. Thus, a black band will appear connecting all points with an orientation that is the same as that of the polarizer (or analyzer). This black band (or fringe) is called an *isoclinic* (constant inclination). If the polarizer and analyzer are moved in unison then a whole family of isoclinics can be obtained. Complete extinction is also obtained whenever the argument of the phase term is a multiple of π; that is,

$$\frac{2\pi}{\lambda}\left(\frac{n_1 - n_2}{2}\right)h = N\pi \quad \text{or} \quad n_1 - n_2 = N\frac{\lambda}{h} \qquad N = 0, 1, 2, \ldots.$$

where N is called the fringe order. For monochromatic light, black lines (fringes) will occur that are loci of constant phase retardation and hence, the constant difference of indices of refraction. These fringes are wavelength dependent, hence in white light bands of constant color are observed. For this reason, these fringes are called *isochromatics*.

It is surprising how small a difference in indices is needed in order for fringes to be observed. For example, suppose one fringe is observed through a thickness of 3 mm, then

$$n_1 - n_2 = N\frac{\lambda}{h} = \frac{1 \times 600 \times 10^{-9}}{3 \times 10^{-3}} = 0.0002$$

If the average index is 1.5 then

$$n_1 = 1.5001 \qquad n_2 = 1.4999$$

It is seen that the polariscope is a very sensitive instrument for detecting small differences in indices of refraction.

The polarizer and analyzer can also be placed parallel to each other. The fringes observed with this arrangement are a half order from those observed in the crossed polariscope. Further, when there is no phase difference ($n_1 = n_2$) the light intensity is $I = E_0^2$, whereas for the crossed arrangement it is $I = 0$. For this reason the respective arrangements are often referred to as light and dark fields, respectively. Examples are shown in Figure 2.54.

The isochromatics and isoclinics appear simultaneously in the plane polariscope; this, at times, can make fringe interpretation difficult. A refinement to the plane polariscope is obtained by inserting quarter-wave ($\lambda/4$) plates before and after the model and oriented at 45° to the polarizer. This arrangement is called a *circular polariscope* and it removes the isoclinics leaving only the isochromatics. It is important to note that a quarter-wave plate works at only one wavelength and an error is introduced if it is used with the wrong light source or with white light to determine other than whole or half-order fringes.

Finally, a word about the use of white light. For a constant thickness model, a given index difference will cause different fringe orders depending on the wavelength used. Complete extinction, however, will occur only at those wavelengths for which

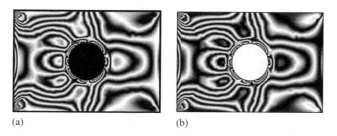

(a)   (b)

Figure 2.54: Isochromatics in dark and light field polariscopes. (a) Dark field gives whole-order fringes. (b) Light field gives half-order fringes.

$(n_1 - n_2)h/\lambda$ is an integer. Therefore, for a white light source, the wavelength that makes the above ratio an integer will be blocked and the color observed at that point will be white minus the particular wavelength. Note that as the retardation gets larger, it is possible to have multiple wavelengths being extinguished (at different orders but the same point). This makes fringe interpretation through colors difficult at the high retardation levels and is the reason this approach is generally restricted to fringe orders less than four or five. Isoclinics are best observed using white light because they appear as black fringes on a colored background.

Optical Constitutive Relations

It is of importance now to establish the relationship between the fringe patterns and the mechanical variables. The fringes are related to the indices of refraction, whereas constitutive relations are written more conveniently in terms of the dielectric tensor. Hence, it is first necessary to relate the dielectric components to the fringe data.

The measured optical data (N, θ) can he written in the form

$$N\frac{\lambda}{h} = n_1 - n_2 = \sqrt{1 + e_1} - \sqrt{1 + e_2}, \qquad \tan 2\theta = \frac{2e_{xy}}{e_{xx} - e_{yy}}$$

Noting that $e_i = n_i^2 - 1$, then we get

$$e_1 - e_2 = n_1^2 - n_2^2 = (n_1 + n_2)(n_1 - n_2) \approx 2n(n_1 - n_2)$$

where n is the average index of refraction of the medium. The basic optical relations can now be written as

$$e_{xx} - e_{yy} = (\epsilon_1 - \epsilon_2) \cos 2\theta = 2n\frac{N\lambda}{h} \cos 2\theta$$

$$2e_{xy} = (e_1 - e_2) \sin 2\theta = 2n\frac{N\lambda}{h} \sin 2\theta$$

or in matrix form

$$\frac{1}{2n} \begin{Bmatrix} e_{xx} - e_{yy} \\ 2e_{xy} \end{Bmatrix} = \frac{N\lambda}{h} \begin{Bmatrix} \cos 2\theta \\ \sin 2\theta \end{Bmatrix}$$

These are the basic equations relating the observed optical data to the dielectric properties of the medium.

Birefringence is observed in a variety of polymeric materials, and these have been used as the basis for a variety of experimental stress analysis tools. References [57, 58, 59, 60, 61] discuss the optical characterization of some of these materials. The structural materials of most common interest are those that are linear elastic, and it is no surprise that the birefringent effect has most of its applications in this area. We will restrict ourselves to that material.

In photoelasticity, the stresses are simply related to the fringes by

$$\frac{1}{2n} \left\{ \begin{matrix} e_{xx} - e_{yy} \\ 2e_{xy} \end{matrix} \right\} = \frac{\lambda}{f_\sigma} \left\{ \begin{matrix} \sigma_{xx} - \sigma_{yy} \\ 2\sigma_{xy} \end{matrix} \right\} = \frac{N\lambda}{h} \left\{ \begin{matrix} \cos 2\theta \\ \sin 2\theta \end{matrix} \right\}$$

where f_σ is called the (photoelastic) *material fringe value* (also sometimes called the fringe-stress coefficient) and is obviously a function of the wavelength. This coefficient is experimentally obtained and will be different for different materials. This relation may be written alternatively as

$$N\frac{f_\sigma}{h} = \sigma_1 - \sigma_2 = \sqrt{(\sigma_{xx} - \sigma_{yy})^2 + 4\sigma_{xy}^2}, \qquad \tan 2\theta = \frac{2\sigma_{xy}}{\sigma_{xx} - \sigma_{yy}} \qquad (2.21)$$

which is usually referred to as the stress-optic law.

The isochromatics and isoclinics thus have simple interpretation in terms of the stresses. The isoclinics give the loci of points whose principal stress orientation coincides with that of the polarizer/analyzer combination. By rotating the polarizer/analyzer combination in increments, a whole family of isoclinics may be obtained. The isoclinics are usually recorded in increments of 5°. When a monochromatic light source is used, black fringes labeled 0, 1, 2, ... are observed. Along a free edge (where one of the principal stresses is zero), the greater number of fringes occur in regions of high stress. If the polarizer and analyzer are put in a parallel arrangement (light field), then the black fringes correspond to $N = \frac{1}{2}, 1\frac{1}{2}, 2\frac{1}{2}, \ldots$ and this allows intermediate data to be obtained. Reference [83] gives a thorough listing of the properties of isoclinic and isochromatic fringes.

Since photoelasticity is a model method, each photoelastic material must be calibrated to obtain the material fringe value f_σ. The simplest calibration model is the uniaxial specimen, and for this

$$\sigma_1 = \frac{P}{Wh}, \qquad \sigma_2 = 0$$

where W is the specimen width. This gives the relationship between load and fringe order as

$$P = f_\sigma WN$$

Therefore, f_σ can be obtained from the slope of the load versus fringe graph. Note that the thickness of the model does not figure in this formula. Other calibration schemes are discussed later.

Table 2.3: Properties of some photoelastic materials.

material:		f_σ [kN/m]	(lb/in.)	E [MPa]	(ksi)	ν
Epoxy:	(Araldite, Epon)	11.0	(60.0)	3300	(430)	0.37
Polycarbonate:	(PSM-1, Makrolon)	7.0	(40.0)	2600	(360)	0.38
PMMA:	(Plexiglas, Perspex)	130	(700.)	2800	(400)	0.38
Polyester:	(Homolite 100)	24.0	(140.0)	3900	(560)	0.35
Polyurethane:	(Hysol)	0.2	(1.0)	3	(0.5)	0.46

The properties of some common photoelastic materials are given in Table 2.3. Because of the variability inherent in plastics, these property values are to be taken as nominal values, and calibration for a particular batch of materials is always recommended. Note that the PMMA material is essentially nonbirefringent and therefore often used as part of a built-up or sandwich model.

Extensions of Photoelasticity

Many extensions have been made to the basic photoelastic method and References [43, 48] give a good survey of them. Just three of these are mentioned here. Recent advances in techniques based on digital methods are discussed in References [137, 144].

I: Stress Freezing

Polymeric materials are composed of long chain molecules. Some of these are well bonded in a 3-D network of primary bonds, but a large number are only loosely bonded into shorter secondary chains. At room temperature, in the glassy state, both sets of bonds resist the deformation. As the temperature of the polymer is increased, in particular, when it goes past the critical or glass transition temperature, the secondary bonds break down and the deflections that the primary bonds undergo are quite large. When the temperature is lowered back to room temperature (while the load is maintained) the secondary bonds re-form and lock the long chains in their extended state. Thus, when the load is removed the deformation is not recovered.

After stress freezing, a photoelastic model can be cut or sliced without disturbing the character of the deformation or the fringe pattern. This makes it suitable for creating 3-D models that are appropriately sliced and analyzed to get the full stress distributions.

The primary materials used for stress freezing are the epoxy resins. The epoxies consist of a base monomer that is polymerized using either an amine or anhydride curing agent. Generally, the anhydrides are preferred because of the large exothermic reactions of the amines. A typical epoxy would have properties, at the effective elevated temperature, of

Temp: 120 °C   250 °F
f_σ: 350 N/m   2.0 lb/in.
E: 17 MPa   2500 psi

When analyzing 3-D slices, the fringes are related to the principal stress difference in the plane of the slice. Generally, supplementary information is required in order to get the full stress distribution. This can be obtained either from multiple oblique views or by using the shear difference method. Both of these procedures effectively reduce the photoelastic method down to a point by point method, and hence are not really advanced here.

II: Integrated Photoelasticity

A potentially powerful 3-D photoelastic method is that of integrated photoelasticity [1, 2, 55]. Here, the complete 3-D model is analyzed without the destructive slicing of the stress-freezing method. Since the rays pass through the complete model, the resulting birefringent effect include the accumulated or integrated effect of all the stresses along the path.

The birefringent effect along each ray of light is governed by the differential system of Equation (2.19) derived from Maxwell's equations for (optically) inhomogeneous media [1, 2, 55]. While integrating these is difficult enough, another complication is that because of the changing index of refraction, the ray path is no longer straight. A simplification made is that (to first approximation) the rays are straight and the accumulated birefringence is simply

$$N = \frac{1}{f_\sigma} \int (\sigma_1^* - \sigma_2^*) \, dz$$

where σ_1^* and σ_2^* are the secondary principal stresses in the plane perpendicular to the direction of propagation. The method poses challenges in recording enough information as well as in performing the inverse problem.

III: Birefringent Coatings

The greatest disadvantage of the classical photoelastic method is that it is a model method—it cannot be applied to the prototype directly. A model (of another material) must be used and thus imposes restrictions on the geometry and materials of prototypes suitable for analysis.

The birefringent coating method overcomes this disadvantage by bonding a thin photoelastic coating to the prototype. The coating is mirrored at the interface so that when observed in a reflection polariscope, the optical effect is essentially at the surface of the structure. The advantages of the method are: it is a whole-field method, it is nondestructive, it may be used on the prototype, and it is not prototype material dependent.

The coating material should have many of the following characteristics: high optical coefficient K to maximize the number of fringes and minimize the coating thickness;

low modulus of elasticity to minimize reinforcing effects; linear strain optical response for ease of data reduction; high proportional limit to maximize the measuring range; and exhibit little stress relaxation. Some typical coating materials are mentioned below: Polycarbonate, which are available only as flat sheets; Epoxy, which is good for contouring because it can be cast; and Polyurethane, because it is suitable for measuring large strains.

For curved surfaces, the procedure is to cast a slow curing sheet, and when it is in the rubbery state, contour it by hand to fit the curved surface of the structure. When the polymerization is completed, the sheet becomes a rigid contoured shell, which may then be bonded to the specimen. It may be necessary to check the variation of thickness before bonding.

The reflection polariscope has the same elements as the regular polariscope but here the light goes through the model, reflects at the back surface, goes through the coating again, and is then observed through the analyzer. The optical response of the coating is therefore

$$\frac{N\lambda}{2h} \begin{Bmatrix} \cos 2\theta \\ \sin 2\theta \end{Bmatrix} = K \begin{Bmatrix} \epsilon_{xx} - \epsilon_{yy} \\ \gamma_{xy} \end{Bmatrix}$$

where K is the strain optic coefficient for the coating and can be obtained from a calibration experiment.

Experiment: Calibration of Photoelastic Materials

There are a variety of calibration tests that can be used to obtain the photoelastic material fringe value f_σ. Two of the more common ones are the uniaxial test and the disk under diametral compression. Both of these are discussed in Reference [48].

For 3-D stress freezing where the model is made by casting, the calibration specimen must be made and tested at the same time as the model itself; that is, the calibration specimen must go through the same temperature cycle as the model. A convenient test for this purpose is the four-point bend test.

Figure 2.55 shows a dark-field fringe pattern for a stress-frozen specimen made of PLM-4 epoxy [177]. More details on the experiment and the application can be found in References [155, 156]. The center portion is under pure bending (as witnessed by the almost horizontal fringes) achieved by a whippletree arrangements of links.

The overall load (including links) is $P = 15.4\,\mathrm{N}\,(3.46\,\mathrm{lb})$; this produces a moment of $M = WP/2$, which in turn causes a bending stress given by

$$\sigma = \frac{My}{I} = \frac{6PWy}{Bh^3}$$

Consequently, the fringe distribution in the uniform section should obey

$$N = \left(\frac{1}{f_\sigma}\right)\frac{6PWy}{h^3}$$

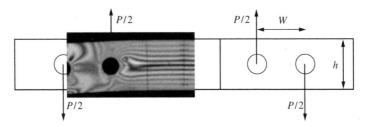

Figure 2.55: Calibration experiment. Fringe pattern for beam in four-point bending.

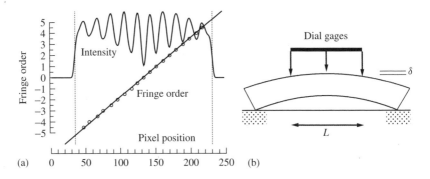

Figure 2.56: Calibration experiment. (a) Fringe order plot. (b) Scheme to determine Young's modulus.

The material fringe value is obtained from the slope as

$$f_\sigma = \frac{1}{\text{slope}} \frac{6PW}{h^3}$$

The fringes in Figure 2.55 do not have very good contrast; however, the intensity plot of Figure 2.56(a) makes it very easy to identify the whole and half-order fringes. (The intensity plot is taken slightly to the right of center away from the hash marks.) The fringe order plot is remarkably linear.

The edge of the specimen is sometimes hard to detect, again, the intensity plot makes the task easier. The slope is determined from the extrapolated edge values as slope = $(N_u - N_l)/h = 10.446$ fringe/in. from which the material fringe value is determined as

$$f_\sigma = 0.574\,P = 350 \text{ N/m} \quad (1.99 \text{ lb/in.})$$

It is interesting to compare this to the nominal room temperature value of 10.5 kN/m (60. lb/in.); almost a factor 30 difference in sensitivity.

A slight curvature of the beam can be noticed in Figure 2.55. This is because a permanent deformation is caused through the stress-freezing process. A very useful attribute of the four-point bending specimen is that this curvature can be used to estimate the Young's modulus. With reference to Figure 2.56(b), the radius and strain are estimated as, respectively,

$$R \approx \frac{L^2}{8\delta}, \qquad \epsilon = \frac{y}{R} = \frac{8\delta y}{L^2}$$

Combine this with the stress relation to get the Young's modulus

$$E = \frac{\sigma}{\epsilon} = \frac{3PWL^2}{4Bh^3\delta} = 16.8\,\text{MPa} \ (2430\,\text{psi})$$

This compares to the nominal room-temperature value of 3300 MPa (450 ksi). The epoxy softens considerably at the stress-freezing temperature.

3

Inverse Methods

As outlined in the introduction, the key to solving ill-posed inverse problems is to incorporate some supplementary information. How this can be done in a rational way is the subject of this chapter.

We begin with a discussion of modeling and the use of experimental data. The connection between model and data can be rather remote and give rise to some significant computational challenges. The essential idea that emerges is that of regularizing the solution; that is, rather than trying to have a best-fit of the data, we instead negotiate between satisfying the data and satisfying levels of smoothness based on *a priori* information about the solution. As illustrated in Figure 3.1, the "best" solution is a compromise between both requirements.

The main ideas presented in this chapter come from References [143, 172]; interesting additional background material can be found in References [165, 166]. Reference [127]

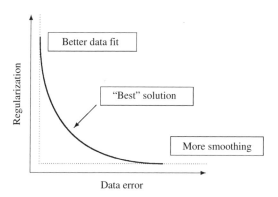

Figure 3.1: The trade-off curve where the "best" solution is a compromise between satisfying the data and satisfying levels of smoothness.

Modern Experimental Stress Analysis: completing the solution of partially specified problems.   James Doyle
© 2004 John Wiley & Sons, Ltd   ISBN 0-470-86156-8

gives a more extensive mathematical treatment of regularization and its relationship to least squares solution of problems; it also has an extensive bibliography.

3.1   Analysis of Experimental Data

Experimental data from measurements have inherent variations caused by any number of reasons depending on the particular experimental setup. Further, when multiple measurements must be made, this too can cause a variation in the final calculated result. It is important to understand how the variations in the individual measurements contribute to the accuracy of the overall result, because this can have a profound influence on the design of the experiment; for example, it is wasteful to purchase an expensive piece of apparatus to perform/obtain a reading, the influence of which on the final result is negligible.

There are basically two forms of variations that can occur in readings:

- Systematic variations, as can occur because of, say, a wrongly calibrated gauge.
- Random variations.

It is the latter type of variation that is of concern here. The former can be prevented only by the proper attention to detail, and by knowing the performance characteristics of the equipment.

Noise and Statistics

We need to distinguish a number of related concepts that play a role in our later developments. The statistical analysis of data characterizes the population of data in terms of its average, spread, and so on. Error analysis of data clarifies the factors contributing to the spread of the data. The modeling of data divines the underlying law (or model) governing the central behavior of the data. Sometimes this last operation is simply the curve fit of noisy data, but as we will see, can at times give rise to some significant issues that must be addressed.

A uniform deviate is a set of random numbers that lie within a range (0 to 1, say) with any one number as likely to occur as another. There are other deviates such as numbers drawn from a normal (Gaussian) distribution that has a specified mean and standard deviation. Figure 3.2 shows the histograms (the number of times a particular number occurred) for both the uniform and Gaussian distributions. These plots were made by taking 100,000 random samples of each and counting the number of times the sample fell into amplitude bins discretized at 0.01. The code for generating these deviates is taken from Reference [143].

The Gaussian distribution is described by

$$p(u) = e^{-\left(\frac{u - \bar{u}}{4\sigma}\right)^2}$$

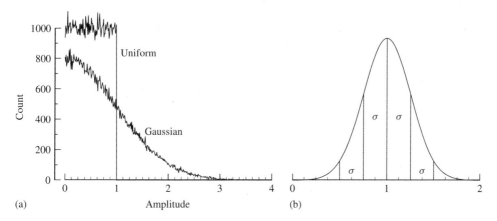

(a)   Amplitude   (b)

Figure 3.2: Random distributions. (a) Comparison of uniform and Gaussian deviates. (b) Parameters of a Gaussian distribution.

where $\bar{u}$ is the mean and σ is the standard deviation—these terms will be explained presently. Note that the distribution is symmetric and the amplitudes can be up to four times that of the uniform deviates. In later chapters, when noise is added to the synthetic data to make it somewhat more realistic, we will use Gaussian noise.

Statistics of Random Variations

Suppose we have a set of measurements u_i with $i = 1, 2, \ldots, N$. Rather than deal with the individual numbers, it is sometimes desirable to represent them by a reduced set of numbers, their statistics. To make the example concrete, suppose the extensional properties of a piece of polymer are being tested by measuring the change in length of a scribed grid (nominally 28 mm in length) to an applied load. This is done by measuring the grid length before and after the application of a dead weight load. The results for 10 specimens are given in Table 3.1.

There is obviously a variation in the data. This could be due to scribing the grids, measuring the spacing or even variations in the material properties itself, to name a few. The question, however, is, Can representative conclusions about the polymer be drawn from this collection of numbers?

The first step is to display the results graphically by plotting a *Histogram*, that is, a plot of the frequency of occurrence that values within certain fixed ranges have. The histograms are plotted in Figure 3.3. It is noted that the data exhibits a central tendency or seem to be clustered around an "average" value. A measure of this average is called the *mean* and is defined as

$$\text{Mean:} \qquad \bar{u} \equiv \frac{1}{N} \sum_{i=1}^{N} u_i$$

Table 3.1: Measurements made on polymer specimens.

Lengths

Specimen#	1	2	3	4	5	6	7	8	9	10
Initial:	26.1	28.2	30.5	26.1	24.6	27.1	28.1	27.0	29.0	27.7
Final:	36.7	37.5	36.0	37.1	36.3	34.3	35.1	39.5	37.2	35.1
Difference:	10.6	9.3	5.5	11.0	11.7	7.2	7.0	12.5	8.2	7.4

Counts

Initial:	Range	24–25	25–26	26–27	27–28	28–29	29–30	30–31	31–32
	Count	1	0	2	3	2	1	1	0
Final:	Range	33–34	34–35	35–36	36–37	37–38	38–39	39–40	40–41
	Count	0	1	2	3	3	0	1	0
Difference:	Range	5–6	6–7	7–8	8–9	9–10	10–11	11–12	12–13
	Count	1	0	3	1	1	1	2	1

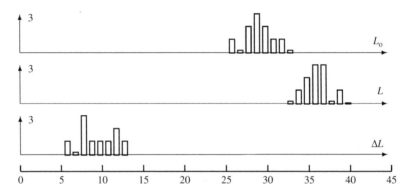

Figure 3.3: Histograms for polymer deformations.

giving for our data

$$\bar{u}_{\text{initial}} = 27.44 , \qquad \bar{u}_{\text{final}} = 36.48 , \qquad \bar{u}_{\text{difference}} = 9.04$$

An additional parameter of value would be one that describes the "spread" of the data. One measure that does this is called the *standard deviation* defined as

$$\text{Standard deviation:} \qquad \sigma_u \equiv \left[\frac{1}{N-1} \sum_1^N (u_i - \bar{u})^2 \right]^{1/2}$$

giving for our data

$$\sigma_{\text{initial}} = 1.66 , \qquad \sigma_{\text{final}} = 1.49 , \qquad \sigma_{\text{difference}} = 2.33$$

It is possible for different samples to have the same mean but different spreads, and so giving the mean and the standard deviation is a shorthand way of characterizing all the data, that is, rather than giving all the individual data points, it is often more convenient to summarize the data in the form

$$\bar{u} \pm 2\sigma_u$$

By taking twice the standard deviation, about 98% of the data collected will be represented, whereas using only one standard deviation would include about 68% of the data; this is illustrated in Figure 3.2(b). These numbers presuppose that the variations fall approximately in a normal or Gaussian distribution. If the distribution has some other shape, then additional statistics are used.

Representing the data in this fashion, it is interesting to observe that the limits on the difference are not simply the sum of the worst limits on the length measurements, that is, 2.33 versus $1.66 + 1.49 = 3.15$. In fact, they are less. The reason for this is considered next.

Adding Variations

Consider the calculation of strain when using an electrical resistance strain gage. The formula for this is

$$\epsilon = \frac{1}{S_g} \frac{\Delta R}{R}$$

Let S_g (supplied by the manufacturer) and $\Delta R / R$ (measured by the experimenter) have variations in their measurement by the amounts

$$S_g = 2 \pm 0.01, \qquad \frac{\Delta R}{R} = 520 \pm 10 \frac{\mu\Omega}{\Omega}$$

These variations could be judged from previous experience or by noting the range within which the results of nominally the same readings fall.

An estimate for the uncertainty in the strain calculation can be obtained by taking the worst possible variations. While the nominal value is given by

$$\epsilon_{nom} = \frac{520}{2.00} = 260.00\,\mu\epsilon$$

the limits are estimated as

$$\epsilon_{max} = \frac{520 + 10}{2.00 - 0.01} = 266.33\,\mu\epsilon, \qquad \epsilon_{min} = \frac{520 - 10}{2.00 + 0.01} = 253.73\,\mu\epsilon$$

Thus, the confidence in the reported strain value is

$$\epsilon = 260.00 \pm 6.33\,\mu\epsilon$$

It is worth emphasizing that determining such a variation also helps in deciding the number of significant figures to be reported. The strain above should be reported as

$$\epsilon = 260 \pm 6\,\mu\epsilon$$

since clearly we cannot have a confidence in the variation ($\pm 6.33\,\mu\epsilon$) greater than in the basic quantity.

This simple test is severe in that it is unlikely that the resistance measurements will be high (say) precisely when the gage factor is low. Note the use of the word "unlikely"; while it is possible for such a situation to occur, it is known from experience that most cases will fall somewhere in between, and consequently give lower bounds. When there are many elements in the experimental setup, the simple approach can give bounds that are too large to be of practical value. Nonetheless, this simple method can be of great value as a rough check, because if the experimental results deviate by more than the amount indicated, then before proceeding any further, both the experimental setup and the data itself must be scrutinized very closely for errors.

A more precise estimate will obviously be obtained by somehow incorporating the probability that a variation will occur. Suppose the quantity of interest Q is to be calculated from the measured quantities $a_1, a_2, \ldots, a_N$ by the equation

$$Q = Q(a_1, a_2, \ldots, a_N)$$

Let there be variations in the readings so that it is $a_j + \Delta a_j$ and not a_j that is actually measured. This, consequently, causes a variation in Q. If the Δ's are small, then by a Taylor's series expansion, the calculated Q is given approximately as

$$Q + \Delta Q = Q(a_1, a_2, \ldots, a_N) + \frac{\partial Q}{\partial a_1}\Delta a_1 + \frac{\partial Q}{\partial a_2}\Delta a_2 + \cdots$$

and the variation is

$$\Delta Q = \frac{\partial Q}{\partial a_1}\Delta a_1 + \frac{\partial Q}{\partial a_2}\Delta a_2 + \cdots$$

A severe measure of the total variation is given as

$$\Delta Q = \left|\frac{\partial Q}{\partial a_1}\Delta a_1\right| + \left|\frac{\partial Q}{\partial a_2}\Delta a_2\right| + \cdots$$

This is the simple method of error analysis introduced earlier.

Another approach is to consider many measurements being made and that ΔQ_i is the deviation for each reading set. The standard deviation over many readings is therefore given as

$$\sigma_Q = \left[\frac{1}{(N-1)}\sum_{i=1}^{N}[\Delta Q_i]^2\right]^{1/2}$$

Substituting for ΔQ_i and multiplying out gives

$$\sigma_Q^2 = \frac{1}{(N-1)} \sum_i \left[\frac{\partial Q}{\partial a_1} \Delta a_1 + \frac{\partial Q}{\partial a_2} \Delta a_2 + \cdots \right]^2$$

$$= \frac{1}{(N-1)} \left[\left(\frac{\partial Q}{\partial a_1} \right)^2 \sum_i \Delta a_1^2 + 2 \left(\frac{\partial Q}{\partial a_1} \right) \left(\frac{\partial Q}{\partial a_2} \right) \sum_i \Delta a_1 \Delta a_2 + \cdots \right.$$

$$\left. \left(\frac{\partial Q}{\partial a_2} \right)^2 \sum_i \Delta a_2^2 + 2 \left(\frac{\partial Q}{\partial a_2} \right) \left(\frac{\partial Q}{\partial a_3} \right) \sum_i \Delta a_2 \Delta a_3 + \cdots \right]$$

Consider the sums $\sum_i \Delta a_1 \Delta a_2$, $\sum_i \Delta a_1 \Delta a_3$, and so on; since any term is as likely to be positive as negative (because they have independent variations), these sums must be close to zero. Hence, we have the approximation

$$\sigma_Q^2 \approx \frac{1}{(N-1)} \left[\left(\frac{\partial Q}{\partial a_1} \right)^2 \sum_i \Delta a_1^2 + 2 \left(\frac{\partial Q}{\partial a_2} \right)^2 \sum_i \Delta a_2^2 + \cdots \right]$$

But $\sum_i \Delta a_j^2/(N-1)$ is the standard deviation σ_j in the measurement of parameter a_j, therefore, the standard deviation in the calculated result is

$$\sigma_Q^2 = \left(\frac{\partial Q}{\partial a_1} \right)^2 \sigma_1^2 + \left(\frac{\partial Q}{\partial a_2} \right)^2 \sigma_2^2 + \cdots \quad \text{or} \quad \sigma_Q = \sqrt{\left(\frac{\partial Q}{\partial a_1} \right)^2 \sigma_1^2 + \left(\frac{\partial Q}{\partial a_2} \right)^2 \sigma_2^2 + \cdots}$$

$$(3.1)$$

In comparison with the simple approach, this gives a smaller variation because the square root of the sum of the squares is smaller than the sum of the components. In this context, we will refer to $\partial Q/\partial a_j$ as the sensitivity of the measurement Q to the parameter a_j.

Let us reconsider the case of the extension of the polymer. The change of length is given as

$$\Delta L = L_f - L_i$$

Making the identifications

$$Q \longrightarrow \Delta L = L_f - L_i, \quad a_1 \longrightarrow L_f, \quad a_2 \longrightarrow L_i$$

gives the sensitivities

$$\frac{\partial Q}{\partial a_1} \longrightarrow 1, \quad \frac{\partial Q}{\partial a_2} \longrightarrow -1$$

and

$$\sigma_{\Delta L} = \sqrt{[1\sigma_f]^2 + [-1\sigma_i]^2} = \sqrt{[1.66]^2 + [-1.49]^2} = 2.23$$

This is very close to the number (2.33) actually obtained.

Now reconsider the strain gage results. The association of variables is

$$Q \longrightarrow \epsilon = \frac{1}{S_g} \frac{\Delta R}{R}, \qquad a_1 \longrightarrow S_g, \qquad a_2 \longrightarrow \frac{\Delta R}{R},$$

$$\frac{\partial Q}{\partial a_1} \longrightarrow -\frac{1}{S_g^2} \frac{\Delta R}{R}, \qquad \frac{\partial Q}{\partial a_2} \longrightarrow \frac{1}{S_g}$$

and this leads to the deviation

$$\sigma_\epsilon = \sqrt{\left(\frac{\Delta R}{R}\right)^2 \frac{\sigma_1^2}{S_g^4} + \frac{\sigma_2^2}{S_g^2}} = \sqrt{16900\,\sigma_1^2 + .25\,\sigma_2^2}$$

Let $\sigma_1 = 0.01$, $\sigma_2 = 10$. Then,

$$\sigma_\epsilon = \sqrt{1.69 + 25} = 5.17\,\mu\epsilon$$

As expected, this value is slightly smaller than the value (6.37) previously obtained. One of the things this exercise shows is that the variation in $\Delta R/R$ dominates the strain measurement error. This, perhaps, is not surprising since the manufacturer controls the quality of the gage, and hence the variation of the gage factor; it is expected that this is controlled within narrow limits.

3.2 Parametric Modeling of Data

We often need to condense or summarize a set of data by fitting it to a model with adjustable parameters. Sometimes the modeling is simply a curve-fit of functions (such as polynomials o r sinusoids), and the fit determines the coefficients. Other times, the model's parameters come from some underlying theory that the data are supposed to satisfy; an example is the Young's modulus from stress/strain data and the assumption that the material behavior is linear elastic. We also encounter situations in which a few data points are to be extended into a continuous function, but with some underlying idea of what that function should look like. This is like curve-fitting except that we bias the fit.

The approach is to choose an error or cost function that measures the agreement between the data and the model. The parameters are then adjusted to achieve a minimum in the error function, yielding the best-fit parameters. This is essentially a problem in minimization in many dimensions.

It must be kept in mind that data are not exact, they are subject to measurement uncertainties and therefore never exactly fit the model, even when that model is correct. Consequently, we must be able to assess whether the model is appropriate. This entails testing the goodness-of-fit against some statistical standard.

The main ideas in this section are covered in great depth in Reference [143].

Linear Chi-Square Fitting

Suppose that we are fitting N data points (x_i, d_i), $i = 1, N$, (where d_i are the measurements at the independent positions x_i) to a model that has M adjustable parameters a_j, $j = 1, \ldots, M$. We can write the functional relationship underlying the model as

$$u(x) = u(x; a_1, \ldots, a_M)$$

Now suppose that each data point has a known standard deviation σ_i; then an estimate of the model parameters is obtained by minimizing the quantity

$$\chi^2 = \sum_{i=1}^{N} \left[\frac{d_i - u(x_i; a_1, \ldots, a_M)}{\sigma_i} \right]^2 \tag{3.2}$$

This quantity is called the "chi-square" error function.

Taking the derivative of Equation (3.2) with respect to the parameters a_k, yields equations that must hold at the chi-square minimum,

$$0 = \sum_i \left[\frac{d_i - u(x)}{\sigma_i^2} \right] \frac{\partial u(x_i, \ldots, a_k \ldots)}{\partial a_k}, \qquad k = 1, \ldots, M$$

This is, in general, a set of M nonlinear equations for the M unknown a_k. The various procedures described subsequently in this section derive from this and its specializations.

In most cases, the individual uncertainties associated with a set of measurements are not known. We can get an estimate of the error bar, however, by assuming that all measurements have the same standard deviation, $\sigma_i = \sigma$, proceeding by assigning an arbitrary constant σ to all points, and then fitting for the model parameters by minimizing χ^2, and finally recomputing

$$\sigma^2 = \frac{1}{(N - M)} \sum_i [d_i - u(x)]^2$$

When the measurement error is not known, this approach gives an error estimate to be assigned to the points.

I: Fitting Data to a Straight Line

We begin by considering the relatively simple problem of fitting a set of N data points (x_i, d_i) to a straight-line model

$$u(x) = u(x; a, b) = a + bx$$

In this case, the chi-square error estimate is

$$\chi^2(a, b) = \sum_i \left[\frac{d_i - a - bx_i}{\sigma_i} \right]^2$$

Now minimize this equation to determine a and b. At the minimum, derivatives of $\chi^2(a, b)$ with respect to a, b are zero to give the two equations

$$\frac{\partial \chi^2}{\partial a} = -2 \sum_i \frac{d_i - a - bx_i}{\sigma_i^2} = 0, \qquad \frac{\partial \chi^2}{\partial b} = -2 \sum_i x_i \frac{d_i - a - bx_i}{\sigma_i^2} = 0$$

This is a $[2 \times 2]$ system of equations, which can be written in the form

$$aS + bS_x = S_d$$

$$aS_x + bS_{xx} = S_{xd} \tag{3.3}$$

where we have introduced the sums:

$$S \equiv \sum_i \frac{1}{\sigma_i^2}, \quad S_x \equiv \sum_i \frac{x_i}{\sigma_i^2}, \quad S_d \equiv \sum_i \frac{d_i}{\sigma_i^2}, \quad S_{xx} \equiv \sum_i \frac{x_i^2}{\sigma_i^2}, \quad S_{xd} \equiv \sum_i \frac{x_i d_i}{\sigma_i^2}$$

The solution of these two equations for the best-fit model parameters is calculated as

$$a = (S_{xx} S_d - S_x S_{xd})/\Delta, \qquad b = (S S_{xd} - S_x S_d)/\Delta, \qquad \Delta \equiv S S_{xx} - S_x S_x$$

The analysis is completed by estimating the probable uncertainties in the estimates of a and b; that is, we will think of a and b as functions of the data. Recall that the variance σ_{f}^2 in the value of any function will be

$$\sigma_{\mathrm{f}}^2 = \sum_i \sigma_i^2 \left(\frac{\partial f}{\partial d_i} \right)^2$$

The derivatives of a and b with respect to d_i can be directly evaluated from the solution (noting that only S_d and S_{xd} depend on d_i and do so linearly) to give

$$\frac{\partial a}{\partial d_i} = \frac{S_{xx} - S_x x_i}{\sigma_i^2 \Delta}, \qquad \frac{\partial b}{\partial d_i} = \frac{S x_i - S_x}{\sigma_i^2 \Delta}$$

Squaring, summing over all the data, and simplifying, leads to the simple expressions

$$\sigma_a^* = \sqrt{S_{xx}/\Delta}, \qquad \sigma_b^* = \sqrt{S/\Delta}$$

Note that these variances in the estimates of a and b do not contain the data; this will be added after we have estimates of the standard deviations in the data. Actually, these estimates are a special case of a more general result. If we write the system of Equation (3.3) as

$$[\ \alpha\ ]\{a\} = \{\beta\}$$

then the inverse matrix $[C_{jk}] = [\alpha_{jk}]^{-1}$ is closely related to the probable uncertainties of the estimated parameters a_j according to

$$\sigma^2(a_j) = C_{jj}$$

that is, the diagonal elements of $[\ C\ ]$ are the variances (squared uncertainties) of the fitted parameters a_j.

Figure 3.4 shows an example of linear curve-fitting. The data were generated for the line

$$u = 2 + 0.5x, \qquad d = 2 + 0.5x \pm R$$

where R are random numbers; these data are also shown in Figure 3.4. We do not know the standard deviation of each data point, so we take it as unity, that is, setting $\sigma_i = 1$, gives the intermediate calculations

$$S = 11.0, \qquad S_x = 55.0, \qquad S_d = 50.1, \qquad S_{xx} = 385.0, \qquad S_{xd} = 306.1$$

This leads to estimates for the coefficients as

$$a = 2.017, \qquad b = 0.507, \qquad \Delta = 1210.0$$

This is shown plotted as the heavy line in Figure 3.4.

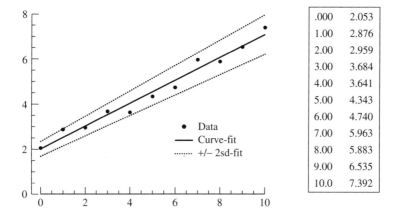

.000	2.053
1.00	2.876
2.00	2.959
3.00	3.684
4.00	3.641
5.00	4.343
6.00	4.740
7.00	5.963
8.00	5.883
9.00	6.535
10.0	7.392

Figure 3.4: Results for linear curve-fitting.

The χ^2 error for the fit is computed as

$$\chi^2 = \sum_i [d_i - a - bx_i]^2 = \sum_i [d_i - 2.017 - 0.507x_i]^2 = 0.749$$

An estimate of the variances of the data (on the assumption that the curve-fit is good) is now

$$\sigma = \sqrt{\chi^2/(N-2)} = 0.288$$

The variances in the coefficients are estimated as

$$\sigma_a = \sigma\sigma_a^* = \sigma\sqrt{S_{xx}/\Delta} = 0.163, \qquad \sigma_b = \sigma\sigma_b^* = \sigma\sqrt{S/\Delta} = 0.028$$

The top and bottom dashed lines in Figure 3.4 correspond to plotting

$$u_+(x) = [a + 2\sigma_a] + [b + 2\sigma_b]x, \qquad u_-(x) = [a - 2\sigma_a] + [b - 2\sigma_b]x$$

respectively, which correspond to the worst case scenarios of twice the standard deviations. These estimates do a good job in bracketing the data.

II: Uncertainties in Both Dimensions

If the experimental data are subject to uncertainties in both the measurements d_i and the positions x_i, then the task of curve-fitting is much more difficult. We can write the χ^2 error function as

$$\chi^2(a, b) = \sum_i \frac{[d_i - a - bx_i]^2}{\sigma_{di}^2 + b^2\sigma_{x_i}^2}$$

where σ_{xi} and σ_{di}, are respectively the x and u standard deviations for the ith point. The form of variances in the denominator arises as the variance between each data point and the line with slope b, and also as the variance of the linear combination $d_i - a - bx_i$ of two random variables x_i and d_i.

The occurrence of b in the denominator of χ^2 makes the equation for the slope $\partial\chi^2/\partial b = 0$ nonlinear. The corresponding condition for the intercept, $\partial\chi^2/\partial a = 0$ is still linear and yields

$$a = \sum_i \frac{[d_i - bx_i]}{\sigma_{di}^2 + b^2\sigma_{x_i}^2} \Big/ \sum_i \frac{1}{\sigma_{di}^2 + b^2\sigma_{x_i}^2}$$

A reasonable strategy is to minimize χ^2 iteratively with respect to b, while using this equation for a (updated at each stage) to ensure that the minimum with respect to b is also minimized with respect to a.

III: General Linear Least Squares

A generalization of the previous section is to fit a set of data points (x_i, d_i) to a model that is a linear combination of any M specified functions of x. The general form of this kind of model is

$$u(x) = \sum_k a_k \phi_k(x)$$

where $\phi_1(x), \ldots, \phi_M(x)$ are specified functions of x, called the *basis functions*. For example, the functions could be $1, x, x^2, \ldots, x^{M-1}$, in which case their general linear combination,

$$u(x) = a_1 + a_2 x + a_3 x^2 + \cdots + a_M x^{M-1}$$

is a polynomial of degree $M - 1$. The functions, alternatively, could be sines and cosines, in which case their general linear combination is a harmonic series. In fact, it can be shown that a Fourier series is a least squares curve-fit of a function.

Define an error function

$$\chi^2 = \sum_i \left[\frac{d_i - \sum_k a_k \phi_k(x_i)}{\sigma_i} \right]^2$$

and pick as best parameters those that minimize χ^2. The minimum of χ^2 occurs where the derivatives with respect to all M parameters a_k are zero simultaneously. This yields the M equations

$$\sum_j \alpha_{kj} a_j = \beta_k \qquad \text{or} \qquad [\,\alpha\,]\{a\} = \{\beta\}$$

where

$$\alpha_{kj} \equiv \sum_i \frac{\phi_j(x_i)\phi_k(x_i)}{\sigma_i^2}, \qquad \beta_k \equiv \sum_i \frac{d_i \phi_k(x_i)}{\sigma_i^2}$$

In this, $[\,\alpha\,]$ is an $[M \times M]$ matrix, and $\{\beta\}$ a vector of length M.

These equations are called the *normal equations* of the least squares problem. They can be solved for the vector of parameters $\{a\}$ by the standard methods for systems of equations. The solution of a least squares problem directly from the normal equations is rather susceptible to roundoff error. These difficulties can be alleviated by using singular value decomposition (SVD); indeed, it is the recommended technique [143] for all but "easy" least squares problems. We will return to this issue later.

Nonlinear Chi-Square Fitting

Consider curve-fitting the apparently simple model

$$u(x) = A e^{-Bx}$$

where A and B are the unknown parameters. This is an example of a nonlinear curve-fit since B does not appear in a linear fashion. Note that this particular case could be linearized by using

$$\log[u] = \log[A] - Bx$$

and treating $\log[A]$ as one of the unknowns. Here, however, we wish to consider directly the problem of nonlinear curve-fitting.

Let the model depend nonlinearly on the set of M unknown parameters a_k, $k = 1, 2, \ldots, M$. We use the same approach as in the previous subsection, namely, define a χ^2 error function and determine best-fit parameters by its minimization; that is,

$$\chi^2(a_k) = \sum_i \frac{1}{\sigma_i^2} [d_i - u(x_i; a_k)]^2$$

With nonlinear dependencies, however, the minimization must proceed incrementally/iteratively, that is, given trial values for the parameters, we develop a procedure that improves the trial solution. The procedure is then repeated until χ^2 stops decreasing.

I: Brute Force Method

The simplest method of finding the parameters of a model is to scan the possible space of the parameters and evaluate the error function to find which combination of parameters gives the minimum error. In two dimensions, the approach is enhanced by displaying the results in the form of contour plots.

Figure 3.5 shows contours for the function

$$u(x) = Ae^{-Bx}, \qquad A = 0.6, \qquad B = 1.4$$

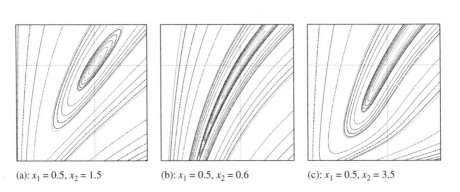

(a): $x_1 = 0.5, x_2 = 1.5$ (b): $x_1 = 0.5, x_2 = 0.6$ (c): $x_1 = 0.5, x_2 = 3.5$

Figure 3.5: Brute force function evaluations over the range $A[0, 1]$ horizontal, $B[0, 2]$ vertical, for different separations of measurement points x_1 and x_2. (a) Reasonable separation. (b) Too close. (c) Too remote.

for different measurement points x_1, x_2. The error is evaluated as

$$\chi^2 = [d_1 - Ae^{-Bx_1}]^2 + [d_2 - Ae^{-Bx_2}]^2$$

The topology of the contours is of a single valley. Clearly, when the measurement points are good, this brute force scheme correctly identifies the minimum point and hence the correct parameters. It has two serious drawbacks, however. First, it has terrible scaling as the number of parameters M increases; the number of function evaluations scales as N^M, where N is the number of evaluations in a typical dimension. Second, to get precise parameter values requires globally changing the resolution N, thus further exacerbating the computational cost.

There are obviously many refinements that could be made to improve this basic scheme; indeed, that is the point of the following subsections. But this example highlights a very crucial issue which will recur frequently throughout this book: irrespective of the scheme of finding the minimum of a function, the topology of the search space (and hence the success of the method) is profoundly affected by the choice of measurements. Devising schemes that are robust with respect to the measurements is one of our preoccupations.

II: Steepest Descent Method

Suppose we have an initial (approximate) guess for the parameters and we wish to improve them; looking at an error function topology such as Figure 3.5(a) immediately suggests a scheme—simply keep moving in the direction of the steepest slope and eventually the bottom will be met.

With reference to the one-dimensional plot of Figure 3.6(a), suppose we are nowhere near the minimum, then it is best to treat the function as linear (and not quadratic which is the case at the actual minimum) and take a step down the gradient; that is,

$$a_{\text{next}} = a^o - h \frac{\partial \chi^2}{\partial a}\Big|_{a^o}$$

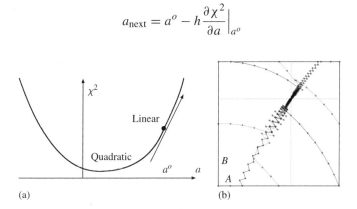

| (a) | (b) |

Figure 3.6: The steepest descent scheme. (a) Away from the minimum, assume the function is nearly linear. (b) Convergence performance for different initial guesses on the perimeter of the search space $A[0, 1]$, $B[0, 2]$.

where h is the step size (discussed further on). After the step is taken, χ^2 is then evaluated at the new point: if it got smaller, the new point is accepted; if it did not get smaller, then a smaller step size is taken.

For the multidimensional case, the gradient of χ^2 with respect to the parameters a_k, has the components

$$\frac{\partial \chi^2}{\partial a_k} = -2\sum_i \frac{1}{\sigma_i^2} [d_i - u(x_i; a_k)] \frac{\partial u(x_i; a_k)}{\partial a_k} \equiv -\beta_k , \qquad k = 1, 2, \ldots, M$$

leading to the updating relation as

$$a_k = a_k^o + h_k \beta_k \qquad \text{or} \qquad \{a\} = \{a^o\} + \{h\beta\}$$

The choice of the step size is important for the efficiency of this scheme. A reasonable choice is to take it as some fraction of the search space. However, as the minimum is achieved, the derivative tends to zero, the increments in parameter will become very small (for fixed h) and convergence will be excruciatingly slow. An alternative is to use the derivative purely for directional information, that is, the updating is done according to

$$a_k = a_k^o + h_k \hat{e}_k , \qquad \hat{e}_k = \frac{h_k \beta_k}{\sqrt{(h_1 \beta_1)^2 + (h_2 \beta_2)^2 + \cdots}}$$

The advantage is that the step size can be related to the global search space.

This scheme was applied to the situation of Figure 3.5(a). The derivative functions are

$$\frac{\partial u}{\partial A} = e^{-Bx} , \qquad \frac{\partial u}{\partial B} = -x A e^{-Bx}$$

The results for different initial guesses are shown in Figure 3.6(b). It is clear that the scheme can find the valley; but it then bogs down as the absolute minimum is approached.

Sophistications can be applied to improve the stepping scheme, but for now we conclude that the algorithm should be modified once the minimum is close and take into account the quadratic form of the function. This is discussed next.

III: Least-Squares Newton–Raphson Method

Assume we have a guess a_k^o for the parameters, then a Taylor series expansion gives

$$u(x) = u(x; a_1 \ldots a_M) \approx u(x; a_k^o + \Delta a_k) \approx u(x; a_k^o) + \sum_j \frac{\partial u(x; a_k^o)}{\partial a_j} \Delta a_j + \cdots$$

The direct application of the Newton–Raphson method would obtain the parameter increments by solving

$$\left[\frac{\partial u(x; a_k^o)}{\partial a_j} \right] \{\Delta a_j\} = \{u(x) - u(x, a_k^o)\}$$

But we actually have discretized values $\{d_i\}$ at positions x_i and therefore the system is overdetermined. Substituting the expansion into the error function

$$\chi^2 = \sum_i \left[d_i - u_i^o - \sum_j \frac{\partial u}{\partial a_j} \Delta a_j \right]^2, \qquad u_i^o \equiv u(x_i, a_k^o)$$

and minimizing with respect to the parameters yields the M equations

$$\sum_j \alpha_{kj} \Delta a_j = \beta_k \qquad \text{or} \qquad [\ \alpha\ ]\{\Delta a\} = \{\beta\}$$

where

$$\alpha_{kj} \equiv \sum_i \frac{\partial u_i}{\partial a_k} \frac{\partial u_i}{\partial a_j}, \qquad \beta_k \equiv \sum_i [d_i - u_i^o] \frac{\partial u_i}{\partial a_k}$$

In this, $[\ \alpha\ ]$ is an $[M \times M]$ matrix, and $\{\beta\}$ a vector of length M. Starting with a guess for a_k^o, we compute u_i^o, solve for Δa_k, update the value of a_k and repeat the process until convergence. Convergence is assumed to be achieved when χ^2 no longer changes significantly. One possibility is

$$\left| \frac{\chi^2 - \chi^{o2}}{\chi^2_{\text{ref}}} \right| < \epsilon, \qquad \epsilon = 1.0 \times 10^{-10}$$

where χ^2_{ref} could be taken as the value at the initial guess.

Before we do an example, it is of value to show the connection between the Newton–Raphson method and the direct use of the χ^2 error function as was done with the steepest descent method.

Consider a Taylor series expansion about the current estimated state

$$\chi^2(a_k) = \sum_i \frac{1}{\sigma_i^2} [d_i - u(x_i; a_k)]^2 \approx \chi^2(a_k^o + \Delta a_k)$$

$$\approx \chi^2(a_k^o) + \sum_k \frac{\partial \chi^2}{\partial a_k} \Delta a_k + \frac{1}{2} \sum_{k,m} \frac{\partial^2 \chi^2}{\partial a_k \partial a_m} \Delta a_k \Delta a_m + \cdots$$

The gradient of χ^2 with respect to the parameters a_k, has the components

$$\frac{\partial \chi^2}{\partial a_k} = -2 \sum_i \frac{1}{\sigma_i^2} [d_i - u(x_i; a_k)] \frac{\partial u(x_i; a_k)}{\partial a_k} \equiv -2\beta_k, \qquad k = 1, 2, \ldots, M$$

The second partial derivative is

$$\frac{\partial^2 \chi^2}{\partial a_k \partial a_m} = 2 \sum_i \frac{1}{\sigma_i^2} \left[\frac{\partial u}{\partial a_k} \frac{\partial u}{\partial a_m} - [d_i - u(x_i; a_k)] \frac{\partial^2 u}{\partial a_k \partial a_m} \right]$$

Look at the term $[d_i - u(x_i; a_j)]$ multiplying the second derivative; for a good fit, this term should just be the random-measurement uncertainties and are as likely to be positive as negative. Therefore, the second derivative terms tend to cancel out when summed over i. Hence, we use as an approximation

$$\frac{\partial^2 \chi^2}{\partial a_k \partial a_l} \approx 2 \sum \frac{1}{\sigma_i^2} \left[\frac{\partial u(x_i; a_j)}{\partial a_k} \frac{\partial u(x_i; a_j)}{\partial a_l} \right] \equiv 2\alpha_{kl}$$

and thus avoid computing second derivatives of the curve-fitting functions. We therefore have for the expansion

$$\chi^2(a_k) \approx \chi^2(a_k^o k) - \sum_k 2\beta_k \Delta a_k + \frac{1}{2} \sum_{k,m} 2\alpha_{km} \Delta a_k \Delta a_m + \cdots$$

Minimizing this allows the increments of parameters to be obtained from the set of linear equations

$$\sum_m \alpha_{km} \Delta a_m = \beta_k \qquad \text{or} \qquad [\, \alpha \,]\{\Delta a\} = \{\beta\}$$

The equations turn out to be the same as the least squares Newton–Raphson method and we conclude that (in comparison to the steepest descent method) the Newton–Raphson method utilizes the quadratic form of the error function.

This scheme was applied to the situation of Figure 3.5(a). The derivative functions are the same as the previous example.

The results for different initial guesses are shown in Figure 3.7. It is clear that the scheme can find the valley quickly, but it also proceeds quickly to find the minimum. The inset table shows the rapid convergence typically attained.

As the example shows, this scheme usually works very well. However, when the number of unknown parameters is large (and unrelated to each other) then there can be a problem in finding good initial guesses. This is addressed next.

IV: Levenberg–Marquardt Method

We motivate the solution by considering a one-parameter problem. With reference to Figure 3.8, consider the two situations: first is when we are nowhere near the minimum,

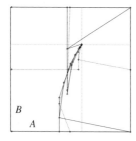

Iter	A	B	χ^2
0	0.500000	0.200000	0.153587E-01
1	0.476992	0.668307	0.122130E-01
2	0.524141	1.12104	0.580532E-03
3	0.585408	1.36342	0.862764E-05
4	0.599706	1.39957	0.677030E-08
5	0.600000	1.40000	0.622455E-15

Figure 3.7: Newton–Raphson iterations. Convergence performance of the Newton–Raphson method for different initial guesses taken on the perimeter of the search space $A[0, 1]$, $B[0, 2]$. Also shown is the typical convergence performance against iteration.

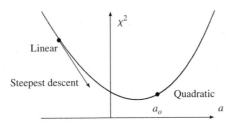

Figure 3.8: Estimating the minimum for the error function.

and second is when we are close to the minimum. The shape of the χ^2 function is different in each regime. Both of these situations will then be combined into a single algorithm.

Suppose we are nowhere near the minimum, then it is best to treat the function as linear and take a step down the gradient, as in the steepest descent method; that is,

$$a_{\text{next}} = a_o - \text{step} \times \frac{\partial \chi^2}{\partial a}\Big|_{a_o} \qquad \text{or} \qquad \Delta a = \text{step} \times \beta$$

where the value for the step will be discussed shortly.

On the other hand, suppose we are close to the minimum, we then expect the χ^2 function to be reasonably approximated by the quadratic form; that is,

$$\chi^2(a) = \chi^2(a_o + \Delta a) \approx \chi_o^2 + \frac{\partial \chi^2}{\partial a}\Delta a + \frac{1}{2}\frac{\partial^2 \chi^2}{\partial a^2}\Delta a^2$$

where the derivatives are evaluated at the current value of $a = a_o$. Choose Δa to make χ^2 a minimum

$$\frac{\partial \chi^2}{\partial \Delta a} = 0 = \frac{\partial \chi^2}{\partial a} + \frac{\partial^2 \chi^2}{\partial a^2}\Delta a \qquad \text{or} \qquad [\,\alpha\,]\Delta a = \beta$$

Now contrast the two methods

$$\Delta a = \text{step} \times \beta \qquad \text{versus} \qquad [\,\alpha\,]\Delta a = \beta$$

Both methods are quite similar except that the quadratic method has a natural step size determined from the derivatives of the error function. We will now connect both methods by using an adaptable step size.

The quantity χ^2 is nondimensional, hence, the constant of proportionality between β and Δa must therefore have the dimensions of a^2. An obvious quantity with these dimensions is $1/\alpha$. But this quantity as a scale might be too big, hence divide it by some (nondimensional) factor γ, with the possibility of setting $\gamma \gg 1$ to cut down the step; that is, replace the steepest descent equation with

$$\Delta a = \frac{1}{\gamma\alpha} \times \beta \qquad \text{or} \qquad [\gamma\alpha]\Delta a = \beta$$

Both methods can now be united because both left-hand sides depend on α. We will do it for the multiparameter case.

Reference [143] discusses the very elegant Levenberg–Marquardt implementation of the above idea for solving nonlinear least squares problems. In this method, the extreme approaches are combined by defining a new matrix $[\,\alpha'\,]$ such that

$$\alpha'_{jj} = \alpha_{jj}(1 + \gamma)\,; \qquad \alpha'_{jk} = \alpha_{jk} \quad (j \neq k)$$

and then replacing both of the equations by

$$\sum \alpha'_{kl}\Delta a_l = \beta_k \tag{3.4}$$

When γ is very large, the matrix $[\,\alpha'\,]$ is forced into being diagonally dominant, and we get the steepest descent method. On the other hand, as γ approaches zero, we get the quadratic form. It is necessary that all α_{kk} be positive, and this is automatically achieved with the above forms for $[\,\alpha_{ij}\,]$.

Given an initial guess for the set of fitted parameters a_k, the recommended Levenberg–Marquardt algorithm works as follows:

Step 1: Compute $\chi^2(a_k^o)$.
Step 2: Pick a modest value for γ, say $\gamma = 0.001$.
Step 3: Solve the linear Equations (3.4) for Δa_k.
Step 4: Evaluate $\chi^2(a_k^o + \Delta a_k)$.
    If $\chi^2(a_k^o + \Delta a_k) > \chi^2(a_k^o)$, increase γ by a factor of 10 and go back to Step 3.
    If $\chi^2(a_k^o + \Delta a_k) < \chi^2(a_k^o)$, decrease γ by a factor of 10 and go to Step 5.
Step 5: Update the trial solution $a_k = a_k^o + \Delta a_k$, and go back to Step 3.

Once the minimum has been found, set $\gamma = 0$ and compute the matrix

$$[\,C\,] \equiv [\,\alpha\,]^{-1}$$

which is the estimated covariance matrix of the standard errors in the fitted parameters a_k.

Reference [143] gives the source code for this routine. This Levenberg–Marquardt method works very well in practice and has become the standard for solving nonlinear least squares systems of equations. We will show its application in the next section.

3.3   Parameter Identification with Extrapolation

One of the most difficult identification problems is that of using remote data to infer parameters by extrapolation. A simple example is using the fringe data of Figures 2.38 and 2.40 to estimate the maximum fringe value at the notch root.

When the gradient is not steep and there is sufficient data, this can usually be accomplished accurately. The more challenging case is when the gradient is steep and sufficient data are not available. To proceed requires the imposition of some underlying model on the data. We use the example of fracture mechanics to illustrate some of the issues and difficulties.

Photoelastic Stress Analysis of Cracks

In a linear elastic fracture mechanics analysis (LEFM), there are three regions surrounding the crack-tip. With reference to Figure 3.9, they are as follows

    I: Nonlinear Zone—large deformation, nonlinear material behavior.

    II: Linear Zone—LEFM is valid, singular stresses dominate.

    III: Remote Region—boundaries dominate.

We will collect the experimental data from the linear zone and extrapolate to the crack-tip area.

Figure 3.10 shows simulated isochromatics near a crack tip under Mode I and Mixed-mode loading. The effect of the shearing is to rotate the figure-of-eight fringes about the crack tip; the effect of σ_o is to tilt the fringes. The purpose of a photoelastic stress analysis of cracks is to use this fringe information to determine the stress intensity factors. While the combination of isochromatics and isoclinics could be used for this purpose, only isochromatics will be used in this discussion.

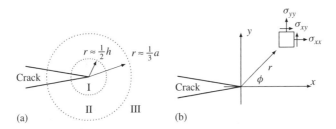

Figure 3.9: Stressed crack. (a) Three regions around the crack tip. (b) Stressed element close to the crack tip.

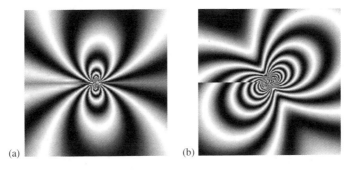

Figure 3.10: Light-field isochromatics near a crack tip. (a) Mode I with $K_1 = 2$, $\sigma_o = 0.1$. (b) Mixed mode with $K_1 = 2$, $K_2 = 2$, $\sigma_o = 0.1$.

On the basis of linear elastic fracture mechanics [48], the stress distribution in Region II close to the crack tip can be written as

$$\sigma_{xx} = \frac{K_1}{\sqrt{2\pi r}} \left[\cos\phi \cos\frac{3}{2}\phi + \frac{1}{2}\sin\phi \sin\frac{3}{2}\phi \right]$$

$$+ \frac{K_2}{\sqrt{2\pi r}} \left[-2\cos\phi \sin\frac{3}{2}\phi + \frac{3}{2}\sin\phi \cos\frac{3}{2}\phi \right] - \sigma_o$$

$$\sigma_{yy} = \frac{K_1}{\sqrt{2\pi r}} \left[\cos\phi \cos\frac{3}{2}\phi + \frac{3}{2}\sin\phi \sin\frac{3}{2}\phi \right] + \frac{K_2}{\sqrt{2\pi r}} \left[\frac{1}{2}\sin\phi \cos\frac{3}{2}\phi \right]$$

$$\sigma_{xy} = \frac{K_1}{\sqrt{2\pi r}} \left[\frac{1}{2}\sin\phi \cos\frac{3}{2}\phi \right] + \frac{K_2}{\sqrt{2\pi r}} \left[\cos\phi \cos\frac{3}{2}\phi + \frac{1}{2}\sin\phi \sin\frac{3}{2}\phi \right] \quad (3.5)$$

The stress distribution is characterized by two types of parameters: K_1 and K_2 are the stress intensity factors that reflect the level of loading, and σ_o is an additional stress to account for boundaries and extend the range of the near-field solution. In the present analysis, the boundaries are assumed to be sufficiently far away that the single parameter σ_o is adequate for characterizing their effect.

The photoelastic data in the vicinity of the crack tip are given by

$$\left(\frac{f_\sigma}{h} \right) N \begin{Bmatrix} \cos 2\theta \\ \sin 2\theta \end{Bmatrix} = \begin{Bmatrix} \sigma_{xx} - \sigma_{yy} \\ 2\sigma_{xy} \end{Bmatrix} = -\frac{(K_1 \sin\phi + 2K_2 \cos\phi)}{\sqrt{2\pi r}} \begin{Bmatrix} +\sin\frac{3}{2}\phi \\ -\cos\frac{3}{2}\phi \end{Bmatrix}$$

$$+ \frac{K_2 \sin\phi}{\sqrt{2\pi r}} \begin{Bmatrix} \cos\frac{3}{2}\phi \\ \sin\frac{3}{2}\phi \end{Bmatrix} - \begin{Bmatrix} \sigma_o \\ 0 \end{Bmatrix} \quad (3.6)$$

The fringe order (isochromatics) in the vicinity of the crack is obtained by squaring the first and second of the equations above and adding to give

$$\left(\frac{f_\sigma}{h} N \right)^2 = (\sigma_{xx} - \sigma_{yy})^2 + (2\sigma_{xy})^2$$

$$= \frac{1}{2\pi r}[K_1 \sin\phi + 2K_2 \cos\phi]^2 + \frac{1}{2\pi r}[K_2 \sin\phi]^2 \quad (3.7)$$

$$+ \frac{2\sigma_o}{\sqrt{2\pi r}}[K_1 \sin\phi + 2K_2 \cos\phi] \sin\frac{3}{2}\phi - \frac{2\sigma_o}{\sqrt{2\pi r}}[K_2 \sin\phi] \cos\frac{3}{2}\phi + \sigma_o^2$$

This shows that as r goes to zero, the fringe order N goes to infinity, which implies that it is impossible to get data very close to the crack tip.

The objective is to determine K_1, K_2, and σ_o having measured many values of $N(r_i, \phi_i)$. This is a nonlinear identification problem.

Parameter Extraction

One scheme often adopted to handle the nonlinear relations is to replace them with a linearized approximation; that is, if we assume that K_1 is dominant and that most of the data are collected perpendicular to the crack, then the fringe can be written approximately as

$$\left(\frac{f_\sigma}{h}\right) N \approx \frac{1}{\sqrt{2\pi r}}(K_1 \sin\phi + 2K_2 \cos\phi) + \sigma_o \sin\frac{3}{2}\phi + \cdots$$

These equations are amenable to a linear analysis. Because of the $\sqrt{r}$ in the denominator, the data very close to the crack tip would have inadvertent high weighting. It is usual to rearrange the equations as

$$\left(\frac{f_\sigma}{h}\right) \sqrt{2\pi r} N = K_1 \sin\phi + 2K_2 \cos\phi + \sigma_o \sin\frac{3}{2}\phi\sqrt{2\pi r} \qquad (3.8)$$

A plot of $\sqrt{r}N$ against $\sqrt{r}$ for data taken along a radial line will be linear with an intercept related to the stress intensity factor(s). This is the scheme used in Reference [158] and in Reference [74] for an error analysis of parameter extraction.

Our interest here is to show the nonlinear modeling. Rewrite the nonlinear fringe relation as

$$\overline{N}^2 = f_{11}K_1^2 + f_{12}K_1K_2 + f_{22}K_2^2 + f_{10}K_1\sigma_o + f_{20}K_2\sigma_o + f_{00}\sigma_o^2$$

where f_{ij} are functions of (r, ϕ) given by

$$f_{11} = \sin^2\phi, \quad f_{12} = 4\sin\phi\cos\phi, \quad f_{22} = 3\cos^2\phi + 1, \quad f_{00} = 2\pi r$$
$$f_{10} = [2\sin\phi\sin\tfrac{3}{2}\phi]\sqrt{2\pi r}, \quad f_{20} = [4\cos\phi\sin\tfrac{3}{2}\phi - 2\sin\phi\cos\tfrac{3}{2}\phi]\sqrt{2\pi r}$$

and $\overline{N} = Nf_\sigma 2\pi r/h$. Define the error function as

$$\mathcal{A} = \sum_i \left[\left(\frac{f_\sigma}{h}\right)^2 2\pi r N^2(x_i, y_i) - \overline{N}^2(K_1, K_2, \sigma_o)\right]^2$$

The derivative vector is

$$\frac{\partial\mathcal{A}}{\partial K_1} = 2f_{11}K_1 + f_{12}K_2 + f_{10}\sigma_o$$

$$\frac{\partial\mathcal{A}}{\partial K_2} = 2f_{22}K_2 + f_{12}K_1 + f_{20}\sigma_o$$

$$\frac{\partial\mathcal{A}}{\partial\sigma_o} = f_{10}K_1 + f_{20}K_2 + 2f_{00}\sigma_o$$

so that the simultaneous equations become

$$\left\lfloor\frac{\partial\mathcal{A}}{\partial K_1}, \frac{\partial\mathcal{A}}{\partial K_2}, \frac{\partial\mathcal{A}}{\partial\sigma_o}\right\rfloor \left\{\begin{matrix}\Delta K_1 \\ \Delta K_2 \\ \Delta\sigma_o\end{matrix}\right\} = \left\{\left(\frac{f_\sigma\sqrt{2\pi r}}{h}\right)^2 N^2 - \overline{N}_o^2\right\}^2 \qquad (3.9)$$

This is written for each data point and the least squares procedure is then used to reduce it to normal form. Starting with a guess for K_1, K_2, and σ_o, the equations are solved iteratively to get improvements on the guesses. This is the scheme used in Reference [50], which also showed how the formulation can be made more general by including the crack-tip positions as additional unknowns.

Experimental Studies

An epoxy specimen was cut to the dimensions shown in Figure 3.11(a) and an artificial crack was machined using a very fine blade. The specimen was then loaded as described in Reference [74] and sent through a stress-freezing cycle.

A calibration specimen was also stress frozen at the same time. This had an eccentric load (as indicated in Figure 3.11(b)) such that a slight variation of stress across the specimen was achieved. After stress freezing, a dark-field white light isochromatic image was recorded using an Olympus E20 digital camera. Intensity scans of the color separations allowed the fringe order at the center to be estimated; the material fringe value was then obtained as $f_\sigma = P/WN$ as described in Section 2.9. The three values are given in Figure 3.11(b); note that the alignment with color wavelength is only approximate and not important since the same camera and polariscope will be used to analyze the cracked specimen.

Figure 3.12(b) shows the RGB separations for a region near the crack tip. The scale was taped to the specimen. Intensity scans were taken along a vertical ($\phi = 90°$) line through the crack tip and Figure 3.12(a) shows the whole and half-order data plotted as indicated in Equation (3.8). The expected value for the stress intensity is $K_1/K_o = 1.37$ where $K_o = \sigma_\infty \sqrt{\pi a}$.

There is a good deal of variability among the data but the estimations are reasonably robust. The statistics for the three extrapolations are shown in Table 3.2. The blue

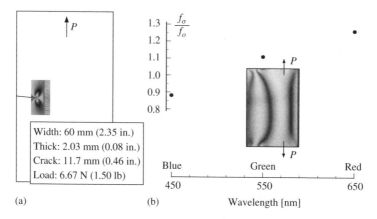

(a)   (b)   Wavelength [nm]

Figure 3.11: Crack experiment. (a) Specimen with dimensions. (b) Calibration for the color separations where $f_o = 183\,\text{N/m}\,(1.00\,\text{lb/in.})$.

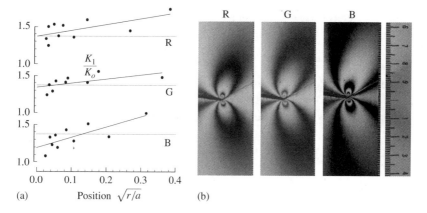

(a)   Position $\sqrt{r/a}$ (b)

Figure 3.12: Extrapolation method. (a) Data from a vertical scan. (b) The three color separations.

Table 3.2: Stress intensity parameters results.

Parameters:	K_1/K_o	1.37	1.34	1.19	1.26
	K_2/K_o	0.00	0.00	0.00	−0.18
	σ_o^*	0.77	0.54	1.34	0.35
Deviations:	σ_{K_1}	0.05	0.04	0.05	0.06
	σ_σ	0.31	0.26	0.39	0.08

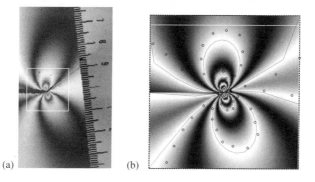

(a)   (b)

Figure 3.13: Crack parameter extraction treated as a mixed-mode problem. (a) Experimental dark-field fringe pattern for green separation. (b) Reconstruction with superposed data.

extrapolation is lower than the other two by more than two standard deviations; it is not clear why.

The crack, as machined, is at a small angle to the edge and therefore there is a small amount of mixed-mode behavior. Figure 3.13(a) shows the green separation with the image oriented with respect to the crack. The mixed-mode behavior is evident in the inner fringes. We will attempt to estimate the amount of K_2.

As shown in Reference [74], in order to get robust estimates of mixed-mode parameters, it is necessary to have an angular spread in the data. To accommodate this, we will use discrete data; the half-order fringes inside the white box of Figure 3.13(a) were discretized and are shown as circles in Figure 3.13(b).

The Levenberg–Marquardt algorithm was used in conjunction with Equation (3.9) to estimate the crack parameters. The results are given in Table 3.2. The associated reconstructions are shown in Figure 3.13(b). Convergence was usually within the first four or so iterations.

The expected stress intensities for a crack at a small angle of $\theta \approx -6°$ is given approximately by

$$K_1 \approx 1.37\, K_o \cos^2 \theta = 1.36\, K_o\,, \qquad K_2 \approx 1.37\, K_o \sin \theta \cos \theta = -0.14\, K_o$$

By and large, the reconstructions give good estimates.

Discussion of Photoelasticity and LEFM

This photoelastic study of cracks is a reasonable example of nonlinear modeling. However, from a fracture mechanics point of view, it suffers from the serious drawback that the assumed model may be incorrect. For example, the region where the expansion for the stresses is valid is not known *a priori*; hence, data from outside this region may inadvertently be used. Reference [48] discusses how this can be overcome by the use of higher-parameter expansions of the near-field stresses. The drawback is that the data must now be capable of resolving the additional parameters.

On the other hand, the LEFM equations themselves are not valid very close to the crack tip because of various nonlinearities. For these two reasons, this approach is not very satisfactory.

The crux of the matter is that, for practical problems, the LEFM equations are too restrictive as a model—we require a more general-purpose modeling of the problem. This, of course, must come from FEM. We will revisit this crack problem beginning in Chapter 4 after we have established the means of combining experimental data with FEM analyses.

3.4 Identification of Implicit Parameters

The foregoing sections dealt with the identification of model parameters using experimental data. The parameters of the model appeared explicitly since the choice and/or construction of the model is at the discretion of the modeler. A closely related topic concerns the identification of parameters that appear only implicitly in the model. This situation usually arises because the model was constructed with some purpose in mind (such as structural analysis) other than the modeling of data. This raises a number of issues that will bear directly on some of the later chapters.

To elaborate, as shown in Chapter 1, the finite element analysis of a linear structural problem reduces to solving the system of equations

$$[K]\{u\} = \{P\}$$

The stiffness matrix $[K]$ is constructed by assembling many elements of different shapes and properties. Suppose the properties of one (or many) of these elements is unknown and we wish to use measurements to identify them. Clearly, it would be very difficult to rearrange the governing equation so that the thickness (say) of these elements appears explicitly. The FEM model was constructed so as to make structural analyses convenient and turns out to be unsuitable for identification problems.

With some imagination, it is possible to conceive of a way to make the parameters explicit in the above FEM analysis. But suppose the problem is nonlinear, then this rearrangement (if possible at all) would require complete access and control of the inner workings of the FEM program. This is not something normally available and therefore we need to approach the problem differently. The basic idea is presented next and will be elaborated on in the later chapters.

Sensitivity Response Method

Assume we have the ability to compute responses based on the model, and the model has a number of parameters to be identified; that is,

$$u = u(a_M)$$

For example, for FEM modeling, we can write

$$\{u(a_M)\} = [K]^{-1}\{P\} = [K(a_M)]^{-1}\{P(a_M)\}$$

where the parameters could be thicknesses, Young's moduli, or even shape. We will treat the model as a black box in the sense that for a completely specified problem it can compute the response but we do not have access to its inner workings.

Assume we have a guess a_k^o for the parameters; a Taylor series expansion about this guess gives

$$u(x_i) \approx u(x_i; a_p^o + \Delta a_p) \approx u(x_i; a_p^o) + \sum_j \frac{\partial u(x_i; a_p^o)}{\partial a_j} \Delta a_j + \cdots$$

In this context, the derivatives are called *sensitivities*, that is,

$$S_j(x_i; a_p^o) \equiv \frac{\partial u(x_i; a_p^o)}{\partial a_j}$$

gives an indication of how the solution changes when the parameter a_j changes.

Let the measurements be $\{u_i\}$ at positions x_i and form an error function between the measurements and the results from analysis that used the parameters a_p^o, that is,

$$\chi^2 = \sum_i \left[u_i - u_i^o - \sum_j S_j(x_i; a_p^o)\Delta a_j \right]^2, \qquad u_i^o = u(x_i; a_p^o)$$

Minimizing with respect to the parameters yields the M equations

$$\sum_j \alpha_{kj} \Delta a_j = \beta_k \qquad \text{or} \qquad [\ \alpha\ ]\{\Delta a\} = \{\beta\}$$

where

$$\alpha_{kj} \equiv \sum_i S_k(x_i; a_p^o) S_j(x_i; a_p^o), \qquad \beta_k \equiv \sum_i [u_i - u_i^o] S_k(x_i; a_p^o)$$

This system of equations is then solved iteratively with the parameters updated by

$$a_p = a_p^o + \Delta a_p$$

on each iteration.

As presented, the method is almost identical to the Least Squares Newton–Raphson method presented earlier. The key difference here is that we cannot obtain the sensitivities directly, that is, in the foregoing section, this derivative could be performed explicitly in terms of the parameters because we knew the explicit form of the model. Here we must do it implicitly. The simplest way to do this is to run the analysis with a specified set of parameters (the initial guesses) and then rerun the analysis, changing each parameter in turn by a small amount. For each run, the sensitivity is estimated as

$$S_k(x_i; a_p^o) \approx \frac{u(x_i; a_p^o, a_k^o + d\, a_k) - u(x_i; a_p^o)}{d\, a_k}$$

Clearly, there are issues dealing with appropriate parameter increments $d\, a_k$ in order to get reasonable estimates of the sensitivities; these will be considered in detail as particular problems are considered. When the sensitivities are obtained as above, we will refer to this method as the *sensitivity response method* (SRM).

Example

Consider the situation shown in Figure 3.14 where an array of detectors, I_i, are used to measure the intensity (specularly) reflecting off a moving object O; the objective is to use the intensity information to infer the position of the object.

The source and sensors are aligned at a distance D from the origin, and the object is at position (x, y). For simplicity, suppose the intensity obeys an inverse power law so that it diminishes as $1/r$ for the total distance travelled irrespective of reflections. A

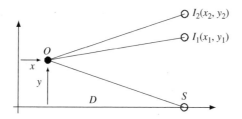

Figure 3.14: Detectors I_1 and I_2 measuring intensity reflected from an object. The objective is to determine the position (x, y).

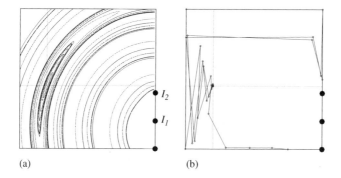

(a)   (b)

Figure 3.15: Locating the position of an object. (a) Contours of the error function. (b) Convergence for initial guesses on the rim of the search space.

detector receives the intensity

$$I_i = \frac{S}{\sqrt{(D - x)^2 + y^2} + \sqrt{(D - x_i)^2 + (y - y_i)^2}}$$

We wish to use multiple measurements to infer the (x, y) position, that is, x and y are the parameters we wish to identify.

Form the χ^2 error function as

$$\chi^2 = [I_1 - \overline{I}(x_1, y_1; x^o, y^o)]^2 + [I_2 - \overline{I}(x_2, y_2; x^o, y^o)]^2$$

where $\overline{I}(x_i; x^o, y^o)$ is the model prediction using the current estimates (x^o, y^o) of the unknown parameters. Figure 3.15(a) shows contours of χ^2 over the search space for (x, y), where two sensors are used and located as indicated in the figure. There is a definite minimum to the error function. For the iterative scheme, the sensitivities were computed as

$$S_x(x_i; x^o, y^o) \approx \frac{\overline{I}(x_i; x^o + \mathrm{d}x, y^o) - \overline{I}(x_i; x^o, y^o)}{\mathrm{d}x}$$

$$S_y(x_i; x^o, y^o) \approx \frac{\overline{I}(x_i; x^o, y^o + \mathrm{d}y) - \overline{I}(x_i; x^o, y^o)}{\mathrm{d}y}$$

so that for each guess (x^o, y^o) three function evaluations of I are performed. Figure 3.15(b) shows the convergence for different initial guesses starting on the corners of the search space. Convergence is assumed to be achieved when χ^2 no longer changes significantly. The stopping criterion was taken as

$$\left| \frac{\chi^2 - \chi^{o2}}{\chi_{\text{ref}}^2} \right| < \epsilon, \qquad \epsilon = 1.0 \times 10^{-10}$$

where χ_{ref}^2 was the value at the initial guess. In each case, convergence occurred within six or seven iterations. This was fairly insensitive to the finite difference approximation of the sensitivity; that is, the results did not change much when Δa_k of $1/100$, $1/10$, $1/2$ of the search space was used.

A final point worth considering is that the iterated values were always constrained to be inside the search space. This explains why some of the iterates lie on the boundaries in Figure 3.15(b).

3.5   Inverse Theory for Ill-Conditioned Problems

In the foregoing sections, the model parameters were obtained by the process of minimizing the errors between the data and the model. This process is highly ill-conditioned in the sense that small deviations of input data can cause large deviations of output parameters. This can also be observed in the topology of the search space as depicted in Figures 3.5 and 3.15(a)—the more ill-conditioned the problem, the less well defined is the minimum error.

What we address in this section is the use of regularization as a way of making ill-conditioned problems better conditioned or more "regular". But first we give some background on problems involving inversion.

Problems Involving Inversion

Reference [172] introduces the scenario of a zookeeper with a number of animals, and on the basis of the number of heads and feet, we try to infer the number of animals of each type. Let us continue with this scenario and suppose the zookeeper has just lions (L) and tigers (T), then the number of heads (H) and paws (P) of all his animals can be computed as

$$\begin{aligned} H &= L + T \\ P &= 4L + 4T \end{aligned} \qquad \text{or} \qquad \left\{ \begin{matrix} H \\ P \end{matrix} \right\} = \begin{bmatrix} 1 & 1 \\ 4 & 4 \end{bmatrix} \left\{ \begin{matrix} L \\ T \end{matrix} \right\}$$

This is a well-stated forward set of equations; that is, given the populations of lions and tigers (the input), and number of heads and paws per each animal (the system), we can compute precisely the total number of heads and paws (the output).

Let us invert the problem, that is, given the number of heads and paws, determine the populations of lions and tigers. The answer is

$$\begin{Bmatrix} L \\ T \end{Bmatrix} = \begin{bmatrix} 1 & 1 \\ 4 & 4 \end{bmatrix}^{-1} \begin{Bmatrix} H \\ P \end{Bmatrix}$$

Unfortunately, the indicated matrix inverse cannot be computed because its determinant is zero. When the determinant of a system matrix is zero, we say that the system is *singular*. It is important to realize that our data are precise and yet we cannot get a satisfactory solution. Also, it is incorrect to say we cannot get a solution; in fact, we get an infinite number of solutions that satisfy the given data.

Thus, it is not always true that N equations are sufficient to solve for N unknowns: the equations may be consistent but contain insufficient information to separate the N unknowns. Indeed, we can even overspecify our problem by including the number of eyes as additional data and still not get a satisfactory solution. The origin of the problem is that our choice of features to describe the animals (heads, paws, eyes) does not distinguish between the animals—"lions and tigers have the same number of heads and paws."

Let us choose the weight as a distinguishing feature and say (on average) each lion weighs 100 and each tiger weighs $100 + \Delta$. Then the heads and weight totals are

$$\begin{Bmatrix} H \\ W \end{Bmatrix} = \begin{bmatrix} 1 & 1 \\ 100 & 100 + \Delta \end{bmatrix} \begin{Bmatrix} L \\ T \end{Bmatrix}, \qquad \begin{Bmatrix} L \\ T \end{Bmatrix} = \frac{1}{\Delta} \begin{bmatrix} 100 + \Delta & -1 \\ -100 & 1 \end{bmatrix} \begin{Bmatrix} H \\ W \end{Bmatrix}$$

As long as (on average) the animals have a different weight ($\Delta \neq 0$), we get a unique (single) solution for the two populations. The heads are counted using integer arithmetic and we can assume it is precise, but assume there is some variability in the weight measurement, that is, $W \rightarrow W \pm \delta$. Then the estimated populations of lions and tigers are

$$L^* = L \mp \frac{\delta}{\Delta}, \qquad T^* = T \pm \frac{\delta}{\Delta}$$

where L and T are the exact (true) values, respectively. We notice two things about this result. First, for a given measurement error δ, the difference in weight between the animals acts as an amplifier on the errors in the population estimates; that is, the smaller the Δ is, the larger is the δ/Δ. When Δ is small (but not zero) we say the system is *ill-conditioned*—small changes in data (δ) cause large changes in results. If there were no measurement errors, then we would not notice the ill-conditioning of the problem, but the measurement errors do not cause the ill-conditioning—this is a system property arising from our choice of weight as a purported distinguishing feature between the animals.

The second thing we notice about the estimated populations is that the errors go in opposite directions. As a consequence, if the estimated populations are substituted back into the forward equations, then excellent agreement with the measured data is obtained even though the populations may be grossly in error. The sad conclusion is that excellent agreement between model predictions and data is no guarantee of a good model if we suspect that the model system is ill-conditioned.

What we will be concerned with for the remainder of this section is developing methods for obtaining robust solutions to ill-conditioned problems. It is clear from the previous paragraph that the data must be supplemented, somehow, with additional or outside information to effect these solutions—this is the topic of regularization.

A word about terminology. Strictly, an *inverse problem* is one requiring the inverse of the system matrix. Thus, all problems involving simultaneous equations (which means virtually every problem in computational mechanics) are inverse problems. As a simple example, in Chapter 1 dealing with FEM analyses, we represented the solution of a linear elastic problem as

$$[K]\{u\} = \{P\} \qquad \text{or} \qquad \{u\} = [K^{-1}]\{P\}$$

Certainly in these problems, there are issues of ill-conditioning—if the number of elements is increased, then a stage would be reached when the inversion of the stiffness matrix would be meaningless. But we do not want to refer to these FEM problems as "inverse" problems; indeed, we actually want to refer to them as "forward" problems.

Consequently, we shall refer, collectively, to methods, techniques, algorithms, and so on, for the robust inversion of matrices as *Inverse Methods* and give looser meanings to *forward* and *inverse* problems based on their cause and effect relationship. Thus, a forward problem (irrespective of whether a matrix inversion is involved or not) is one for which the system is known, the inputs are known, and the responses are the only unknowns. An inverse problem is one in which something about the system or the input is unknown and knowledge of something about the response is needed in order to effect a solution. Thus, the key ingredient in inverse problems is the use of response data and hence its connection to experimental methods. Parenthetically, the antonyms of "forward" are "backward" and "reverse"—neither of which sounds like edifying problems to study.

Ill-Conditioning of Least-Squares Solutions

Consider the situation shown in Figure 3.16 in which two detectors, I_1 and I_2, are used to determine the strength of two sources, S_1 and S_2.

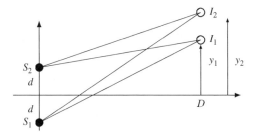

Figure 3.16: Two detectors measuring emissions from two sources.

Suppose the intensity obeys an inverse power law so that it diminishes as $1/r$, then the detectors receive the intensities

$$\left\{ \begin{matrix} I_1 \\ I_2 \end{matrix} \right\} = \begin{bmatrix} 1/r_{11} & 1/r_{12} \\ 1/r_{21} & 1/r_{22} \end{bmatrix} \left\{ \begin{matrix} S_1 \\ S_2 \end{matrix} \right\}, \qquad r_{ij} = \sqrt{D^2 + (y_i + d_j)^2}$$

where the subscripts on r are (sensor, source). We wish to investigate the sensitivity of this system to errors.

Figure 3.17 shows the reconstructions of the sources (labelled as direct) when there is a fixed 0.001% of source error in the second sensor. The sources have strengths of 1.0 and 2.0, respectively. There is a steady increase in error as the sensors become more remote (D increases for fixed d). A similar effect occurs as d is decreased for fixed D

Let us now use least squares to establish the system of equations. This leads to

$$\begin{bmatrix} 1/r_{11} & 1/r_{21} \\ 1/r_{12} & 1/r_{22} \end{bmatrix} \begin{bmatrix} 1/r_{11} & 1/r_{12} \\ 1/r_{21} & 1/r_{22} \end{bmatrix} \left\{ \begin{matrix} S_1 \\ S_2 \end{matrix} \right\} = \begin{bmatrix} 1/r_{11} & 1/r_{21} \\ 1/r_{12} & 1/r_{22} \end{bmatrix} \left\{ \begin{matrix} I_1 \\ I_2 \end{matrix} \right\}$$

The results are also shown in Figure 3.17(a). The least squares solution shows very erratic behavior as D/d is increased. It is sometimes mistakenly thought that least squares will "cancel" errors in the data and thereby give the correct (true or smoothed) answer. Least squares will average errors in the reconstructions (of the data) but in all likelihood the computed solution will not be smooth.

Closer inspection shows that when one solution is up the other is down, and *vice versa*. What we are seeing is the manifestation of *ill-conditioning*. It should also be

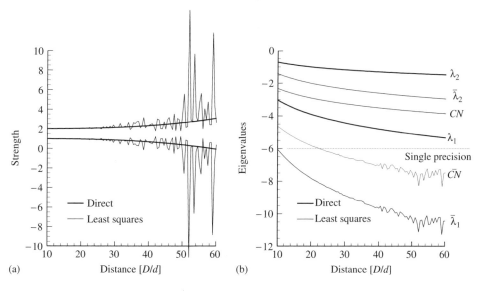

Figure 3.17: Sensitivity of system to errors. (a) Reconstructions of sources. (b) Eigenvalues of system matrix.

emphasized that these erratic values of S_1 and S_2 reconstruct the measurements, I_1 and I_2, very closely (i.e., within the error bounds of the measurements).

Ill-conditioning is a property of the system and not the data; the effect of errors in the data is to manifest the ill-conditioning. To get a deeper insight into this, Figure 3.17(b) shows the eigenvalues λ_1, λ_2 and $\bar{\lambda}_1$, $\bar{\lambda}_2$ of the direct and least squares system matrices, respectively. In both cases, the first eigenvalue goes to zero indicating a near singular matrix; the significant difference is that the least squares does it much sooner and more rapidly (with respect to the system parameters) than the direct method.

The condition number of a matrix is the magnitude of the ratio of its largest eigenvalue to its smallest. A matrix is singular if the condition number is infinite and ill-conditioned when its reciprocal is less than the machine precision (approximately 10^{-6} for single precision and 10^{-12} for double precision). Figure 3.17(b) shows the reciprocal of the condition number for both matrices.

As it happens, the condition number of the matrix product $[A^{\mathrm{T}}][A]$ is the square of that of the matrix $[A]$ (a factor of 2 on the log scale) and this is why least squares solutions are more apt to be ill-conditioned. Reference [143] strongly recommends, therefore, that all least squares problems be solved using singular value decomposition. The power of SVD is that it can automatically determine the number of independent solutions the system is capable of producing—in the present case, this would mean determining the number of sources the system of equations can identify.

In this book, we take a different tact but with the same underlying idea of reducing the number of unknowns. The basic idea is suggested by Figure 3.17(a): if we somehow "average" the wild oscillations between the two solutions, then the result will better reflect the expected solution. In other words, we will impose on the solution some *a priori* belief in the smoothness or regularity of the solution.

Minimizing Principle with Regularization

To effect a robust solution to ill-conditioned problems, we must incorporate (in addition to the measurements) some extra information either about the underlying system, or about the nature of the response functions, or both.

To elaborate on this point, consider the fanciful situation when we are outside a room and overhear the partial sentence (Figure 3.18),

"... COW ... OVER ..."

We wonder to ourselves as to what the complete sentence could be.

Many people might jump to the conclusion that the sentence is "the COW jumped OVER the moon". This conclusion is not warranted by the data but is a result of an outside bias or what we will refer to as supplementary information. For example, if we notice that the room is a child's bedroom and recognize the voice as that of an adult, then the nursery rhyme assumption/conclusion is more reasonable. On the other hand, if we know/recognize/assume the room to be a TV room and the voice to be that of

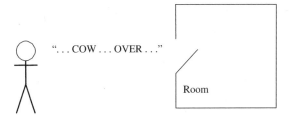

Figure 3.18: Interlocutor outside a room overhears a partial conversation.

Table 3.3: Some solutions to the given data.

Solution	Supplementary information
The COW jumped OVER the moon	Child's bedroom
The proletarian revolution in MosCOW is OVER	Newscast
He sang his COWboy songs all OVER the land	Music show
She spilled the coCOA on her bib OVERalls	Play school
Noel COWaRd wrote many plays	Theater talk show
Blah blah blah COW blah OVER blah blah	Cadence
COW OVER	None

a newscaster, then a possibility is "the proletarian revolution in MosCOW is OVER". Some other possibilities are given in Table 3.3, but the total number is countless.

Although the ideas in this section are much more general than curve-fitting, to help fix ideas, we will use curve-fitting as another simple example. Suppose we have the data set

$$(x_i, d_i): \qquad (3, 2.5) \quad (6, 2.0) \quad (9, 1.0) \quad (12, 1.5)$$

and wish to know what is happening at each point between 1 and 14. We could set this up as a system of simultaneous equations

$$\begin{bmatrix} 0 & 0 & 1 & 0 & 0 & 0 & 0 & 0 & 0 & 0 & 0 & 0 \\ 0 & 0 & 0 & 0 & 0 & 1 & 0 & 0 & 0 & 0 & 0 & 0 \\ 0 & 0 & 0 & 0 & 0 & 0 & 0 & 0 & 1 & 0 & 0 & 0 \\ 0 & 0 & 0 & 0 & 0 & 0 & 0 & 0 & 0 & 0 & 0 & 1 \end{bmatrix} \begin{Bmatrix} u_1 \\ u_2 \\ u_3 \\ \vdots \\ u_{14} \end{Bmatrix} = \begin{Bmatrix} d_1 \\ d_2 \\ d_3 \\ d_4 \end{Bmatrix}$$

or $[A]\{u\} = \{d\}$. Clearly, we cannot solve for the 14 unknown u_k having only 4 data d_i. We obviously must supplement this system with additional information.

In the previous section, we determined the parameters of a model by minimizing the χ^2 functional. With this in mind, suppose that u_k is an "unknown" vector that we plan to determine by some minimizing principle.

Suppose that our functional $\mathcal{A}(u_k)$ has the particular form of χ^2,

$$\mathcal{A}(u) = \left\{ \{d\} - [\ A\ ]\{u\} \right\}^{\mathrm{T}} \left\{ \{d\} - [\ A\ ]\{u\} \right\}$$

If $[\ A\ ]$ has fewer rows (N, data) than columns (M, unknowns) (as in our example), then minimizing $\mathcal{A}(u)$ will give the correct number of equations (M); however, it will not give a unique solution for $\{u\}$ because the system matrix will be *rank deficient*. (A matrix is rank-deficient when two or more rows are not linearly independent; as a consequence, the determinant is zero and the matrix cannot be inverted.)

Suppose, instead, we choose a nondegenerate quadratic form $\mathcal{B}(u)$, for example,

$$\mathcal{B}(u) = \{u\}^{\mathrm{T}}[\ H\ ]\{u\}$$

with $[\ H\ ]$ being a $[M \times M]$ positive definite matrix (the unit matrix, say), then minimizing this will lead to a unique solution for $\{u\}$. Of course, the solution will not be very accurate since none of the data were used.

Here is the key point: if we add any multiple γ times $\mathcal{B}(u)$ to $\mathcal{A}(u)$ then minimizing $\mathcal{A}(u) + \gamma \mathcal{B}(u)$ will lead to a unique solution for $\{u\}$, that is, when a quadratic minimizing principle is combined with a quadratic constraint, only one of the two need be nondegenerate for the overall problem to be well posed.

Thus, the essential idea in our inverse theory is the objective to minimize

$$\mathcal{A} + \gamma \mathcal{B} = \left\{ \{d\} - [\ A\ ]\{u\} \right\}^{\mathrm{T}} \lceil\ W\ \rfloor \left\{ \{d\} - [\ A\ ]\{u\} \right\} + \gamma \{u\}^{\mathrm{T}}[\ H\ ]\{u\}$$

where $\lceil\ W\ \rfloor$ is a general weighting array although we will take it as a diagonal. We minimize the error functional for various values of $0 < \gamma < \infty$ along the so-called trade-off curve of Figure 3.1. When $M \gg N$ (many more unknowns than data), then $\{u\}$ will often have enough freedom to be able to make $\mathcal{A}$ quite very small. Increasing γ pulls the solution away from minimizing $\mathcal{A}$ in favor of minimizing $\{u\}^{\mathrm{T}}[\ H\ ]\{u\}$. We then settle on a "best" value of γ by some criterion ranging from fairly objective to entirely subjective. The most appropriate criteria are specific to particular problems and we consider some in the later chapters.

To summarize, there are two positive functionals, $\mathcal{A}$ and $\mathcal{B}$. The first measures the agreement of the model to the data. When $\mathcal{A}$ by itself is minimized, the agreement becomes very good but the solution becomes unstable, often oscillating wildly, reflecting that $\mathcal{A}$ alone is usually a highly ill-conditioned minimization problem. The second functional measures the "smoothness" of the desired solution, or sometimes a quantity reflecting *a priori* judgments about the solution. This function is called the *regularizing function*. Minimizing $\mathcal{B}$ by itself gives not only a solution that is smooth but also a solution that has nothing at all to do (directly) with the measured data.

The concept of regularization is a relatively recent idea in the history of mathematics; the first few papers seem to be those of References [123, 139, 171] although Reference [165] gives citations of some earlier literature in Russian. A readable discussion of regularization is given in Reference [143] and a more extensive mathematical treatment can be found in Reference [127].

The minimization problem is reduced to a linear set of normal equations by the usual method of differentiation to yield

$$[[A]^{\mathrm{T}} \lceil W \rfloor [A] + \gamma [H]]\{u\} = [A]^{\mathrm{T}} \lceil W \rfloor \{d\}$$

where $\lceil W \rfloor$ are relative weights on the data. These equations can be solved by standard techniques such as $[LU]$ decomposition. In comparison to the standard least squares form, these equations are not ill-conditioned because of the role played by γ and the regularization term.

For future reference, our standard statement of the problem entails a set of data relations

$$[A]\{\tilde{P}\} = \{B\} \tag{3.10}$$

where $\{B\}$ is related to the data $\{d\}$ and $\{\tilde{P}\}$ is related to the unknown parameters. The least squares approach, in conjunction with regularization, reduces to solving the set of simultaneous equations

$$[[A]^{\mathrm{T}} \lceil W \rfloor [A] + \gamma [H]]\{\tilde{P}\} = [A]^{\mathrm{T}} \lceil W \rfloor \{B\} \tag{3.11}$$

In the subsequent sections and chapters, we will therefore concentrate on establishing the appropriate contents of Equation (3.10).

3.6 Some Regularization Forms

We now discuss some particular forms of regularization. The regularization method we use is generally called *Tikhonov [8, 116, 165] regularization*. Typically, the functional B involves some measures of smoothness that derive from first or higher derivatives.

Linear Regularization

Consider a distribution represented by $u(x)$ or its discretization $\{u\}$, which we wish to determine from some measurements. Suppose that our *a priori* belief is that, locally, $u(x)$ is not too different from a constant, then a reasonable functional to minimize is associated with the first derivative

$$B \propto \int \left[\frac{\mathrm{d}u}{\mathrm{d}x} \right]^2 \mathrm{d}x \propto \sum_{m=1}^{M-1} [u_m - u_{m+1}]^2 = ([D]\{u\})^2$$

since it is nonnegative and equal to zero only when $u(x)$ is constant. We can write the second form of B as

$$B = \{u\}^{\mathrm{T}}[D]^{\mathrm{T}}[D]\{u\} = \{u\}^{\mathrm{T}}[H]\{u\}$$

where [D] is the [$(M - 1) \times M$] first difference matrix

$$[D] = \begin{bmatrix} -1 & 1 & 0 & \cdots \\ 0 & -1 & 1 & \cdots \\ 0 & 0 & -1 & \cdots \end{bmatrix}$$

and [H] is the [$M \times M$] symmetric matrix

$$[H] = [D]^T[D] = \begin{bmatrix} 1 & -1 & 0 & \cdots \\ -1 & 2 & -1 & \cdots \\ 0 & -1 & 2 & \cdots \\ \cdots & & -1 & \end{bmatrix}$$

Note that [D] has one less rows than columns. It follows that the symmetric matrix [H] is degenerate; it has exactly one zero eigenvalue. (It is easy to verify, for example, that when $M = 3$ the determinant of [H] is zero.) However, the data equations will help make the overall system determinate.

The above choice of [D] is only the simplest in an obvious sequence of derivatives. If our *a priori* belief is that a linear function is a good local approximation to $u(x)$, then minimize

$$\mathcal{B} \propto \int \left[\frac{d^2 u(x)}{dx^2} \right]^2 dx \propto \sum_{m=1}^{M-2} [-u_m + 2u_{m+1} - u_{m+2}]^2$$

This gives for the matrices [D] and [H]

$$[D] = \begin{bmatrix} -1 & 2 & -1 & 0 & \cdots \\ 0 & -1 & 2 & -1 & \cdots \\ 0 & 0 & -1 & 2 & \cdots \end{bmatrix}$$

and

$$[H] = [D]^T[D] = \begin{bmatrix} 1 & -2 & 1 & 0 & 0 & 0 & \cdots \\ -2 & 5 & -4 & 1 & 0 & 0 & \cdots \\ 1 & -4 & 6 & -4 & 1 & 0 & \cdots \\ 0 & 1 & -4 & 6 & -4 & 1 & \cdots \\ \cdots & & & & & 0 & \end{bmatrix}$$

This time note that [D] has two fewer rows than columns. Reference [143] gives explicit forms for the matrices [D] and [H] for some higher order smoothing.

A special case of the above that is worth noting is called *zeroth-order regularization*. Here, our *a priori* belief is that there is no connection between the points of $u(x)$; that is, our unknowns bear no spatial relation to each other. This minimizes

$$\mathcal{B} \propto \int [u(x)]^2 dx \propto \sum_{m=1}^{M} [u_m]^2$$

and gives for the matrices $[D]$ and $[H]$

$$\lceil D \rfloor = \begin{bmatrix} 1 & 0 & 0 & 0 & \cdots \\ 0 & 1 & 0 & 0 & \cdots \\ 0 & 0 & 1 & 0 & \cdots \end{bmatrix}, \qquad \lceil H \rfloor = \lceil D \rfloor^{\mathrm{T}} \lceil D \rfloor = \begin{bmatrix} 1 & 0 & 0 & 0 & \cdots \\ 0 & 1 & 0 & 0 & \cdots \\ 0 & 0 & 1 & 0 & \cdots \end{bmatrix}$$

This time note that $\lceil D \rfloor$ has as many rows as columns. This is useful when the unknowns are not actually related to each other, but we, nonetheless, want to regularize the solution.

It might seem that first-order regularization, for example, is no different than linear curve-fitting. But linear curve-fitting is a global operation; indeed as will be seen in the example, this is achieved in the above when γ is very large; however, for medium values of γ we get more of a local linear fit.

There are other forms of regularization that we will consider duly.

Estimating the Amount of Regularization

An apparent weakness of regularization methods is that the solution procedure does not determine the amount of regularization to be used. This is not exactly true, and we now look at the effect regularization has on our original curve-fitting problem to examine some of the issues involved.

Figure 3.19 shows results for first- and second-order regularization on our curve-fit problem. A few points are worth noting. First, of course, is that a solution is obtained even though there are more unknowns than data. Second, the estimated points fall into two groups: those that are interpolated between the data, and those that are extrapolated to outside the data range. As γ is decreased, the interpolated curve-fit matches the data more closely. However, irrespective of the value of γ, outside the data range the curve-fit follows our underlying regularization assumption be it constant or linear. Third, it is only when γ is very large that we see a global trend in the curve-fitting. This reinforces the

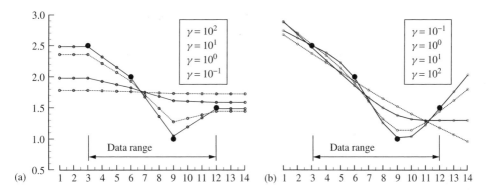

Figure 3.19: Curve-fitting with regularization. (a) First order. (b) Second order.

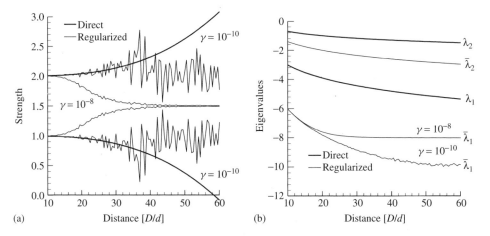

Figure 3.20: Zeroth-order regularization of the sensor array problem with different values of γ. (a) Reconstructions. (b) Eigenvalues of the system matrices.

idea that regularization is a local effect. As γ is driven to zero the curve-fit shows wild oscillations.

Figure 3.20 shows results for zero-order regularization on the sensor array problem. The two values of γ, $(10^{-8}, 10^{-10})$, are close to giving a good solution. Too much smoothing forces them to have the same value but note that this is only if ill-conditioning is a problem. Too little smoothing gives erratic behavior, but it has a mean in the vicinity of the direct value (at least up to $D/d \approx 40$). Figure 3.20(b) shows the eigenvalues of the regularized system. The achievement of regularization is to put a lower limit to the lower eigenvalue, and therefore sets a limit to the condition number. It is also worth noting that the regularization has no noticeable effect on the larger eigenvalue.

The final point to be discussed is the value of γ to be used. First, be clear that there is no "correct" value; the basis of the choice can range from the purely subjective ("the curve-fit looks reasonable") to somewhat objective, depending on the variances of the data. A reasonable beginning value of γ is to try

$$\gamma = \mathrm{Tr}[[\ A\ ]^{\mathrm{T}}[\ A\ ]]/\mathrm{Tr}[\ H\ ]$$

where Tr is the trace of the matrix computed as the sum of diagonal components. This choice will tend to make the two parts of the minimization have comparable weights; adjustments (in factors of 10) can be made from there.

Figure 3.21 shows the trade-off curve for our simple curve-fitting example using second-order regularization. Equal logarithmic increments of γ were used. The vicinity of $\gamma \approx 1.00$ is where there is about equal balance of data fitting and smoothing. It is interesting to compare this value with the corresponding plot in Figure 3.19(b); it is only by having outside information about the underlying process that we could judge which of the curve-fits is "best". In that sense, the trade-off curve is not especially useful, and in the later chapters we will rely more on knowledge about the problem in order to choose a value for γ.

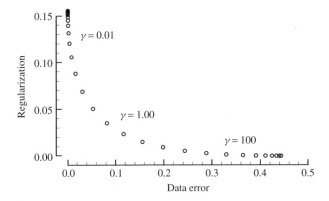

Figure 3.21: Trade-off curve for simple curve-fitting example.

To emphasize this last point, the purpose of regularization is to give us a mechanism to enforce on the solution some of our *a priori* expectations of the behavior of the solution. Therefore, its determination cannot be fully objective in the sense of only using the immediate measured data. We will see in later chapters that other judgments can and should come into play. For example, suppose we are determining a transient force history; before the event begins, we would expect the reconstructed force to show zero behavior, and if it is an impact problem we would expect the force to be of only one sign. Subjectively judging these characteristics of the reconstructions can be very valuable in assessing the amount of regularization to use.

Other Regularizations

The regularization methods just discussed involved various levels of smoothness associated with derivatives. It is worth noting that regularization can also be done with respect to a variety of constraints. We will mention just a couple here.

Suppose we do a Moiré analysis of a deflected beam and have deflection values at many points. Further, let us be interested in the distribution of bending stresses that are obtained from the displacements by differentiation

$$\sigma = \frac{1}{2} E h \frac{d^2 v}{dx^2}$$

Unfortunately, double differentiation of experimental data is very susceptible to small variations in the data and the results would, generally, be highly suspect showing great irregularities. We therefore want to regularize the solution. One possibility is to use the displacement data to find strain values, and then curve-fit (with regularization) the strain values as if they are new data points.

Alternatively, we can keep the displacement distribution as the basic unknown but use higher levels of smoothing, and, further, choose the level on the basis of the governing differential equation. For example, the governing differential equation for a beam with distributed loading is

$$EI\frac{d^4 v}{dx^4} = q(x)$$

where $q(x)$ is the applied distributed load. We could assume that the applied load, locally, is not too different from a constant and therefore choose as our regularizing function

$$\mathcal{B} \propto \int \left[\frac{d^5 v}{dx^5}\right]^2 dx \propto \sum_{m=1}^{M-5} [u_m, u_{m+1}, u_{m+2} \cdots]^2$$

We can then proceed as before. The outstanding advantage is that we add a significant degree of inherent smoothness to the second derivative of the data. Furthermore, if we so desire, we can allow the system to even tell us the best values for the boundary conditions. This can be significant if the beam ends are encased, say, in another elastic material.

The same could also be done in multidimensions but this is not something we will pursue here. By regularizing with respect to a differential equation, we are actually solving the differential equation using finite differences. This is reasonable for some simple problems such as the beam but is not the method of choice for general problems in solid and structural mechanics—we would prefer to use the finite element method. The next few chapters will show how that is done.

It may well be that, based on previous solutions of similar problems, we have a general idea of what the solution should look like. It is then possible to regularize with respect to this expectation. We choose

$$\mathcal{B} = \{u - f\}^{\mathrm{T}}\{u - f\}$$

where $f(x)$ is our expected behavior of the solution $u(x)$. This leads to the interesting optimized solution

$$\{u\} = [A^{\mathrm{T}}A + \gamma I]^{-1}\{A^{\mathrm{T}}d + \gamma f\}$$

where $[\ I\ ]$ is the identity matrix. The regularization parameter appears in two places. In the inverse term, it regularizes the solution as if it is zero-order regularization and would be present even if $f(x) = 0$. But it also contributes to the effective data in such a way that as γ is made very large the predicted solution becomes the bias solution. This is a clear example of how our *a priori* information/bias can be made part of the solution.

3.7   Relocation of Data onto a Grid Pattern

Consider the situation in which we have some spatially distributed data. It may be uniformly or randomly distributed, but we want interpolated values on a particular grid. The data may have arisen from one of the optical methods discussed in Chapter 2 and the grid could be the node points of an FEM mesh. This section considers some ways of achieving this objective.

Relocation to Nearest Node

Consider the case where we wish to relocate the data onto the nearest node. There are a number of ways to approach this problem depending on the amount of data and density of mesh points. The scheme illustrated here finds the triangle surrounding the data point and then distributes the data equally among the three nodes. Since multiple data can appear in a triangle, a count is made (for each node) of the number of contributions it received and then the total is divided by this number.

Determining if a data point is inside a triangle is quite straightforward. The area coordinates $h_1(x_i, y_i)$, $h_2(x_i, y_i)$, and $h_3(x_i, y_i)$, of Equation (1.32), are computed relative to each triangle. The triangle for which

$$h_1 \quad \text{or} \quad h_2 \quad \text{or} \quad h_3 < 0$$

does not contain the data point.

Figure 3.22 shows the result of relocating some data points. For simplicity, the element data values range from 1.00 at the bottom (Element 1) to 6.00 at the top (Element 6). The inserted table shows the nodal values computed.

Clearly, more sophisticated methods would be needed if the mesh were coarse or the data changed more rapidly. This is not likely to be an issue; however, since the meshes will be FEM meshes and its density will be dictated by the mechanics of the problem. Consequently, this will be our method of choice when only a limited number of data points need to be relocated.

Area Distributions and Spatial Regularization

Consider the situation in which we have spatially distributed data as before, but this time we want interpolated values on every node of a particular grid. This is different from the

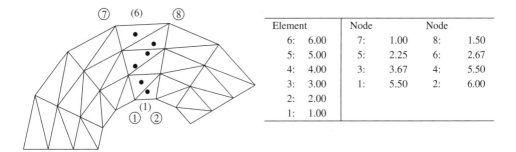

Element		Node		Node	
6:	6.00	7:	1.00	8:	1.50
5:	5.00	5:	2.25	6:	2.67
4:	4.00	3:	3.67	4:	5.50
3:	3.00	1:	5.50	2:	6.00
2:	2.00				
1:	1.00				

Figure 3.22: Relocation to nearest node. The element values are averaged at each node.

previous case for two reasons. First, there may be no data near some nodes and hence gross extrapolation/interpolation must be used. Second, there may be many more nodal points than data points. This is essentially the 2-D equivalent of the curve-fitting problem considered earlier.

To make the example specific, consider the situation shown in Figure 3.23(a) where the black dots represent the data point locations. Note that there are some triangles with multiple data, and some triangles with none. A simple relocation scheme such as assigning the node value as that of the nearest data point would not be very accurate. A scheme based on interpolation as used for one-dimensional data would be very difficult to implement. We will use a regularization method because there are more nodal points than data points.

Let the collection of data be $\{d\}$ and the collection of (unknown) nodal point values be $\{w\}$. The minimizing principle is

$$E(w) = \{d - Qw\}^{\mathrm{T}} \lceil\ W\ \rfloor \{d - Qw\} + \gamma\, \{w\}^{\mathrm{T}} [\ H\ ]\{w\}$$

where $\lceil\ W\ \rfloor$ is a general weighting array and $[\ H\ ]$ is our regularization array based on derivatives. The matrix $[\ Q\ ]$ depends on our local assumptions about the distributions, which we will discuss shortly. We minimize the error with respect to $\{w\}$ to get

$$[Q^{\mathrm{T}} W Q + \gamma\, H]\{w\} = [\ Q\ ]^{\mathrm{T}} [\ W\ ]\{d\}$$

It remains now to determine $[\ Q\ ]$ and the form of regularization $[\ H\ ]$.

Let the area be discretized using triangular regions. Then the distribution within each triangle is given by the linear interpolations

$$w(x, y) = \sum_{i}^{3} h_i(x, y) w_i$$

where w_i are the desired nodal values. The row of the $[\ Q\ ]$ matrix for data d_i is

$$\lfloor Q \rfloor_i = \lfloor \ldots, 0, h_1(x_i, y_i), h_2(x_i, y_i), h_3(x_i, y_i), 0, \ldots \rfloor_i$$

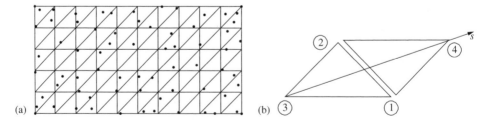

(a)   (b)

Figure 3.23: Relocating random data onto a triangular grid. (a) Distributed data to be interpolated onto every node of the grid. (b) Node numbering scheme for the second derivative of two neighboring triangular elements.

This is constructed for each data point. The nonzero entries are not necessarily contiguous since they relate to the node numbers.

The derivatives needed for first-order regularization are

$$\frac{\partial w}{\partial x} = \sum_i^3 \frac{\partial h_i}{\partial x} w_i = \sum_i^3 b_i w_i, \qquad \frac{\partial w}{\partial y} = \sum_i^3 \frac{\partial h_i}{\partial y} w_i = \sum_i^3 c_i w_i$$

Introducing the derivative matrices for each triangle

$$\lfloor D_x \rfloor = \lfloor 0, \ldots, b_1, b_2, b_3, \ldots, 0 \rfloor, \qquad \lfloor D_y \rfloor = \lfloor 0, \ldots, c_1, c_2, c_3, \ldots, 0 \rfloor$$

Each row corresponds to an element and the nonzero entries are not necessarily contiguous. The regularization matrix is formed as

$$\mathcal{B} = \{w\}^T [D_x]^T [D_x]\{w\} + \{w\}^T [D_y]^T [D_y]\{w\}$$

$$= \{w\}^T [H_x]\{w\} + \{w\}^T [H_y]\{w\} = \{w\}^T [H]\{w\}$$

Constructed in this way, the matrix [H] is positive definite.

Approximation for the second derivative requires the involvement of at least two elements. One possible scheme is shown in Figure 3.23(b) where the direction of the derivative is indicated by the arrow. The derivative is approximated as

$$\frac{\partial^2 w}{\partial s^2} \approx w_3 - 2(w_1 + w_2)\frac{1}{2} + w_4 = -w_1 - w_2 + w_3 + w_4$$

(Note that, for the purposes of regularization, it is not necessary to involve the distances in the derivatives.) This derivative can be formed for each contiguous element (to a maximum of three); they could be arranged row-wise as three per element, but it is slightly more efficient to arrange them row-wise with the first node as a superposition. This gives for the matrices [D] and [H]

$$\lfloor D \rfloor = \lfloor 0, \ldots, -1, -1, 1, 1, \ldots 0 \rfloor, \qquad [H] = [D]^T [D]$$

It is possible to derive higher-order smoothings, but it does not seem to be justified.

Some results are shown in Figure 3.24. The data came from a random sampling of the function $w(x, y) = \sin(2\pi x) \sin(\pi y)$ and relatively large random errors of 10% were added. The amount of regularization used was the suggested value of

$$\gamma = \text{Tr}[Q^T W Q]/\text{Tr}[H]$$

Clearly, the second-order regularization outperforms the first-order regularization. The dots in Figure 3.24(b) are the exact values (without noise) of the contours.

Depending on the density of data, and fineness of the mesh, it is generally a good idea to use an interpolation mesh that is finer than the required discretization. This overcomes

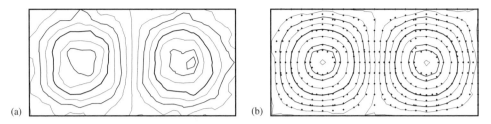

Figure 3.24: Contours of interpolated data. (a) First-order regularization. (b) Second-order regularization.

the limitation of linear interpolation. The results shown are for a mesh that is twice as fine (in both directions) than that shown in Figure 3.23(a).

The purpose of this example was to obtain good interpolated values from a spatially random set of data, and the results show that this can be done effectively. A word of warning, however; these interpolated values are not suitable for further processing that involve derivatives. For example, if the data were from a shadow Moiré analysis of the out-of-plane displacements, it might be tempting to double differentiate the interpolated data to get the strains. Poor results would be obtained, especially along the boundaries where the maximum strains occur. The next few chapters will show more appropriate ways of handling this processed data.

Experiment: Interpolating Fringe Data

Figure 3.25 shows a fringe pattern from a Moiré experiment. We will discuss this experiment in more detail in Section 4.2; here we are concerned with the issue of taking a discretized version of a fringe pattern such as this and interpolating the data to every nodal point of the indicated mesh. The reason this is needed is because the initial fringe pattern must be subtracted from the deformed pattern and this must be done at common points (chosen here to be the mesh nodes).

The pattern is first discretized at a number of points and the half-order (black fringes) identified. These data are shown as the circles in Figure 3.26(a). The data are relocated

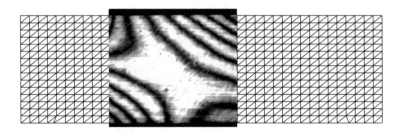

Figure 3.25: Moiré u-displacement fringe data.

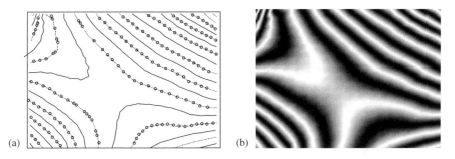

Figure 3.26: Interpolated Moiré *u*-displacement fringe data.

to the node point using the scheme just discussed; second-order regularization was used. The resulting interpolated pattern is shown in Figure 3.26(a) as the full (whole order) and dashed (half order) lines.

Note that the objective was not to curve-fit the data but to interpolate it onto the nodes of the mesh. Only because this was achieved could the synthetic fringe pattern of Figure 3.26(b) be reconstructed. This fringe pattern is to be compared to the fringes in Figure 3.25.

Whole-field Phase Unwrapping

An interesting use of the previous developments could be the phase unwrapping of whole-field fringe data. Let the intensity plot of a fringe pattern be represented by

$$I(x, y) = A(x, y) + B(x, y) \cos[\phi(x, y)]$$

The phase $\phi(x, y)$ is related to the mechanical information (e.g., displacement from Moiré or principal stress difference from photoelasticity) and hence our first task is to determine it at the grid points.

The unknowns in this equation are $A(x, y)$, $B(x, y)$, and $\phi(x, y)$, and their relation to the phase is nonlinear. Using some iterative scheme, we can either interpolate all the data $I(x, y)$ to the nodal points or we can use interpolation on triangles to determine the coefficients at the data points. Either way, we establish a set of simultaneous equations that can be solved using least squares plus regularization.

The underlying assumption of this approach is that the phase and coefficients change smoothly relative to the intensity data and, therefore, a reasonable amount of regularization will keep the iterations within bounds.

The task of finding phase information from intensity data has received a great deal of attention because of its practical uses. For example, in medical X-ray tomography [14], the intensity detected at a sensor is related to the density of the body along the light path; by scanning a series of such paths in a plane it is possible to reconstruct the density distribution in the plane and thereby "see into" the body. A somewhat similar situation arises in astronomy, holography, and X-ray diffraction [77, 78, 85] but here the intensity recordings are in the image and diffraction planes. The Gerchberg–Saxton algorithm [85] bounces back and forth between the image domain where *a priori* information such as nonnegativity is applied and the Fourier domain where the measured diffraction data are applied. Reference [28] gives a good summary of these iterative methods and has an extensive bibliography.

Since unwrapping the phase is not an end in itself but a step toward solving the stress analysis problem, it is reasonable to attempt to use the intensity data directly in the inverse method. In this way, supplementary information about the problem can be utilized. This will be considered in the next few chapters.

3.8 Discussion

Our inverse theory, as we formulated it, has the objective to minimize

$$\mathcal{A} + \gamma \mathcal{B} = \left\{ \{d\} - [\,A\,]\{u\} \right\}^{\mathrm{T}} \lceil\,W\,\rfloor \left\{ \{d\} - [\,A\,]\{u\} \right\} + \gamma\,\{u\}^{\mathrm{T}}[\,H\,]\{u\}$$

for various values of $0 < \gamma < \infty$. We can have various levels of sophistication in the description of the regularization; as suggested we can even regularize with respect to the governing equation or some *a priori* information about how the solution should behave.

This formulation is missing a number of ingredients that are very important in experimental stress analysis. For example, it often happens that our measurements are related to displacement or strain, but the unknowns are forces or stresses—essentially, we do not know the matrix $[\,A\,]$ in the form as indicated above. In solid and structural mechanics, as shown in Chapter 1, there is a relation between the forces and displacements; in static problems it is the stiffness relation, but for dynamic problems it is a more complicated convolution relation. How to incorporate this supplementary information is problem dependent, and therefore becomes a main concern of the next four chapters.

We have additional challenges. For inverse problems on complex structures, there is not a simple division between known response and unknown inputs; we may know responses at some points (two strain measurements, say) and not at any other point, while at the same time knowing the loads at many points (where it is traction free, say) but not knowing the impact load. The reformulation of the problem to handle these mixed situations is also a primary concern of the next four chapters.

4

Static Problems

Figure 4.1 shows a model of a lug statically loaded. It has some stress concentrations (locations where the stresses are high relative to the mean) as well as stress singularities where the stresses are theoretically infinite—the crack tip and load point are examples. It is our intention to develop methods that will allow determination of the complete solution (applied loads, singularities, concentrations, etc.) of these problems using limited experimental measurements.

This method developed is based on FEM modeling of the structure and can determine multiple unknown forces and parameters using data that can come from point sensors and/or whole-field images. An important issue that arises is the appropriate parameterization of a problem and this is discussed in depth through the use of many examples.

One of the significant side benefits of the FEM-based method is a consistent underlying framework across all problems that, for example, already has the postprocessing

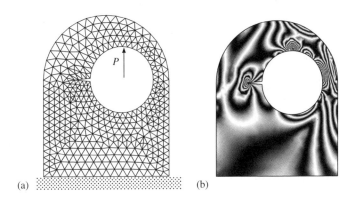

(a)   (b)

Figure 4.1: A model with two stress singularities and a stress concentration around the hole. (a) Mesh. (b) A fringe pattern related to stresses.

Modern Experimental Stress Analysis: completing the solution of partially specified problems. James Doyle
© 2004 John Wiley & Sons, Ltd   ISBN 0-470-86156-8

apparatus in place for obtaining further information as well as for simply presenting the results in the form of contours.

4.1 Force Identification Problems

As discussed in Chapter 1, the governing equations for a linear static system can be discretized using the finite element method as

$$[\,K\,]\{u\} = \{P\} = [\,\Phi\,]\{\tilde{P}\} \tag{4.1}$$

In this equation, $\{u\}$ is the vector of all the free degrees of freedom (DoF) and is of size $\{M_u \times 1\}$, $[\,K\,]$ is the banded $[M_u \times B]$ symmetric stiffness matrix, $\{P\}$ is the $\{M_u \times 1\}$ vector of all applied loads (some of which are zero), $\{\tilde{P}\}$ is the $\{M_p \times 1\}$ vector of the subset of $\{P\}$ that are nonzero-applied loads, and $[\,\Phi\,]$ is the $[M_u \times M_p]$ matrix that associates these loads with the degrees of freedom.

In the forward problem, given the applied loads $\{\tilde{P}\}$, we can solve for the unknown responses directly. In the inverse problem of interest here, both the applied loads and the displacements are unknown, but we have some information about the responses. We wish to use this information to determine a solution. In particular, assume we have a vector of measurements $\{d\}$ of size $\{M_d \times 1\}$, which are related to the structural DoF according to

$$\{d\} \Leftrightarrow [\,Q\,]\{u\}$$

Note that the $[M_d \times M_u]$ matrix $[\,Q\,]$ could be a difference relation as would be the case for strains but, in general, we will treat it simply as a selection matrix. We want to find the forces $\{\tilde{P}\}$ that make the system best match the measurements.

Preliminary Solution

Consider the general least-squares error given by

$$E(u, \tilde{P}) = \{d - Qu\}^{\mathrm{T}} \lceil\, W\, \rfloor \{d - Qu\} + \gamma \{\tilde{P}\}^{\mathrm{T}} [\,H\,]\{\tilde{P}\}$$

Our objective is to find the set of forces $\{\tilde{P}\}$ that minimize this error functional.

As there is a relation between $\{u\}$ and $\{\tilde{P}\}$, (for minimization purposes) we can take either as the basic set of unknowns; here it is more convenient to use the loads because for most problems, the size of $\{\tilde{P}\}$ is substantially smaller than the size of $\{u\}$. We first decompose $[\,K\,]$ and then solve $[\,\Psi\,]$ as the series of back-substitutions

$$[\,K\,][\,\Psi\,] = [U^{\mathrm{T}}DU][\,\Psi\,] = [\,\Phi\,]$$

that is, let $[\,\Phi\,]$ be made up of the load vectors

$$[\,\Phi\,] = [\{\phi\}_1 \{\phi\}_2 \cdots \{\phi\}_{M_p}]$$

Then we find the solutions

$$[K]\{\psi\}_1 = \{\phi\}_1, \qquad [K]\{\psi\}_2 = \{\phi\}_2, \qquad \ldots, \qquad [K]\{\psi\}_{M_p} = \{\phi\}_{M_p}$$

The matrix $[\Psi]$ is the collection of forward solutions for unit loads applied for each of the unknown forces and assembled as

$$[\Psi] = [\{\psi\}_1\{\psi\}_2 \cdots \{\psi\}_{M_p}]$$

We can therefore write the actual forward solution as

$$\{u\} = [\Psi]\{\tilde{P}\}$$

although $\{\tilde{P}\}$ is not known. In this way, the coefficients of $\{\tilde{P}\}$ act as scale factors on the solutions in $[\Psi]$. This idea is shown schematically in Figure 4.2 in which the two-load problem is conceived as the scaled superposition of two separate one-load problems. This superposition idea leads to simultaneous equations to solve for the unknown scales $\{\tilde{P}\}$.

The error functional can now be written only as a function of $\{\tilde{P}\}$

$$E(\tilde{P}) = \{d - Q\Psi\tilde{P}\}^{\mathrm{T}}\lceil W \rfloor\{d - Q\Psi\tilde{P}\} + \gamma \{\tilde{P}\}^{\mathrm{T}}[H]\{\tilde{P}\}$$

We minimize this with respect to $\{\tilde{P}\}$ to get

$$[\Psi^{\mathrm{T}} Q^{\mathrm{T}} W Q \Psi + \gamma H]\{\tilde{P}\} = [Q\Psi]^{\mathrm{T}}\lceil W \rfloor\{d\} \qquad (4.2)$$

This reduced system of equations is of size $[M_p \times M_p]$ and is relatively inexpensive to solve since it is symmetric and small. Once the loads are known, the actual forward

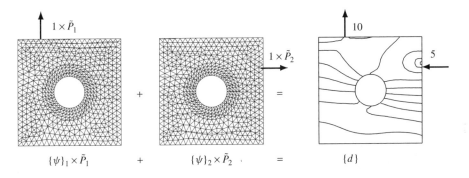

Figure 4.2: The data $\{d\}$ are conceived as synthesized from the superpositions of unit load solutions $\{\psi\}_1$ and $\{\psi\}_2$ modified by scalings $\tilde{P}_1$ and $\tilde{P}_2$.

solution is then

$$\{u\} = [\ \Psi\ ]\{\tilde{P}\}$$

which involves a simple matrix product.

These equations can be solved by standard techniques such as LU decomposition [143]. As discussed in Chapter 3, the purpose of γ and the regularization term is to overcome the ill-conditioning usually associated with solving least-squares systems of equations.

Computer Implementation: Sensitivity Response Method

The above form of solution implies that we have access to the structural stiffness matrix [K]. For relatively simple problems, this could be assembled internally (to the inverse program) having first formulated the element stiffness matrix. However, as the structural complexity increases, requiring more complicated elements (e.g., plates, shells, 3-D solids, and perhaps nonlinear behavior), this would put excessive burden on the programming of the inverse methods. To make our inverse method generally applicable, it is necessary to be able to leverage the power and versatility of commercial FEM codes.

Conceivably, the commercial FEM program could be used to generate [K], but its utilization would require intimate knowledge of its structure; this is something not readily available or is usually not understood even by regular users of FEM programs. The key to understanding our approach to the computer implementation is the series of solutions in the matrix [Ψ]—it is these that can be performed externally (in a distributed computing sense) by a commercial FEM code. The basic relationship is shown in Figure 4.3; StrIDent is the inverse program, GenMesh is an example of a mesh generator program, and StaDyn is an example of an FEM analysis program. One of the roles of StrIDent is to prepare the scripts to run these external programs.

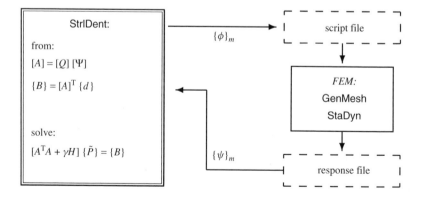

Figure 4.3: Schematic representation of the distributed computing relationship between the inverse program and the FEM programs. The FEM program is run once for each load position.

We now summarize the approach as implemented in the computer code **StrIDent**. The unknown forces are represented as the collection

$$\{P\} = [\{\phi\}_1\{\phi\}_2 \cdots \{\phi\}_{M_p}]\{\tilde{P}\} = [\ \Phi\ ]\{\tilde{P}\} \qquad (4.3)$$

where each $\{\phi\}_m$ is a distribution of forces and $\tilde{P}_m$ is the corresponding scale. As a particular case, $\{\phi\}_m$ could be a single unit force, but it can also represent a uniform or a Gaussian distribution; for traction distribution problems, it has a triangular distribution [41] as will be shown later. We will begin referring to each $\{\phi\}_m$ distribution as a *unit* or *perturbation load*; there are M_p such loads.

Each $\{\phi\}_m$ load causes a response $\{\psi\}_m$ such that the total response (because of the linearity of the system) is

$$\{u\} = [\{\psi\}_1\{\psi\}_2 \cdots \{\psi\}_{M_p}]\{\tilde{P}\} = [\ \Psi\ ]\{\tilde{P}\}$$

The $\{\psi\}_m$ responses are obtained by the finite element method. These responses will have the effects of the complexity of the structure imbedded in them. Consequently, usual complicating factors (for deriving analytical solutions used in inverse methods) such as geometry, material, and boundary conditions are all included in $\{\psi\}_m$, which in turn is handled by the finite element method. Furthermore, note that the responses are not necessarily just displacement; they could be strain or principal stress difference as required by the experimental method, and, consequently, the inverse program does not need to perform differentiation to get strains. In other words, anything particular to the modeling or the mechanics of the structure is handled by the FEM program. We will begin referring to these responses as *sensitivity responses* and our inverse method as the *sensitivity response method* (SRM).

The measured data at a collection of locations are related to the computed responses as

$$\{d\} \Longleftrightarrow [\ Q\ ]\{u\} = [\ Q\ ][\ \Psi\ ]\{\tilde{P}\}$$

where $[\ Q\ ]$ only symbolically plays the role of selecting the subset of $[\ \Psi\ ]$ participating in forming the data.

That is, the array

$$[\ A\ ] = [\ Q\ ][\ \Psi\ ]$$

is established directly from the participating components of $\{\psi\}_m$, and the product $[\ Q\ ]$ $[\ \Psi\ ]$ is never actually performed. The least-squares minimization leads to

$$[A^\mathrm{T} W A + \gamma\, H]\{\tilde{P}\} = \{A^\mathrm{T} W d\} \qquad (4.4)$$

If supplementary information is available, it is incorporated simply by appending to the matrix $[\ A\ ]$ before the least-squares procedure is used.

A final point concerns the amount of regularization. A reasonable beginning value of γ is to try

$$\gamma = \mathrm{Tr}[A^\mathrm{T} W A]/\mathrm{Tr}[\ H\ ]$$

where Tr is the trace of the matrix computed as the sum of diagonal components. This choice will tend to make the two parts of the minimization have comparable weights. It is found, however, that in most experiments this is too large; in other words, we actually have more confidence in the data than in our *a priori* beliefs. But this value is a good starting point and can be changed in the decades that follow. Three cases of regularization are implemented in StrIDent: zeroth, first, and second order.

A Test Example

Consider the uniaxial rod shown in Figure 4.4 that is modeled as three elements with one applied load. From measurements at Nodes 2 and 3, we wish to determine the applied load P at Node 2 and the reaction at Node 4. This example will be used to illustrate some aspects of the matrix formulation of problems; it will also illustrate the small rearrangements necessary when an external FEM program is to be used.

I: Forward Solution

We first demonstrate the solving of the forward problem by determining the responses at the nodes. The reduced element stiffnesses are

$$[k^{12}] = \frac{EA}{L}[\ 1\ \ 1\ ], \qquad [k^{23}] = \frac{EA}{L}\begin{bmatrix} 1 & -1 \\ -1 & 1 \end{bmatrix}, \qquad [k^{34}] = \frac{EA}{L}[\ 1\ \ 1\ ]$$

The assembled system of equations is

$$\frac{EA}{L}\begin{bmatrix} 2 & -1 \\ -1 & 2 \end{bmatrix}\begin{Bmatrix} u_2 \\ u_3 \end{Bmatrix} = \begin{Bmatrix} P \\ 0 \end{Bmatrix} \qquad \text{or} \qquad [\ K\ ]\{u\} = \{P\}$$

This is solved to give the responses

$$\begin{Bmatrix} u_2 \\ u_3 \end{Bmatrix} = \frac{PL}{3EA}\begin{Bmatrix} 2 \\ 1 \end{Bmatrix}$$

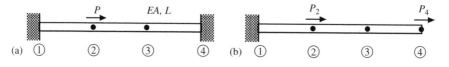

Figure 4.4: Simple rod problem. (a) Rod with fixed ends. (b) Problem modeled with unknown boundary reaction.

The reaction at Node 4 is found from

$$\left\{ \begin{matrix} F_1 \\ F_2 \end{matrix} \right\}^{34} = \frac{EA}{L} \begin{bmatrix} 1 & -1 \\ -1 & 1 \end{bmatrix} \left\{ \begin{matrix} u_2 \\ u_3 \end{matrix} \right\} = \frac{P}{3} \left\{ \begin{matrix} 1 \\ -1 \end{matrix} \right\}, \qquad R_4 = -\frac{1}{3}P$$

II: Internal Inverse Solution

What we mean by internal inverse solution is that the inverse method has access to system matrices such as the stiffness matrix.

For the inverse problem, some of the boundaries can be replaced with unknown tractions. With reference to Figure 4.4(b), we have

$$\{u\} = \{u_2, u_3, u_4\}^{\mathrm{T}}$$

$$\{d\} = \{u_2, u_3\}^{\mathrm{T}} = \begin{bmatrix} 1 & 0 & 0 \\ 0 & 1 & 0 \end{bmatrix} \{u\} = [\ Q\ ]\{u\}$$

$$\{P\} = \{P_2, P_3, P_4\}^{\mathrm{T}} = \begin{bmatrix} 1 & 0 \\ 0 & 0 \\ 0 & 1 \end{bmatrix} \left\{ \begin{matrix} P_2 \\ P_4 \end{matrix} \right\} = [\ \Phi\ ]\{\tilde{P}\}$$

Let the weighting arrays be simply

$$\lceil W \rfloor = \alpha \begin{bmatrix} 1 & 0 \\ 0 & 1 \end{bmatrix}, \qquad [\ H\ ] = \begin{bmatrix} 1 & 0 \\ 0 & 1 \end{bmatrix}$$

The reduced element stiffnesses are

$$[k^{12}] = \frac{EA}{L}[\ 1\ 1\ ], \qquad [k^{23}] = \frac{EA}{L}\begin{bmatrix} 1 & -1 \\ -1 & 1 \end{bmatrix}, \qquad [k^{34}] = \frac{EA}{L}\begin{bmatrix} 1 & -1 \\ -1 & 1 \end{bmatrix}$$

The assembled structural stiffness is

$$[\ K\ ] = \frac{EA}{L}\begin{bmatrix} 2 & -1 & 0 \\ -1 & 2 & -1 \\ 0 & -1 & 1 \end{bmatrix}$$

The $[UDU]$ decomposition of this gives

$$[\ K\ ] \Rightarrow [UDU] = \frac{EA}{L}\begin{bmatrix} 1 & 0 & 0 \\ -\frac{1}{2} & 1 & 0 \\ 0 & -\frac{2}{3} & 1 \end{bmatrix}\begin{bmatrix} 2 & 0 & 0 \\ 0 & \frac{3}{2} & 0 \\ 0 & 0 & \frac{1}{3} \end{bmatrix}\begin{bmatrix} 1 & -\frac{1}{2} & 0 \\ 0 & 1 & -\frac{2}{3} \\ 0 & 0 & 1 \end{bmatrix}$$

The matrix [Ψ] has three rows and two columns and is obtained by solving the system with the columns of the [Φ] matrix as the load vectors, that is,

$$[UDU]\{\psi\}_1 = \begin{Bmatrix} 1 \\ 0 \\ 0 \end{Bmatrix}, \qquad [UDU]\{\psi\}_2 = \begin{Bmatrix} 0 \\ 0 \\ 1 \end{Bmatrix}$$

This then leads to the unit solutions

$$[\Psi] = \frac{L}{EA} \begin{bmatrix} 1 & 1 \\ 1 & 2 \\ 1 & 3 \end{bmatrix}$$

We are now in a position to form the matrix products

$$[Q][\Psi] = \begin{bmatrix} 1 & 0 & 0 \\ 0 & 1 & 0 \end{bmatrix} \frac{L}{EA} \begin{bmatrix} 1 & 1 \\ 1 & 2 \\ 1 & 3 \end{bmatrix} = \frac{L}{EA} \begin{bmatrix} 1 & 1 \\ 1 & 2 \end{bmatrix}$$

leading to

$$[\Psi^{T}Q^{T}WQ\Psi] = \alpha \left(\frac{L}{EA}\right)^2 \begin{bmatrix} 2 & 3 \\ 3 & 5 \end{bmatrix}, \qquad [\Psi^{T}Q^{T}W]\{d\} = \alpha \left(\frac{L}{EA}\right) \begin{Bmatrix} u_2 + u_3 \\ u_3 + 2u_5 \end{Bmatrix}$$

The reduced system of equations to be solved is then

$$\left[\alpha \left(\frac{L}{EA}\right)^2 \begin{bmatrix} 2 & 3 \\ 3 & 5 \end{bmatrix} + \gamma \begin{bmatrix} 1 & 0 \\ 0 & 1 \end{bmatrix} \right] \begin{Bmatrix} P_2 \\ P_4 \end{Bmatrix} = \alpha \left(\frac{L}{EA}\right) \begin{Bmatrix} u_2 + u_3 \\ u_2 + 2u_3 \end{Bmatrix}$$

The two forces are obtained as

$$P_2 = \frac{c_2}{c_1} \frac{(2u_2 - u_3) + (\gamma/c_1)(u_2 + u_3)}{1 + 7(\gamma/c_1) + (\gamma/c_1)^2}$$

$$P_4 = \frac{c_2}{c_1} \frac{(-u_2 + u_3) + (\gamma/c_1)(u_2 + 2u_3)}{1 + 7(\gamma/c_1) + (\gamma/c_1)^2}, \qquad \frac{c_2}{c_1} = \frac{\alpha(L/EA)}{\alpha(L/EA)^2} = \frac{EA}{L}$$

In this particular instance, because the data are adequate for a unique solution, we can get the forces by setting $\gamma = 0$ and the displacements to the measured values to give

$$P_2 = P, \qquad P_4 = -\tfrac{1}{3}P$$

This is the solution that would be obtained if the straight (unregularized) least-squares method had been used.

III: External Solution

What we mean by external solution is that the inverse method does not have access to system matrices such as the stiffness matrix. But a FEM program is available that can provide solutions, given a set of applied loads.

The specification of the problem is the same as before, thus with reference to Figure 4.4(b), we have

$$\{u\} = \{u_2, \, u_3, \, u_4\}^{\text{T}}$$

$$\{d\} = \{u_2, \, u_3\}^{\text{T}} = \begin{bmatrix} 1 & 0 & 0 \\ 0 & 1 & 0 \end{bmatrix} \{u\} = [\, Q\,]\{u\}$$

$$\{P\} = \{P_2, \, P_3, \, P_4\}^{\text{T}} = \begin{bmatrix} 1 & 0 \\ 0 & 0 \\ 0 & 1 \end{bmatrix} \begin{Bmatrix} P_2 \\ P_4 \end{Bmatrix} = [\, \Phi \,]\{\tilde{P}\}$$

Let the weighting arrays be simply

$$\lceil\, W\, \rfloor = \alpha \begin{bmatrix} 1 & 0 \\ 0 & 1 \end{bmatrix}, \qquad [\, H\,] = \begin{bmatrix} 1 & 0 \\ 0 & 1 \end{bmatrix}$$

The unit loads are

$$\{\phi\}_1 = \begin{Bmatrix} 1 \\ 0 \\ 0 \end{Bmatrix}, \qquad \{\phi\}_2 = \begin{Bmatrix} 0 \\ 0 \\ 1 \end{Bmatrix}$$

These are sent to the FEM program, which returns the sensitivity responses

$$\{\psi\}_1 = \frac{L}{EA} \begin{Bmatrix} 1 \\ 1 \\ 1 \end{Bmatrix}, \qquad \{\psi\}_2 = \frac{L}{EA} \begin{Bmatrix} 1 \\ 2 \\ 3 \end{Bmatrix}$$

The matrix [A] is the subset of the responses associated with the sensors, thus

$$[\, A\,] = \frac{L}{EA} \begin{bmatrix} 1 & 1 \\ 1 & 2 \end{bmatrix}, \qquad \{B\} = \begin{Bmatrix} u_2 \\ u_3 \end{Bmatrix}$$

Application of the least squares then leads to

$$[A^{\text{T}} W A] = \alpha \left(\frac{L}{EA} \right)^2 \begin{bmatrix} 2 & 3 \\ 3 & 5 \end{bmatrix}, \qquad [A^{\text{T}} W B] = \alpha \left(\frac{L}{EA} \right) \begin{Bmatrix} u_2 + u_3 \\ u_3 + 2u_5 \end{Bmatrix}$$

This gives the same system of equations and solution as before.

IV: Effect of Regularization

To see the utility of the regularization term, consider the situation in which everything is the same as above except that we have only one measurement, that is, we wish to obtain two forces from a single measurement. The data is represented by

$$\{d\} = \{u_2\} = \begin{bmatrix} 1 & 0 & 0 \end{bmatrix} \{u\} = [\ Q\ ]\{u\}$$

The matrix $[\ \Psi\ ]$ is unchanged giving the product

$$[Q\Psi] = [1, \ 1]$$

The system of equations finally becomes

$$\left[\alpha\left(\frac{L}{EA}\right)^2 \begin{bmatrix} 1 & 1 \\ 1 & 1 \end{bmatrix} + \gamma \begin{bmatrix} 1 & 0 \\ 0 & 1 \end{bmatrix} \right] \begin{Bmatrix} P_2 \\ P_4 \end{Bmatrix} = \alpha\left(\frac{L}{EA}\right) \begin{Bmatrix} u_2 \\ u_2 \end{Bmatrix}$$

Note that if $\gamma = 0$, then the system is singular without a unique solution—this would manifest itself numerically as an underflow or overflow. With nonzero γ, the two forces are obtained as

$$P_2 = \frac{c_2}{c_1} \frac{(\gamma/c_1)u_2}{(2 + (\gamma/c_1))(\gamma/c_1)} = \frac{1}{(2 + \gamma/\alpha(L/EA)^2)} P, \qquad P_4 = P_2$$

Depending on the value of γ, the solutions range from $\frac{1}{3}P$ to $0P$. None of these solutions are "correct." What regularization contributes is determinacy; that is, we will always get a solution even if we have more unknowns than measurements.

Looking at the traces of the arrays, to have equal amounts of regularization and data fitting, we get

$$\alpha\left(\frac{L}{EA}\right)^2 2 = \gamma 2 \qquad \text{or} \qquad \gamma \approx \left(\frac{L}{EA}\right)^2$$

But (EA/L) is the stiffness of a rod of length L, so γ, in general, is on the order of the reciprocal of the stiffness squared when displacement data are used.

Combination of Known and Unknown Forces

The foregoing equation assumed that all forces present were either zero or unknown. There are problems in which there might be a combination of known and unknown forces present. For example, a complex structure may have a definite applied load, but attachment points (in the form of bolted joints or links) are not clearly defined and therefore could alternatively be represented by unknown forces. We will use the principle of superposition to handle these cases.

Let the general system be represented by

$$[\ K\ ]\{u\} = [\ \Phi\ ]\{\tilde{P}\} + \{P_k\}$$

that is, we have a problem in which, in addition to the unknown force-related term $\{\tilde{P}\}$, we also have a known forcing term $\{P_k\}$. Solve the two systems of equations

$$[\ K\ ][\ \Psi\ ] = [\ \Phi\ ], \qquad [\ K\ ]\{\psi_P\} = \{P_k\}$$

The actual response is the linear combination

$$\{u\} = [\ \Psi\ ]\{\tilde{P}\} + \{\psi_P\}$$

The expression for the error becomes

$$E(\tilde{P}) = \{d - Q\Psi\tilde{P} - Q\psi_P\}^{\mathrm{T}}\lceil\ W\ \rfloor\{d - Q\Psi\tilde{P} - Q\psi_P\} + \gamma\{\tilde{P}\}^{\mathrm{T}}[\ H\ ]\{\tilde{P}\}$$

Minimizing with respect to $\tilde{P}$ leads to

$$[\Psi^{\mathrm{T}}Q^{\mathrm{T}}WQ\Psi + \gamma\ H]\{\tilde{P}\} = [Q\Psi]^{\mathrm{T}}[\ W\ ]\{d - Q\psi_P\}$$

This is of the same form as discussed earlier if we take as effective data the term

$$\{d^*\} \equiv \{d\} - [\ Q\ ]\{\psi_F\}$$

In relation to Figure 4.3, the only change is one additional call to the FEM program to obtain $\{\psi_P\}$.

4.2 Whole-Field Displacement Data

We use two whole-field examples. The first is holography to measure the out-of-plane displacements, and the second is Moiré interferometry to determine in-plane displacements. Both examples illustrate the need to relocate the measured data onto a grid.

Double Exposure Holography

Figure 4.5 shows a double exposure holographic fringe pattern for the out-of-plane displacement of a pressure-loaded plate. The plate is made as one end of a pressurized cylinder, and the inner white dots of the figure show the flexing part of the plate. A small change of pressure was applied accurately using a manometer.

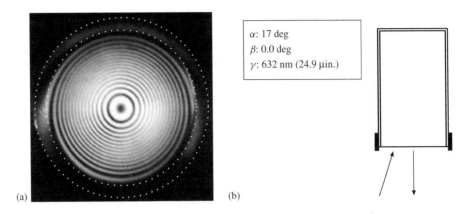

(a)   (b)

Figure 4.5: Holographic experiment of a pressure-loaded plate. (a) Out-of-plane displacement $w(x, y)$ fringe pattern. (b) Cylindrical pressure vessel.

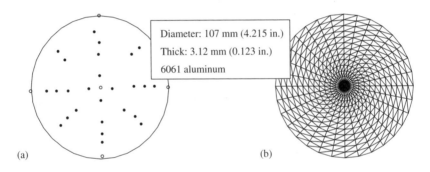

(a)   (b)

Figure 4.6: Pressure-loaded circular plate. (a) Discretized data positions. (b) FEM mesh and properties.

As sometimes happens with holography, the fringe pattern is not of uniform quality. Furthermore, there is a slight asymmetry in the fringe pattern that could be due to a slight overall rigid-body motion of the cylinder.

Figure 4.6(a) shows the limited number of discretized data points used. Because of the expected symmetries in the results, the points were distributed somewhat uniformly so as to provide averaging. Figure 4.6(b) shows the FEM mesh used. The data were relocated onto the nearest nodes of this mesh by the scheme discussed in Section 3.7. The edge of the plate was modeled as being rigidly clamped.

Pressure was taken as the only unknown in formulating the inverse solution. Since the number of data far exceed the number of unknowns, regularization is not required. The nominal measured and computed pressures are given in Figure 4.7. The comparison is very good.

Figure 4.7(a) shows a comparison of the reconstructed fringe pattern (bottom) with that measured (top). Since the pressure comparison is good, it is not surprising that this comparison is also good.

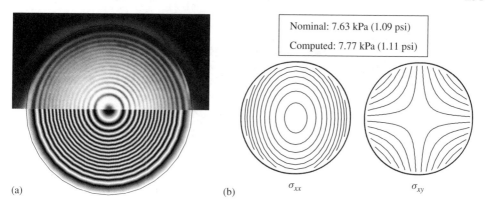

Nominal: 7.63 kPa (1.09 psi)

Computed: 7.77 kPa (1.11 psi)

σ_{xx}

σ_{xy}

(a)   (b)

Figure 4.7: Reconstructions. (a) Displacement fringe pattern. (b) Stress contours.

A theme running through the inverse solutions in this book is that once the unknown forces (or whatever parameterization of the unknowns is used) have been determined, the analytical apparatus is then already available for the complete stress analysis. Figure 4.7(b) shows, for example, the σ_{xx} and σ_{xy} stress distributions. What is significant about these reconstructions is that (in comparison to traditional ways of manipulating such holographic data) the process of directly differentiating experimental data is avoided. Furthermore, this would still be achievable even if the plate was more complicated by having, say, reinforcing ribs (in that case, the strains are not related to the second derivatives of the displacement). Note also that because the experimental data were not directly differentiated, only a limited number of data points needed to be recorded.

Moiré Interferometry

Figure 4.8 shows a u-displacement Moiré fringe pattern for a beam loaded in three-point bending. The data are collected only from the region near the loaded area. It is common in problems like these that the initial (or null-field) fringe pattern is nonzero and must be subtracted from the loaded pattern. To exaggerate this effect, our initial pattern will actually be for a loaded state; this initial pattern is shown in the inset.

Our first task is to subtract these two digitized images. There are a variety of ways of doing this; the scheme adopted here is to first relocate the data to every nodal point of the mesh shown and then subtract the data at each node.

To this end, each pattern is first discretized at a number of points and the half-order (black fringes)is identified. This data is relocated to the nodal point using the scheme discussed in Section 3.7; Figure 3.26 shows an example of the discretized and relocated data when second-order regularization is used. The same procedure was used on the initial fringe pattern, and the resulting difference pattern is shown in Figure 4.9. Data were relocated onto every node of the mesh, but only difference data within the indicated

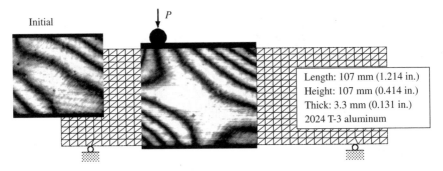

Figure 4.8: Moiré u-displacement fringe data. The inset is the initial fringe pattern.

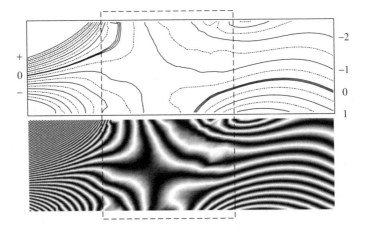

Figure 4.9: Difference data. The dashed box indicates the usable data region and the heavy contour line is the zero-order fringe.

dashed box are suitable for further manipulation, because that is the region of the original discretized data.

The fringe pattern looks reasonable for a beam in three-point bending, but the fringe numbering is off; a zero-order fringe is expected in the central area. The problem is not simply a miscounting of fringes (which could be corrected simply by adjusting by an integer amount) because testing showed that a fractional fringe was required. It was concluded that the fringe pattern has superposed a rigid-body motion on it.

The model used for the inverse solution is shown in Figure 4.10(a); the mesh is the same as shown in Figure 4.8, but since there was an unknown rigid-body motion, this was included in the solution procedure as an additional unknown force; that is, by putting the beam on horizontal rollers and attaching a spring of specified stiffness K, the unknown

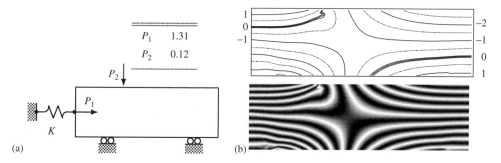

Figure 4.10: Beam in three-point bending. (a) Model with two unknown forces; P_1 is related to the unknown horizontal rigid-body motion. (b) Reconstructions of the whole-field displacements.

horizontal rigid-body displacement is related to the unknown applied load by $u = P_1/K$. The particular value of K is not important, but the value chosen was such that the two forces had comparable magnitudes.

No regularization was used in solving for the two forces; this is because the two forces are unrelated, and the amount of data far exceeds the number of unknowns. The comparison of the computed forces are shown in the inset table of Figure 4.10.

The reconstructions of the u displacements are shown in Figure 4.10. Clearly, the fringe patterns are the same as in Figure 4.9. At this stage, the FEM apparatus can be used to determine other quantities of interest such as stresses or the other displacements. If desired, the rigid-body displacement can be obtained as $u = P_1/K$; in the present case, this was determined as 1.31 fringes.

Discussion of Displacement Methods

Moiré interferometry is a very sensitive method for measuring in-plane displacements. As such, it is a good candidate for use with the inverse methods.

It has two drawbacks that must be taken into account. The first is that the initial null-field is sometimes significant and therefore must be accurately subtracted from the loaded field. The scheme presented seems to do this quite well. The other drawback is the presence of rigid-body motions. It is not clear how these arise; indeed, there may be multiple sources. The inverse method allows these to be determined in a straightforward fashion.

Double exposure holography determines out-plane displacements, which traditionally meant double differentiation (in the case of thin plates) of the experimental data in order to obtain strains. The present approach shows that this is no longer a drawback, and, therefore, it too is a good candidate for use with the inverse methods.

4.3   Strain Gages

Strain gages are excellent for determining the "strain at a point." Suppose, however, that we are interested in a "complete" stress analysis, that is, determining the stress at all significant points. The limitations of strain gages then become immediately apparent: they give information at only a limited number of points, and these points must be chosen wisely to best utilize them. But the significant points may not be known in advance of the analysis. This section shows how a limited number of strain gages in conjunction with a model can indeed render a complete solution.

Utilizing Strain as Data

The inverse method has two possibilities to utilize strain gage data: the first is to stay with the displacement formulation of the last section and modify the $[Q]$ matrix to associate strains with displacement derivatives; the other is to use strain distributions as the sensitivity response functions. We illustrate both approaches, although the latter is more preferable because it is consistent with shifting all of the structural analysis onto the FEM program. It is also the method that generalizes better for more complicated problems.

Let the response solutions obtained from the FEM analysis be that of displacement. What we discuss here is how this can be utilized with strain gage data. There must be a rule that relates the strains to the displacements; within an FEM context, this rule is established through the element interpolation functions. The simplest continuum element to formulate is that of the constant strain triangle (CST) element; we, therefore, use it as an illustration.

Consider a triangle with three nodes; as shown in Chapter 1, the displacements have the interpolations

$$u(x, y) = \sum_i^3 h_i(x, y)u_i , \qquad v(x, y) = \sum_i^3 h_i(x, y)v_i$$

where $h_i(x, y)$ are the linear triangle interpolation functions. The displacement gradients are given by

$$\frac{\partial u}{\partial x} = \sum \frac{\partial h_i}{\partial x}u_i = \frac{1}{2A}\sum_i b_i u_i , \qquad \frac{\partial v}{\partial x} = \sum \frac{\partial h_i}{\partial x}v_i = \frac{1}{2A}\sum_i b_i v_i$$

$$\frac{\partial u}{\partial y} = \sum \frac{\partial h_i}{\partial y}u_i = \frac{1}{2A}\sum_i c_i u_i , \qquad \frac{\partial v}{\partial y} = \sum \frac{\partial h_i}{\partial y}v_i = \frac{1}{2A}\sum_i c_i v_i$$

where the coefficients b_i, and c_i are established with respect to the original configuration of the element. The strains are

$$\epsilon_{xx} = \frac{\partial u}{\partial x}, \qquad \epsilon_{yy} = \frac{\partial v}{\partial y}, \qquad 2\epsilon_{xy} = \frac{\partial u}{\partial y} + \frac{\partial v}{\partial x}$$

which are expressed in matrix form as

$$\left\{ \begin{array}{c} \epsilon_{xx} \\ \epsilon_{yy} \\ 2\epsilon_{xy} \end{array} \right\} = \frac{1}{2A} \left[\begin{array}{cccccc} b_1 & 0 & b_2 & 0 & b_3 & 0 \\ 0 & c_1 & 0 & c_2 & 0 & c_3 \\ c_1 & b_1 & c_2 & b_2 & c_3 & b_3 \end{array} \right] \left\{ \begin{array}{c} u_1 \\ v_1 \\ u_2 \\ v_2 \\ u_3 \\ v_3 \end{array} \right\} \quad \text{or} \quad \{\epsilon\} = [\ B\ ]\{u\}$$

The rows of the matrix $[\ B\ ]$ contribute to the $[\ Q\ ]$ matrix.

As the responses $\{\psi\}_m$ are those of displacement, the data relation is

$$\{\epsilon_{\text{data}}\} = [\ Q\ ]\{\epsilon\} = [\ Q\ ][\ B\ ]\{\psi\}\{\tilde{P}\}$$

where, again, $[\ Q\ ]$ simply selects the participating degrees of freedom.

The problem with this approach is twofold. First, the rule must be established for each type of structural component; for example, double differentiation for beams and plates, but a rather complicated relation for shells. Second, unless a very fine mesh is used, the interpolation functions should be the same as those used in the element formulation. But this is not something necessarily known to an FEM user. We conclude that this approach using displacements is suitable only for relatively simple types of structural problems.

Instead, we will assume that we can access the FEM postprocessor and obtain the strains, that is, for the unit load inputs, we can obtain the responses $\{\psi\}_m$ but they will be those of strain

$$\{\epsilon\} = [\{\psi_\epsilon\}_1 \{\psi_\epsilon\}_2 \cdots \{\psi_\epsilon\}_{M_p}]\{\tilde{P}\} = [\ \Psi_\epsilon\ ]\{\tilde{P}\}$$

The data relation is simply

$$\{\epsilon_{\text{data}}\} = \{d\} \Longleftrightarrow [\ Q\ ]\{\epsilon\} = [\ Q\ ][\ \Psi_\epsilon\ ]\{\tilde{P}\}$$

where $[\ Q\ ]$ simply selects the participating degrees of freedom. This is then made part of the minimum principle as demonstrated before.

A final word about the use of strain gage rosettes. As shown in Chapter 2, at a particular point on the surface of a deformed structure, there are, in general, three unknown strains; either $(\epsilon_{xx}, \epsilon_{yy}, \gamma_{xy})$ or $(\epsilon_1, \epsilon_2, \theta)$. Therefore, three independent pieces of experimental data must be recorded if the strain state is to be specified. However, for the inverse method, we do not actually need a complete rosette; individual gages at various angles and positions are sufficient. To see how the data reduction is achieved, consider a gage at an arbitrary orientation; using the transformation law,

$$\epsilon_\theta = \epsilon_{xx} \cos^2 \theta + \epsilon_{yy} \sin^2 \theta + \gamma_{xy} \sin \theta \cos \theta$$

then each normal component can be written in terms of the components referred to in the $x-y$ reference axes as

$$\{\epsilon_\theta\} = \lfloor \cos^2\theta \ \sin^2\theta \ \cos\theta\sin\theta \rfloor \left\{ \begin{array}{c} \epsilon_{xx} \\ \epsilon_{yy} \\ \gamma_{xy} \end{array} \right\} \qquad \text{or} \qquad \{\epsilon_{\text{data}}\} = [\ C\ ]\{\epsilon_{\text{coord}}\}$$

In this way, the orientation of the gages just factors in as part of the application of the selector $[\ Q\ ]$.

Generalizing the sensitivity responses to include quantities other than displacement is an important step because it allows the use of a wider variety of sensors. For example, transmission interferometry uses the sum of normal stresses, and photoelasticity uses the difference of principal stresses.

Experimental Study I: Tabbed Ring

The first experiment involves a gaged ring. The objective is to determine the complete stress distributions.

Figure 4.11(a) shows a photograph of the gaged ring. Figure 4.11(b) shows the ring dimensions, the position of the gages, and the mesh used in the modeling. The load is applied by a pneumatically controlled loading frame.

Figure 4.12(a) shows the recorded strains for four load increments. The plots are linear and exhibit the expected symmetries, although it is worth pointing out that the data are not exactly symmetric.

As is typical in many experimental setups, the best straight lines through the data do not necessarily go through the $(0, 0)$ point. This can be due to a number of factors,

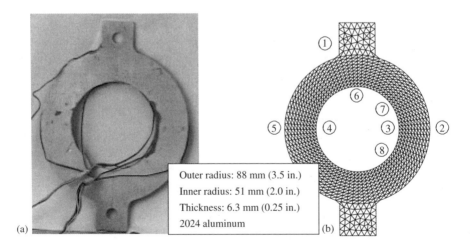

Outer radius: 88 mm (3.5 in.)
Inner radius: 51 mm (2.0 in.)
Thickness: 6.3 mm (0.25 in.)
2024 aluminum

Figure 4.11: Tabbed ring. (a) Photograph of specimen. (b) Mesh and gage positions.

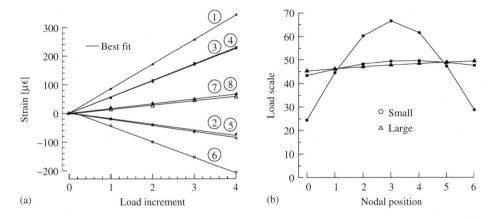

Figure 4.12: Tabbed ring experiment. (a) Recorded strain values. (b) Load distribution at the top of the tab for different amounts of regularization.

Table 4.1: Estimations of applied loads and strain for gages 6, 7, and 8.

	1-load		2-load	Measured
ϵ_2				−78.1
ϵ_3				234.5
ϵ_4				229.9
ϵ_5				−87.6
P	333.7	P_1	163.1	334
		P_2	169.8	
ϵ_6	-210	ϵ_6	-211	−207
ϵ_7	56	ϵ_7	57	66
ϵ_8	49	ϵ_8	49	57

primary among them is the need of the loading apparatus to support the specimen and attachments before applying any strain-producing loads. The offsets were subtracted. Table 4.1 gives the adjusted strain values at the maximum load.

In the first inverse solution, the gages on the horizontal section were used to estimate the load and the strain at gage locations 6, 7, and 8. A single load was assumed to act along the centerline at the top of the model. The results are shown in Table 4.1. The computed load and strains are close to the experimental values.

In the second inverse solution, two loads were determined (the loads were a tab width apart) and these are also shown in Table 4.1. These results indicate a slight asymmetry in the loading as is reflected in the asymmetry of the data.

As a third inverse problem, all top nodes were assumed loaded, and the results are shown in Figure 4.12(b). There are more unknown forces than data, hence some regularization must be used. Figure 4.12(b) shows the effect of the regularization. These plots show that the four gages cannot determine the actual traction distribution; what they show, however, is that the measured strains are insensitive to the actual distribution of traction. This then gives us liberty to model the loaded region in any way that is convenient as long as it is remote from the region of interest. Essentially, this is a confirmation of Saint Venant's principle [56, 167].

Finally, the computed loads were then used to obtain the stress distributions as shown in Figure 4.13. While it may seem obvious that once the loads are obtained, the FEM procedures can easily determine contours and so on, note that the loads come from experimental measurement of strains, and that (conceptually, at least) we could have avoided the discussion of the applied loads and gone directly to the contours. What this highlights is that because the experimental data, inverse methods, and FEM procedures are tightly integrated, producing contours (or whatever other type of information may be of interest) is an almost trivial step.

Experimental Study II: Thin-walled Box Beam

The second experiment is that of a 3-D thin-walled box beam loaded vertically. The objective is to find the value and position of the applied load and to compare them to their nominal values.

Figure 4.14 shows a photograph of the box beam. The vertical loading is applied through a turn buckle that has a force transducer. The dial gauges can be used to locate the shear center (the load position that causes no rotation).

Figure 4.15 shows the geometry and the mesh used in the modeling. The reinforcing steel bars are attached to a thick end plate; the thin aluminum skin is riveted to these bars at every 50 mm (2 in.). The beam is modeled as a 3-D shell. Figure 4.15(b) shows

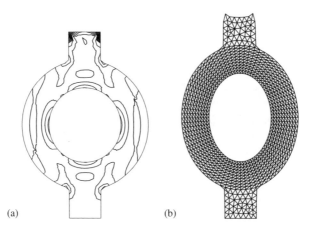

(a)   (b)

Figure 4.13: Examples of postprocessing. (a) Contours of von Mises stress. (b) Exaggerated (x500) deformed shaped.

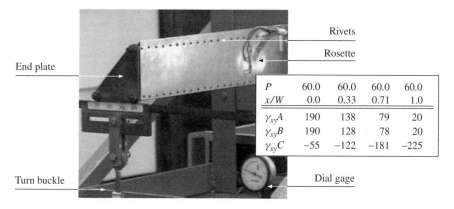

P	60.0	60.0	60.0	60.0
x/W	0.0	0.33	0.71	1.0
$\gamma_{xy}A$	190	138	79	20
$\gamma_{xy}B$	190	128	78	20
$\gamma_{xy}C$	−55	−122	−181	−225

Figure 4.14: Photograph of box beam and experimentally recorded load and strain data.

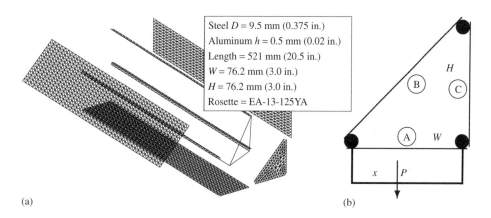

Steel D = 9.5 mm (0.375 in.)
Aluminum h = 0.5 mm (0.02 in.)
Length = 521 mm (20.5 in.)
W = 76.2 mm (3.0 in.)
H = 76.2 mm (3.0 in.)
Rosette = EA-13-125YA

(a)   (b)

Figure 4.15: Box beam geometry. (a) Exploded view of mesh. (b) Dimensions.

the locations of three delta rosettes placed midway along the beam; delta rosettes were used because the shear strain is easily obtained from Equation (2.6) as

$$\gamma_{xy} = \frac{2}{\sqrt{3}}[\epsilon_2 - \epsilon_3]$$

where the ϵ_1 gage is oriented along the x-axis. Figure 4.16 shows the recorded shear strains for six load increments when the nominal load position is $x/W = 0.71$; the data for the other load locations are similar. These data were fitted with straight lines and then the values were adjusted to the origin. The maximum shear strains at the nominal maximum load of $P = 60$ is shown in Table 4.2.

In the first inverse solution, the load is assumed to be applied at the nominal load position. Table 4.2 shows the value of load determined for the four position cases. There is not much spread in the results and each one is quite close to the nominal value.

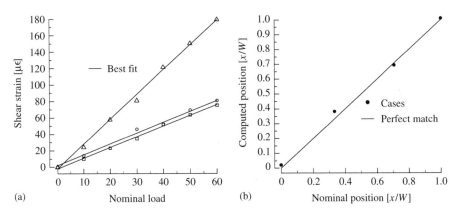

Figure 4.16: Results for the box beam. (a) Experimental shear strains with best-fit lines. (b) Computed position of load.

Table 4.2:   Reconstruction results for the box beam.

P	-58.6	-60.2	-63.1	-60.0
P_1	-58.3	-38.3	-19.3	0.5
P_2	-1.3	-23.8	-43.9	-60.2
$P = P_1 + P_2$	-59.6	-62.0	-63.1	-59.6
Computed x/W	$.02$	0.38	0.70	1.01

In the second inverse solution, both the load and its position are assumed unknown. Rather than treat the position as a variable, the problem is converted instead to a load problem by considering two unknown loads at the lower corners. These two forces are equipollent to the single force at position x and by taking moments about the load point, gives

$$-P_1 x + P_2 (W - x) = 0 \qquad \text{or} \qquad x = \frac{P_2}{P_1 + P_2} W$$

Table 4.2 shows the values of the two forces and the estimated position. Figure 4.16(b) shows a plot of the computed load position against its nominal value. The comparison is quite good.

This experiment shows that an instrumented structure can essentially be its own transducer.

Experimental Study III: Plate with Uniform Pressure

The third experiment is that of a plate with a uniform pressure. In this case, the pressure loading is actually known with a good deal of confidence but because the plate is formed at the end of a pressurized cylinder with bolted clamps, the true boundary conditions

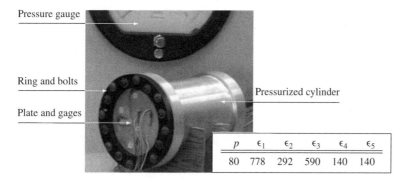

Figure 4.17: Photograph of plate and maximum recorded strains.

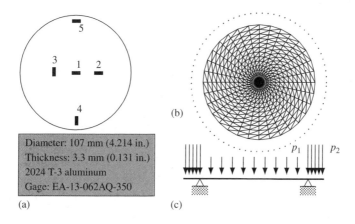

Figure 4.18: Pressure-loaded circular plate. (a) Geometry and gage positions. (b) Center mesh. (c) Clamped boundary modeled as simply supported plate with overhang.

are not rigidly clamped as was originally intended. The objective will be to get a more realistic modeling of the boundary condition.

Figure 4.17 shows a photograph of the plate forming the end of the cylinder. Sixteen bolts along with an aluminum ring are used to clamp the plate to the walls of the cylinder. Note that the cylinder has an increased wall thickness to facilitate this.

Figure 4.18 shows the geometry of the plate and the gage locations. Gages 4 and 5 are not exactly on the edge of the plate, but the gage backing was trimmed so that they could be put as close as possible to the edge. Each gage ended up being approximately 1.8 mm (0.07 in.) from the edge; the meshes, such as those in Figure 4.18(b) were adjusted so as to have a node at the gage positions.

Figure 4.19(a) shows the recorded strains. There is very good linearity with very little offset; consequently, the data was used directly as part of the inverse solutions. The strains at the maximum pressure are given in Figure 4.17.

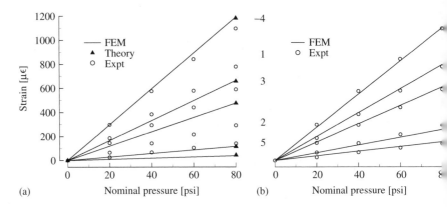

Figure 4.19: Pressurized plate comparisons. (a) Plate with fixed boundary. (b) Plate with flexible boundary.

Good inverse solutions are predicated as having good forward solutions. Figure 4.19(a) shows a comparison of the measured strains with those from an FEM analysis in which the plate is assumed to be rigidly clamped. There are significant differences. Note that there is no systematic difference between the experimental and FEM results, that is, the experimental values are neither consistently higher nor consistently lower than the FEM results; a situation that would imply, for example, an inaccurate pressure measurement or an inaccurate plate thickness. Indeed, it is noted that the experimental value is higher at the center and slightly lower at the fixed edge. This suggests that the FEM modeling does not reflect the true experimental situation.

Also shown in the figure are the exact strains for the problem of a perfectly clamped plate [71]. There is excellent agreement between the FEM and the exact values, and, therefore, it must be concluded that something is amiss with the statement or specification of the problem. The FEM model treats the boundary as fixed, but the actual case has the plate clamped to the end of a cylinder. A reasonable conjecture is that the clamped boundary is somewhat flexible.

We can model this flexible boundary as a spring and use the data to determine the spring stiffness (this would have to be done iteratively as demonstrated later in the chapter). Consistent with our approach (so far) of treating the basic unknowns as loads, we replace instead the unknown boundary conditions with unknown applied loads. Figure 4.18(c) shows one possible modeling. The plate is simply supported on the original diameter, but an overhang with an applied pressure is attached. The inverse problem is now solved with two unknown pressures, that is, the unit solutions correspond to a unit pressure ($p_o = 6.9\,\text{kPa}$, (1.0 psi)) on the inner plate and a unit pressure on the overhang. The results give the pressure scaling on the inner plate as 80.3 and that on the overhang as 2074. Figure 4.19(b) shows a comparison of the strains using the new model; all comparisons show a marked improvement. Because multiple-gage comparisons were improved simultaneously, this gives us the confidence that the new model is indeed more correct.

This example goes to the hearth of the partially specified problems in that, having found a discrepancy between the experimental values and the forward solution, the experimental data can be actively used to provide an improved model. We will leave the issue of determining model parameters until later in the chapter.

4.4 Traction Distributions

Consider the situation in which, for resolution purposes, the FEM mesh has a fine spacing of nodes. The traction distribution would then be represented by a great number of individual forces. As the mesh is made even more fine, the number of forces increases even more, the computational costs increase, and ill-conditioning becomes more troublesome. However, the shape of the distribution of applied tractions does not change with the number of nodes; indeed, it may be a simple curve not requiring very many points for its description. In other words, it would be good to have a traction representation that is independent of the number of nodes used in the FEM discretization. We develop this here.

Space Regularization

As introduced in Chapter 3, the regularization method we use is generally called Tikhonov [8, 116, 165] regularization, where the functional $\mathcal{B}$ involves some measures of smoothness that is derived from first or higher derivatives. For the problem at hand, space regularization connects components in $\{\tilde{P}\}$ across near locations. Zeroth-, first-, and second-order regularization (as discussed in Chapter 3) have been implemented in StrIDent. This regularization cures the ill-conditioning problem, but it does not reduce the computational cost associated with the increased number of forces. That is the objective of the parameterized loading scheme that will be described later.

Consider the plate shown in Figure 4.20(a), which has three sides clamped and the topside free. The Moiré displacement data for a transverse load distributed along the horizontal center line are also shown in the figure. The shape of the distribution is shown in Figure 4.20(b) as the solid line. It is our objective to be able to determine this distribution.

Within the FEM context, the distributed load is represented by individual nodal forces, and for the mesh shown in Figure 4.20, this means that a total of 39 force values must be determined. Because the forces are close to each other, the data will have difficulty resolving between them, and, therefore, ill-conditioning becomes a problem.

Figure 4.20(b) shows the reconstructions for different amounts of regularization using about 70% of the data with no noise. It must be pointed out that even though the data are perfect the results would be meaningless if no regularization was used. The $\gamma = 10^{-12}$ case shows the problem: two neighboring forces, one sightly higher and one sightly lower than the mean, will have only a mean effect on the mostly remote data. The role of the regularization is to impose an *a priori* relationship on the force distribution; in

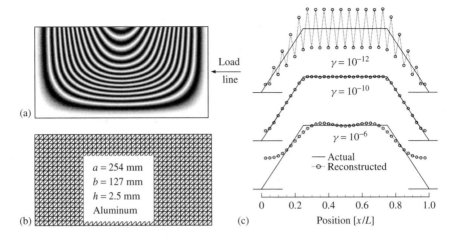

Figure 4.20: Plate, clamped on three sides, with a line loading. (a) Moiré displacement data. (b) Mesh and properties. (c) Reconstructed traction distribution for different amounts of regularization and no noise.

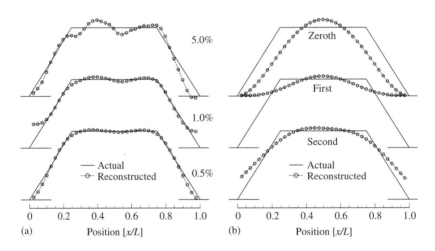

Figure 4.21: Reconstructed traction distributions. (a) Effect of different amounts of noise. (b) Effect of different orders of regularization with 5% noise and excessive γ.

this case essentially, those neighboring forces are not too different from each other. The consequence of this is shown in the other plots.

Figure 4.21(a) shows the effect of noise on the reconstructions. In each case, zero regularization with $\gamma = 10^{-8}$ was used and the noise was Gaussian, with the reported percent being one standard deviation of the average displacement value. The presence of noise can have a significant effect on the distribution but the method itself is robust to the presence of noise.

Generally, the order of regularization has little effect. The difference is seen only when there is excessive regularization; in that case, the effect is one of smoothing and insensitivity to noise. Specifically, zeroth order tends to a zero mean value, first order tends to a flat mean value, and second order tends to a parabolic shape. These are shown in Figure 4.21(b) for 5% noise. This can be quite useful in robustly gauging the overall pattern of a distribution.

Parameterized Loading Scheme

As pointed out, the use of regularization cures the ill-conditioning problem associated with determining many closely related forces, but it does not reduce the computational cost associated with the increased number of forces. That is the objective of parameterizing the load distribution.

The idea is simple: the traction distribution is considered to be comprised of a series of known distributions (with unknown scales) acting at control points as shown in Figure 4.22(b) for triangular distributions; that is,

$$t(x) = \sum_m \tilde{P}_m \phi_m(x)$$

where each $\phi_m(x)$ is a known or specified distribution. The control points need not be equispaced; indeed, they can be adjusted to reflect the gradients (if there is *a priori* information about them) in the tractions.

As a result, the traction distribution has a representation independent of the particular mesh. This distribution is then mapped to nodal forces for a particular mesh. The load vector $\{\phi\}_m$ sent to the FEM program will have multiple loaded nodes, but the treatment of the sensitivity response solution $\{\psi\}_m$ will be the same as in the examples already considered.

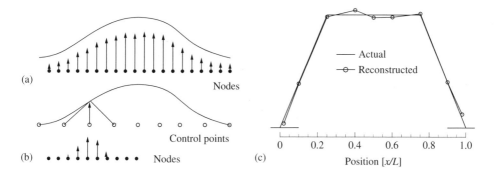

Figure 4.22: Concept of the parameterized loading scheme. (a) A traction distribution is represented as isolated forces acting at every node. (b) Segments of the traction distribution are represented by triangular distributions acting on the control points. These triangular distributions are then mapped to the nearest nodes. (c) Traction distributions for the plate problem.

There are two main advantages in using this method. First, as the method limits the number of unknown forces to a finite number of groups of forces, the computational cost will be controlled. Second, as the method uses only geometry information, a larger variety of traction distributions can be determined. For example, the tractions could form a curve on the model (a 3-D curve if the model is a shell) that does not necessarily coincide with any nodes—the resultant loads are just relocated to the nearest nodes. This method is also easily generalized to allow area distributions of tractions as occurs with pressure loading on a plate.

Figure 4.22(c) shows the reconstructions for the problem of Figure 4.20 when only 9 triangles are used. Clearly, the distribution is captured correctly. Note that the control points are not equispaced.

4.5 Nonlinear Data Relations

Photoelasticity has a nonlinear relationship between the observed fringes and the applied loads. To see this, recall that the fringes are related to the stresses by Equation (2.21)

$$\left(\frac{f_\sigma}{h} N\right)^2 = (\sigma_{xx} - \sigma_{yy})^2 + 4\sigma_{xy}^2$$

From this, we infer that while it is true that a doubling of the load causes a doubling of the stresses, and, therefore, a doubling of the fringe orders, it is not true for nonproportional loading cases; that is, the fringe pattern caused by two loads acting simultaneously is not equal to the sum of the fringe patterns of the separate loads even though the stresses are equal. Since our inverse method is based on superposition, we must modify it to account for this nonlinear relation.

Iterative Formulation

We need to obtain a linear relation between the fringe data and the applied loads. As is usual in nonlinear problems, we begin with a linearization about a known state, that is, suppose we have a good estimate of the stresses as $[\sigma_{xx}, \sigma_{yy}, \sigma_{xy}]_0$, then the true stresses are only a small increment away. Do a Taylor series expansion about this known state to get

$$\left(\frac{f_\sigma}{h} N\right)^2 \approx [(\sigma_{xx} - \sigma_{yy})^2 + 4\sigma_{xy}^2]_0 + 2[(\sigma_{xx} - \sigma_{yy})_0(\Delta\sigma_{xx} - \Delta\sigma_{yy}) + (4\sigma_{xy})_0\Delta\sigma_{xy}]$$

Let the stress increments $(\Delta\sigma_{xx}, \Delta\sigma_{yy}, \Delta\sigma_{xy})$ be computed from the scaled unit loads so that for all components of force, we have

$$\{\Delta\sigma_{xx}, \Delta\sigma_{yy}, \Delta\sigma_{xy}\} = \sum_j \{\psi_{xx}, \psi_{yy}, \psi_{xy}\}_j \Delta\tilde{P}_j$$

The stress sensitivities $\{\psi\}_j$ must be computed for each component of force. The stress-fringe relation becomes

$$\left(\frac{f_\sigma}{h}N\right)^2 \approx [(\sigma_{xx} - \sigma_{yy})^2 \; 4\sigma_{xy}^2]_o$$

$$+ \sum_j [2(\sigma_{xx} - \sigma_{yy})_o(\psi_{xx} - \psi_{yy})_j + 8(\sigma_{xy})_o\psi_{xyj}]\Delta\tilde{P}_j \qquad (4.5)$$

We could now set up an iterative procedure to obtain the increment of force. However, in anticipation of using regularization, we wish to deal directly with the force distributions. Noting that $\Delta\tilde{P} = \tilde{P} - \tilde{P}_o$, we write the approximate fringe relation as

$$\left(\frac{f_\sigma}{h}N\right)^2 \approx \sigma_o^2 - \sum_j G_j\tilde{P}_{oj} + \sum_j G_j\tilde{P}_{oj} \qquad (4.6)$$

with

$$\sigma_o^2 \equiv [(\sigma_{xx} - \sigma_{yy})^2 + 4\sigma_{xy}^2]_o \,, \qquad G_j \equiv [2(\sigma_{xx} - \sigma_{yy})_o(\psi_{xx} - \psi_{yy})_j + 8(\sigma_{xy})_o\psi_{xyj}]$$

Equation (4.6) is our linear relation between the fringes and the applied loads.

Let the data be expressed in the form $\overline{N} \equiv Nf_\sigma/h$ so that we can express the data relation as

$$\overline{N}^2 = Q\sigma_o^2 - Q\sum_j G\tilde{P}_o + Q\sum_j G\tilde{P} \qquad \text{or} \qquad [QG]\{\tilde{P}\} = \{\overline{N}^2 - Q\sigma_o^2 + QG\tilde{P}_o\}$$

where Q plays the role of a selector. Establishing the overdetermined system and reducing then leads to

$$[(QG)^T W QG + \gamma H]\{\tilde{P}\} = [QG]^T[\ W\ ]\{\overline{N}^2 - Q\sigma_o^2 + QG\tilde{P}_o\}$$

This is solved repeatedly using the newly computed $\{\tilde{P}\}$ to replace $\{\tilde{P}\}_o$.

In terms of the computer implementation as shown in Figure 4.23, the sensitivity solution $\{\psi\}$ is determined only once for each load component, but the updating of the stress field $\{\sigma_o\}$ requires an FEM call on each iteration. Both the left- and the right-hand side of the system of equations must be re-formed (and hence decomposed) on each iteration.

Initial Guesses and Regularization

In all iterative schemes, it is important to have a method of obtaining good initial guesses. There are no hard rules for the present type of problem; however, the rate of convergence

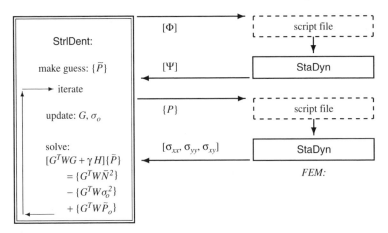

Figure 4.23: Schematic of the distributed computing relationship between the inverse program and the FEM programs when updating is needed.

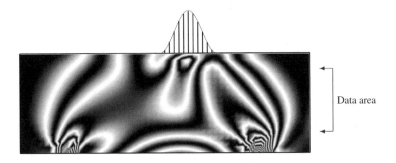

Figure 4.24: Photoelastic fringe pattern with distributed load.

depends on the number of unknown forces; this suggests using fewer forces to obtain an idea of the unknowns to use as starter values. Furthermore, the effect of regularization is sensitive to the number of unknowns, and, therefore, it would be good to obtain the initial guesses without regularization.

Figure 4.24 shows a synthetic fringe pattern for a block with a distributed load. The data used in the analysis is collected in areas away from the loaded and supported edges. Figure 4.25 shows the results of finding force pairs at neighboring nodes at many node locations. No regularization ($\gamma = 0.0$) was used. Shown is the sum and difference of the forces. The sum gives an idea of the resultant force of the actual distribution—for the present distribution, the value should be 6.0 on the scale shown. The force difference gives an idea of the location of the resultant. The separation between the force pairs depends on the ability of the data to resolve the forces without regularization. In the present test, the closest data to the load was a distance of three nodes away and therefore the separation used was four node spacings.

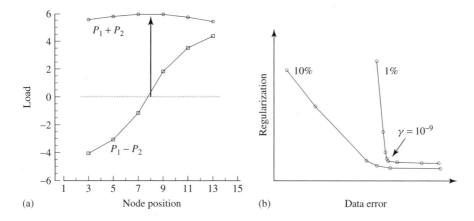

Figure 4.25: Photoelastic results. (a) Force pair scan. (b) Trade-off curves for different amounts of noise.

This information can now be used to make a reasonable initial guess; indeed, using just the resultant and its location (all other forces being zero) is a good initial guess. Convergence occurred in about five iterations for each of these two-force problems.

An alternative way of obtaining an initial guess is to linearize the data relation, that is, assume

$$\sigma_{xx} - \sigma_{yy} = N f_\sigma / h , \qquad \sigma_{xy} = 0$$

This involves additional programming, but in some cases (for example, where no idea of the resultant can be obtained), it gives a reasonable first guess. We will actually use this idea in Chapter 6 when dynamic photoelasticity is considered. It is also noted that if a single force is the only unknown, then the photoelastic problem can be phrased as a linear data relation since the data can be related to the sensitivity responses by

$$\frac{f_\sigma}{h} N = \sqrt{(\psi_{xx} - \psi_{yy})^2 + 4\psi_{xy}^2} \; \tilde{P}$$

We are interested in the multiload cases, but this could be used to obtain initial guesses of resultants.

The next step is to determine the amount of regularization. Whatever supplementary information is available should be used. A useful idea is to treat some knowns as unknowns and make a comparison. For example, suppose we know that the tractions are nonzero over only a limited portion of the surface, then we can also ask for the tractions over a part of the zero traction area. When the solution is good, this portion will have essentially zero tractions, its variation being the effect of the noise.

Figure 4.26 shows the force reconstructions (using 15 nodal force values) for three values of zero-order regularization. Too little regularization shows the effect of the noise, too much shows oversmoothing. A somewhat objective judgment of the optimum can be found from the trade-off curve of Figure 4.25(b); the optimum is at the elbow. A

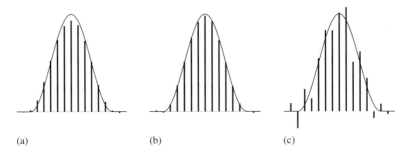

(a)   (b)   (c)

Figure 4.26: Force reconstructions for different amounts of regularization with noise of 1%. (a) Too much regularization. (b) Reasonable amount of regularization. (c) Too little regularization.

subjective judgment of the optimum can be made by looking at the reconstructions outside the nonzero portion. In traction problems like these, the positivity of the tractions could be imposed as an additional constraint.

Experimental Studies

Figure 4.27 shows the loading arrangement for a photoelastic model with a hole. The specimen is essentially under three-point bending. The inset shows the light-field fringe pattern. The fringes were recorded using a diffuse light polariscope with white light; the gray scale image shown is that recorded by the camera. Although there is some washout of the fringes near the concentration points (load and support points plus edge of hole), the fringe patterns are deemed adequate because we will mostly use data away from these points.

The objective is to determine the stresses in the specimen, and in particular, the stresses around the edge of the hole. As an intermediate step, we first determine the

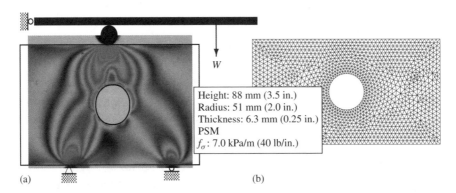

Height: 88 mm (3.5 in.)
Radius: 51 mm (2.0 in.)
Thickness: 6.3 mm (0.25 in.)
PSM
f_σ: 7.0 kPa/m (40 lb/in.)

(a)   (b)

Figure 4.27: Concentrated load experiment. (a) Photoelastic fringe data and dead-weight loading scheme. (b) Mesh and properties.

loadings on the specimen. We do two tests: the first determines isolated forces, the second determines a traction distribution. For the former, the multiple loads correspond to the applied load and the support reaction; for the latter, the applied point load is replaced with a distributed load.

Experiment I: Uncorrelated Forces

Figure 4.27(b) shows the mesh and material properties. Note that the mesh does not exactly coincide with the full length of the actual model; we only mesh the region in the vicinity of the loads since remote regions have essentially zero stress.

The fringe data of Figure 4.27 were discretized in a manner similar to that of the Moiré fringes, that is, the center of the dark fringe was manually identified at a number of points and this data was relocated to the nodes in its vicinity. Only fringes that are clearly visible were discretized. Figure 4.28 shows this discretized data.

When input data is distributed over an area as in the present case, a natural question that arises is the effect of data location on the results. The model was divided into two data regions as shown in Figure 4.28: a region close to the top load, and a region remote from it but close to the bottom supports. Table 4.3 shows the reconstructed

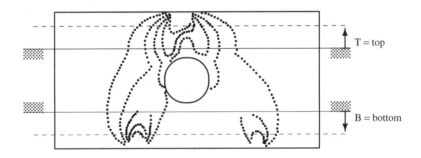

Figure 4.28: Sampling regions supposed on the discretized database.

Table 4.3: Estimations of applied load using different data regions. Left: single top load, Right: two loads.

Region	50	100	150	200	Region	50	100	150	200
T: P_1	80.1	79.2	79.5	78.1	T: P_1	76.0	76.0	77.5	77.3
P_2					P_2	21.3	22.4	24.4	26.2
B: P_1	82.4	84.0	80.2	77.8	B: P_1	83.7	82.3	80.7	78.7
P_2					P_2	29.7	28.7	29.5	29.0
T&B: P_1	80.5	84.0	80.2	77.8	T&B: P_1	80.1	80.2	79.8	78.7
P_2					P_2	29.0	29.3	29.0	28.7
Nom: P_1	80.0	80.0	80.0	80.0	Nom: P_1	80.0	80.0	80.0	80.0

force scale against the amount of data in each region. No regularization was used in the reconstructions. There is a variation in the force but the results show that the method is quite robust.

Rather than set up an experimental arrangement in which two independent forces are applied, we will, instead, take the right boundary support as an additional unknown force. Table 4.3 shows the reconstructed force scale again against different amounts of data. The method gives quite good results for the two forces. We conclude that reasonably good estimates of the loading can be obtained. Furthermore, this need not come from immediately near the load points.

Experiment II: Traction Distribution

Figure 4.29(a) shows the recorded fringe pattern where the load is distributed over an area. It is not the intention here to determine the precise distribution of this load—to do this would require collecting data very close to the distribution. Rather, we are still concentrating on doing a stress analysis in the region close to the hole.

The distributed load was achieved by placing a soft layer between the model and the applied load; consequently, there was no possibility of *a priori* knowing the distribution.

The results to be reported used data taken from both the top and the bottom regions. Figure 4.30(a) shows an estimate of the resultant force as discussed previously. The results clearly indicate the center of the resultant.

The initial guess for the iterative scheme was taken as the resultant load with all others being zero. Figure 4.30(b) shows the reconstructed distribution. The precise distribution depends on the amount of regularization used. A key point, however, is that the solution in the vicinity of the hole does not depend sensitively on the precise nature of the distribution of applied load, that is, all reconstructed distributions give nearly the same solutions in the vicinity of the hole.

Figure 4.29(b) shows the reconstructed fringe pattern corresponding to the distribution of Figure 4.30(b). The patterns close to the hole are almost identical. The patterns near the distribution show a difference that might be due to the presence of a shearing traction.

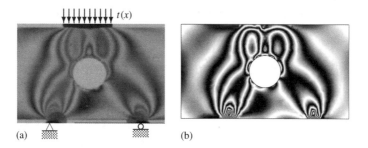

Figure 4.29: Fringe pattern with distributed load. (a) Experimentally recorded. (b) Reconstructed.

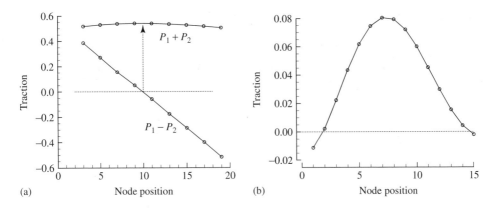

Figure 4.30: Iterative results. (a) Estimation of resultant load. (b) Traction distribution.

Discussion of Photoelastic Results

The photoelastic data bear a nonlinear relation to the mechanical variables, and, therefore, a nonlinear iterative scheme is required when solving the inverse problem. The results just presented show that this can be done robustly.

One aspect not explored, but which has an implication for the general use of photoelasticity, is the ambiguity arising when both a normal and a shear traction are applied and unknown. As seen from the photoelastic relation of Equation (2.21), a σ_{xx} distribution would give the same pattern (in the vicinity of the boundary) as a σ_{xy} distribution of the same shape. If the two stress systems cause different remote stresses (on a beam, for example, the shear would cause bending, while the other would cause uniform axial behavior), the remote measurements could be used to distinguish the two loadings. For the distributed load case discussed, however, the shear would have an overall zero resultant and hence remote measurements would not help.

It is possible that supplementary data such as isoclinics could remove the ambiguity; but in any event, this is an aspect that would warrant further investigation if photoelasticity is to be used to determine general traction distributions.

On the basis of this discussion, it may seem that photoelasticity is not as attractive as other methods. This is perhaps true, but the main reason it is included in this book is as an example of data with a nonlinear relation to the mechanical variables. It may well be that other sensors of the future will also have such a nonlinear relationship, and the algorithms developed here will be able to take advantage of them. For example, there is currently a good deal of activity in developing pressure-sensitive paints, fluorescent paints, and the like; see References [114, 152] for a discussion of their properties and some of the literature. These inexpensive methods, most likely, will have a nonlinear relation between the data and the mechanical variables.

A final point along these lines: in Moiré the relation between fringe intensity and displacement is nonlinear. Thus, using the above iterative approach, it may be possible to use the intensity data directly, and skip the whole process of fringe numbering.

4.6   Parameter Identification Problems

All the previous identification problems focused on loads as the basic unknown to be identified. Even in the case of the pressure-loaded plate of Figure 4.17, the unknown boundary condition was replaced with an unknown load. What we will consider here is the direct determination of structural parameters. Because the basic approach depends on response measurements, we will take the applied loads causing the responses as known.

A Test Example

Consider the simple structures shown in Figure 4.31, which is modeled with two elements. From measurements at various locations, we wish to determine the modulus and height of the member. This example will be used to illustrate some potential pitfalls that might be encountered when multiple parameters are to be identified.

I: Independent Test Data

We begin by solving the forward problem for the rod of Figure 4.31(a). The reduced element stiffnesses are

$$[k^{12}] = \frac{EA}{x_p}[\ 1\ ], \qquad [k^{23}] = \frac{EA}{L - x_p}[\ 1\ ]$$

The assembled system of equations is

$$\frac{EA}{L}\left[\frac{L}{x_p} + \frac{L}{L - x_p}\right]\{u_2\} = \{P_u\} \qquad \text{or} \qquad [\ K\ ]\{u\} = \{P\}$$

This is solved to give the nodal response and displacement distribution as

$$\{u_2\} = \frac{P_u L}{EA}\left\{\frac{L^2}{x_p(L - x_p)}\right\}, \qquad u(x) = \frac{P_u L}{EA}\left[\frac{L^2}{x_p(L - x_p)}\right]\left[\frac{x}{x_p}\right], \qquad x < x_p$$

A similar linear distribution holds to the right of the load where $x > x_p$.

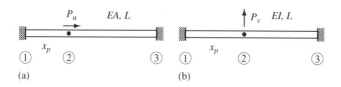

Figure 4.31: Two independent load cases. (a) Member as rod with axial load. (b) Member as beam with transverse load.

Suppose the response at some point is known, $d_i = u(x_i)$, and all structural parameters except E are known; then it is clear that E can be easily determined. Suppose, however, we wish to determine both E and A; it is equally clear that no number of measurements are sufficient because E and A do not appear in an independent way.

Now consider the forward problem for the beam of Figure 4.31(b). The reduced element stiffnesses are

$$[k^{12}] = \frac{EI}{x_p^3} \begin{bmatrix} 12 & -6x_p \\ -6x_p & 4x_p^2 \end{bmatrix}, \qquad [k^{23}] = \frac{EI}{(L - x_p)^3} \begin{bmatrix} 12 & 6(L - x_p) \\ 6(L - x_p) & 4(L - x_p)^2 \end{bmatrix}$$

The assembled system of equations is

$$\frac{EI}{L^3} \left[\frac{L^3}{x_p^3} \begin{bmatrix} 12 & -6x_p \\ -6x_p & 4x_p^2 \end{bmatrix} + \frac{L^3}{(L - x_p)^3} \begin{bmatrix} 12 & -6(L - x_p) \\ -6(L - x_p) & 4(L - x_p)^2 \end{bmatrix} \right] \begin{Bmatrix} v_2 \\ \phi_2 \end{Bmatrix} = \begin{Bmatrix} P_v \\ 0 \end{Bmatrix}$$

This can be solved to give the response; however, as it is too messy to write out explicitly, we will write it symbolically as

$$\begin{Bmatrix} v_2 \\ \phi_2 \end{Bmatrix} = [K]^{-1}\{P\} = \frac{L^3}{EI}[G] \begin{Bmatrix} P_v \\ 0 \end{Bmatrix}$$

The displacement at any point to the left of the load is

$$v(x) = g_3(x)v_2 + g_4(x)x_p\phi_2$$

where $g_i(x)$ are the beam shape functions from Chapter 1. A similar relation holds for the deflection to the right of the load but it is written in terms of $g_1(x)$ and $g_2(x)$. Similar reasoning as for the rod shows that it is not possible to determine both E and I separately no matter how many measurements are available.

Let the depth of the beam be known so that the cross-sectional properties can be written in terms of the height as

$$A = bh, \qquad I = bh^3/12$$

Now we have the important question as to whether measurements from axial and transverse loading cases can give simultaneous evaluation of E and h. Let the two-load points coincide, and let the measurements be taken at the load point; then the two measurements are represented as

$$u_2 = \frac{1}{Eh}\frac{P_u L}{b}\left(\frac{L^2}{x_p(L - x_p)}\right) = \frac{1}{Eh}P_u Q_u, \qquad v_2 = \frac{1}{Eh^3}\frac{P_v 12 L^3}{b}G_{11} = \frac{1}{Eh^3}P_v Q_v$$

Solving simultaneously for E and h gives

$$h = \sqrt{\frac{P_v}{P_u} \frac{Q_v}{Q_u} \frac{u_2}{v_2}}, \qquad E = \frac{P_u Q_u}{u_2} \sqrt{\frac{P_u}{P_v} \frac{Q_u}{Q_v} \frac{v_2}{u_2}}$$

The identification was successful because the thickness entered the responses differently for the two-load cases, that is, in terms of the zookeeper scenario of Section 3.5, "rods and beams have different dependence on the thickness."

To emphasize the point of the need for independent load cases, this particular identification problem can be arranged as two separate independent identification problems: in the first EA is identified, and in the second EI is identified. The separated parameters are then obtained by combining both results

$$h = \sqrt{\frac{12\,EI}{EA}}, \qquad E = \frac{EA}{b} \sqrt{\frac{EA}{12\,EI}}$$

However, for a given loading case, changing the position of the loading or of the measurement does not give independent data as regards E and h.

In the following developments, we will continue to use this simple example but treat it as if the parameters must be identified simultaneously.

II: Implicit Parameters and Sensitivity Responses

What we mean by implicit parameters is that we cannot write an explicit relation that contains the parameters, for example, suppose the load position x_p of Figure 4.31(b) is the unknown parameter; the stiffness of that member is

$$[k(x_p)] = \left[\frac{EI}{x_p^3} \begin{bmatrix} 12 & -6x_p \\ -6x_p & 4x_p^2 \end{bmatrix} + \frac{EI}{(L-x_p)^3} \begin{bmatrix} 12 & -6(L-x_p) \\ -6(L-x_p) & 4(L-x_p)^2 \end{bmatrix} \right]$$

The stiffness is a complicated function of x_p, but x_p appears explicitly in the expression. Suppose now that this member is part of a large frame (with 50 members, say), the assembled structural stiffness matrix $[\,K\,]$ would also be a function of x_p, but it would be very cumbersome to write the explicit function. It would not be impossible, but because of the computer-based assemblage procedure that makes FEM so versatile, it would be very difficult to extract that single member from the rest of the members. We say that the assembled structural stiffness $[\,K\,]$ is an *implicit function* of x_p, and x_p is an *implicit parameter*.

When we do not have explicit access to parameters, we must infer their effect indirectly by looking at their effect on the responses. To this end, we assume that an FEM program is available that can provide solutions (responses) given a set of applied loads; we will then construct sensitivity responses from these solutions.

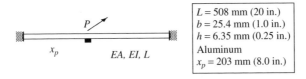

Figure 4.32: Identification of Modulus and height.

Consider the problem shown in Figure 4.32 in which the sensor can measure both u and v displacements and the load is applied at x_p. We will treat this simple problem as if the unknown parameters h and E are implicit.

With properties (including both loads) specified and with assumed parameters h^o and E^o, compute the load point responses

$$u_1^o = \frac{1}{E^o h^o} P_u Q_u, \qquad v_1^o = \frac{1}{E^o h^{o3}} P_v Q_v$$

Let u_1 and v_1 be the measured load point responses. Then the χ^2 error function is

$$\chi^2 = \left[u_1 - u_1^o\right]^2 + \left[v_1 - v_1^o\right]^2$$

Figure 4.33(a) shows contours of χ^2 when both components of force are equal. There are a number of interesting features about this plot. First, there is no discernible minimum; there is a band of minima that is given approximately by $Eh^3 = $ constant. This says that the data cannot distinguish between the different load cases. When both load components are equal, the ratio of the responses is $v/u \approx 1700$. Consequently, the error function is dominated by the transverse displacement.

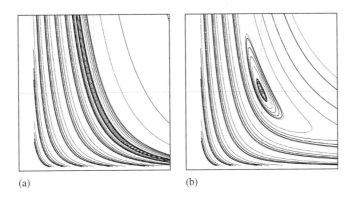

      (a)   (b)

Figure 4.33: Contours of error function over the parameter search space $2h$ (horizontal) and $2E$ (vertical). (a) Both load components are equal giving very unequal responses. (b) Both responses are about equal.

A second feature is that the error function is exceedingly large to the left and to the bottom of the minima—regions where the stiffness is very low. This indicates that in structural analyses, the search should be biased away from the low stiffness quadrant.

Figure 4.33(b) shows contours of χ^2 when the components of force were adjusted so that the two responses (and not the two loads) were nearly equal. Now, there is a definite isolated minimum. In practice, it would probably not be feasible to adjust the loads in this manner, but the same effect can be achieved by applying weighting factors to the data; we will discuss more about this later.

III: Iterative Formulation

The topology of the contours of Figure 4.33(b) (with a single, strong minimum) shows that it should be possible to construct an iterative search method to find the minimum. We will use a variation of the Newton–Raphson method.

The FEM program needs to be run three times. The first time, with the specified properties (including both loads) and assumed parameters h^o and E^o, we compute the responses

$$u_1^o = \frac{1}{E^o h^o} P_u Q_u, \qquad v_1^o = \frac{1}{E^o h^{o3}} P_v Q_v$$

In the second run, h is changed by a specified amount Δh, and we compute the two responses

$$u_1^o + \Delta u_1 = \frac{1}{E^o[h^o + \Delta h]} P_u Q_u \approx u_1^o - \frac{\Delta h}{E^o h^{o2}} P_u Q_u$$

$$v_1^o + \Delta v_1 = \frac{1}{E^o[h^o + \Delta h]^3} P_v Q_v \approx v_1^o - 3\frac{\Delta h}{E^o h^{o4}} P_v Q_v$$

The sensitivities for each response are computed as

$$\psi_{uh}^o = \frac{\Delta u_1}{\Delta h} = -\frac{1}{E^o h^{o2}} P_u Q_u, \qquad \psi_{vh}^o = \frac{\Delta v_1}{\Delta h} = -\frac{3}{E^o h^{o4}} P_v Q_v$$

In the third run, relative to the first run, E is changed by a specified amount ΔE, and we compute the two responses

$$u_1^o + \Delta u_1 = \frac{1}{[E^o + \Delta E]h^o} P_u Q_u \approx u_1^o - \frac{\Delta E}{E^{o2} h^o} P_u Q_u$$

$$v_1^o + \Delta v_1 = \frac{1}{[E^o + \Delta E]h^{o3}} P_v Q_v \approx v_1^o - \frac{\Delta E}{E^{o2} h^{o3}} P_v Q_v$$

The sensitivities for each response are

$$\psi_{uE}^o = \frac{\Delta u_1}{\Delta E} = -\frac{1}{E^{o2} h^o} P_u Q_u, \qquad \psi_{vE}^o = \frac{\Delta v_1}{\Delta E} = -\frac{1}{E^{o2} h^{o3}} P_v Q_v$$

The sensitivities depend on the load level, and, therefore, the load levels must be those of the experiment. The sensitivities also depend on the current values of guessed parameters and therefore must be updated as the guesses of the parameters are updated.

A comparison with the measured data can now be made

$$d_1 = u_1 \longleftrightarrow u_1^o + \Delta u_1 = u_1^o + \psi_{uh}^o \Delta h + \psi_{uE}^o \Delta E$$

$$d_2 = v_1 \longleftrightarrow v_1^o + \Delta v_1 = v_1^o + \psi_{vh}^o \Delta h + \psi_{vE}^o \Delta E$$

The χ^2 error function is

$$\chi^2 = \left[d_1 - u_1^o - \psi_{uh}^o \Delta h - \psi_{uE}^o \Delta E \right]^2 + \left[d_2 - v_1^o - \psi_{vh}^o \Delta h - \psi_{vE}^o \Delta E \right]^2$$

This is reduced to normal form by minimizing with respect to the parameter increments to give the system of simultaneous equations

$$\begin{bmatrix} \psi_{uh}^o & \psi_{vh}^o \\ \psi_{uE}^o & \psi_{vE}^o \end{bmatrix} \begin{bmatrix} \psi_{uh}^o & \psi_{uE}^o \\ \psi_{vh}^o & \psi_{vE}^o \end{bmatrix} \left\{ \begin{matrix} \Delta h \\ \Delta E \end{matrix} \right\} = \begin{bmatrix} \psi_{uh}^o & \psi_{vh}^o \\ \psi_{uE}^o & \psi_{vE}^o \end{bmatrix} \left\{ \begin{matrix} d_1 - u_1^o \\ d_2 - v_1^o \end{matrix} \right\}$$

After solving for the increments, the parameters are updated, new values of responses and sensitivities are computed, and the process is repeated until convergence.

As could be inferred from the contours of Figure 4.33(b), convergence is quite robust as regards initial guess. However, as shown in Figure 4.34, convergence is affected by the approximation of the sensitivities—the percentages are for Δh and ΔE as a fraction of the full search space $[2h, 2E]$. (The initial guesses were taken at 20% of the true value.) If the discretization is too fine at the beginning, convergence is slow; on the other hand, if the discretization is too coarse at the end, then convergence is never achieved.

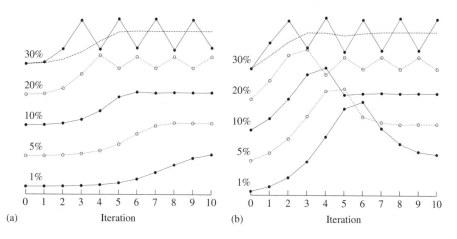

(a) Iteration   (b) Iteration

Figure 4.34: Convergence performance for different approximations of the sensitivities. (a) Thickness *h*. (b) Modulus *E*.

The heavy dashed line for the 30% case used the update rule

$$\text{new} \longleftarrow [(\text{old} + \Delta) + \text{old}]/2$$

which averages the new estimate with the old updated value; the result is somewhat similar (but not identical) to the 15% case. Clearly, there are possibilities for devising a strategy that adaptively improves the sensitivities.

IV: Other Iterative Formulations

The above scheme is essentially the sensitivity response method introduced in Section 3.4; however, it is not the only scheme to make the implicit parameters visible. Reference [105] suggests computing the sensitivity from the equilibrium relation, that is, treating "a_j" as a typical parameter, and then by differentiation

$$[K]\{u\} = P \quad \Longrightarrow \quad [K]\left\{\frac{\partial u}{\partial a_j}\right\} + \left[\frac{\partial K}{\partial a_j}\right]\{u\} = 0$$

Finite difference is used on the stiffness derivative, so the sensitivity is computed from

$$[K]\left\{\frac{\partial u}{\partial a_j}\right\} = -\left[\frac{\partial K}{\partial a_j}\right]\{u\} \quad \text{with} \quad \left[\frac{\partial K}{\partial a_j}\right] \approx \frac{[K(a_j + \Delta a_j)] - [K(a_j)]}{\Delta a_j}$$

Reference [105] refers to this as the *direct differentiation method* (DDM).

Figure 4.35 shows the convergence behavior for the same cases as in Figure 4.34. The character of the convergence is different, but most importantly, it does not show the oscillation at the high value of Δa. This is a significant attribute because in a general case, with many unknown parameters, it may not always be possible to know in advance what a suitable search space should be.

The disadvantage of the method is that it is necessary to have access to the stiffness matrix. In Reference [105], this is not a problem because it is dealing with modification of FEM codes. As discussed earlier in this chapter, however, this is not something regularly available to a user of a commercial FEM package. Another disadvantage is that the product

$$\{\phi\}_j = \left[\frac{\partial K}{\partial a_j}\right]\{u^o\}$$

must be formed. For the product to be accomplished efficiently, it is necessary that the sparsity properties of the stiffness matrix be taken into account. In any event, the result is then a load applied to every node.

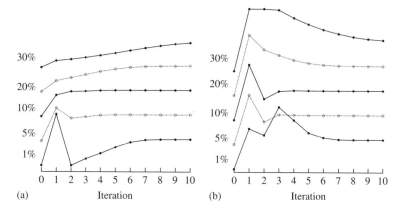

Figure 4.35: Convergence performance for different approximations of the sensitivities using the direct differentiation method. (a) Thickness h. (b) Modulus E.

Both drawbacks suggest that it might be beneficial to find alternative ways of computing $\{\phi\}_j$. We will now pursue this, and in the process, get a better understanding of the iterative method. Expand the finite difference approximation to

$$\{\phi\}_j \approx \frac{[K(a_j^o + \Delta a_j)] - [K(a_j^o)]}{\Delta a_j}\{u^o\}$$

$$= \frac{1}{\Delta a_j}\Big[[K(a_j^o + \Delta a_j)]\{u^o\} - [K(a_j^o)]\{u^o\}\Big] = \frac{1}{\Delta a_j}\Big[\{F\} - \{P\}\Big]$$

The vector $\{P\}$ is the actual applied load that was used in conjunction with $[K(a_j^o)]$ to obtain $\{u^o\}$. The vector $\{F\}$ is a set of nodal loads associated with the new structure $[K(a_j^o + \Delta a_j)]$ with the displacements $\{u^o\}$. We see that $\{\phi\}_j$ are essentially a set of imbalance forces caused by the mismatch of the approximate and exact structure. The form is reminiscent of the loading equation for nonlinear structures as covered in Chapter 1 and as we will see again in Chapter 7.

With this view of $\{F\}$, an alternative way to compute $\{\phi\}_j$ suggests itself: postprocess the FEM solution $\{u^o\}$ to get the element nodal forces, but use the structure $[K(a_j^o + \Delta a_j)]$ instead of $[K(a_j^o)]$. In this way, the inverse solution need not know the inner storage mechanism used by the FEM program.

The advantages of the DDM over the SRM are not compelling to warrant the extra computational effort. By contrast, the SRM has the compelling advantage that precisely the same method is used irrespective of whether the problem is linear or nonlinear, static or dynamic. Consequently, this will be the method implemented.

Computer Implementation of the SRM

We will generalize the programming to the case of many measurements and parameters (and hence the possible need for regularization). Since the parameters can be of very different types, we will also include scaling of the data and the parameters.

Let us have a set of guesses, a_j^o, for the unknown parameters, and solve the system

$$[\ K\ ]_0\{u\}_0 = \{P\}$$

In turn, change each of the parameters (one at a time) by the definite amount $a_j \longrightarrow a_j + da_j$ and solve

$$[\ K\ ]_j\{u\}_j = \{P\}$$

The sensitivity of the solution to the change of parameters is constructed from

$$\{\psi\}_j \equiv \left\{\frac{u_j - u_0}{da_j}\right\} \bar{a}_j$$

where $\bar{a}_j$ is a normalizing factor, typically based on the search window for the parameter. The small change da_j should also be based on the allowable search window; typically it is taken as 5% but in some circumstances it could be chosen adaptively.

The response solution corresponding to the true parameters is different from the guessed parameter solution according to

$$\{u\} = \{u\}_0 + \sum_j \left\{\frac{u_j - u_0}{da_j}\right\} \Delta a_j$$

$$= \{u\}_0 + \sum_j \left\{\frac{u_j - u_0}{da_j}\right\} \bar{a}_j \frac{\Delta a_j}{\bar{a}_j} = \{u\}_0 + \sum_j \{\psi\}_j \Delta \tilde{P}_j$$

It remains to determine Δa_j or the equivalent scales $\Delta \tilde{P}_j$.

Replacing $\Delta \tilde{P}_j$ with $\tilde{P}_j - \tilde{P}_j^o$, and using subscript i to indicate measurements d_i, we can write the error function as

$$\mathcal{A} = \sum_i W_i \left\{d_i - Qu_{0i} + \sum_j Q\psi_{ij}\tilde{P}_j^o - \sum_j Q\psi_{ij}\tilde{P}_j\right\}^2$$

where Q acts as a selector. This is reduced to normal form by minimizing with respect to $\tilde{P}_j$ and by adding regularization to give

$$\left[[Q\Psi]^T \lceil\ W\ \rfloor[Q\Psi] + \gamma[\ H\ ]\right]\{\tilde{P}\} = [Q\Psi]^T \lceil\ W\ \rfloor\{d - Qu_0 + Q\psi\tilde{P}^o\}$$

The system of equations is solved iteratively with the new parameters used to compute updated responses.

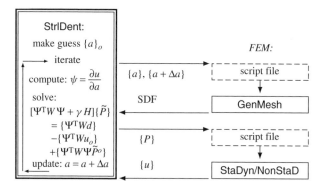

Figure 4.36: Implementation of the sensitivity response method (SRM) for parameter identification. Schematic representation of the distributed computing relationship between the inverse program and the FEM programs when updating is needed.

An implementation of this approach is shown in Figure 4.36: **GenMesh**, the FEM modeler program, produces the changed structure data files (SDF) that are then used in the FEM programs. Its significant advantage is that it is independent of the underlying problem (linear/nonlinear, etc.). There is a lot in common with the force identification scheme of Figure 4.23 when we make the associations $[\ G\ ] \Leftrightarrow [\ \Psi\ ]$. This is not surprising since both are working with a nonlinear data relationship; the difference is that in Figure 4.23, the nonlinear relationship is known explicitly and hence the functional form for $[\ G\ ]$ is also known explicitly. This is not so for $[K(a_j)]$.

Examples

We look at two examples. The first uses synthetic data and tests the ability of the method to determine multiple unknown parameters. The second example revisits the experimental problem of Figure 4.17 and treats the boundary elasticity as the unknown parameter.

I: Layered Materials

As an application, consider a multilayered material cut into the shape of a beam and loaded in a three-point bending as shown in Figure 4.37. The objective is to determine the six moduli distributed through the thickness. These moduli range in value from 100% on the outside, to 60% on the inside with the distribution shown in the figure; the 100% value corresponds to aluminum.

One advantage of the whole-field optical methods is that they can provide an excess of data—this is a case in which it can be put to good use. Figure 4.38(a) shows synthetic Moiré fringes for the beam. The load was purposely put off center.

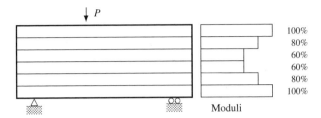

Figure 4.37: Layered beam in a three-point bending.

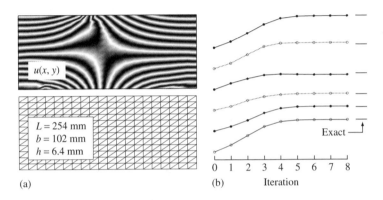

Figure 4.38: Layered beam in three-point bending. (a) Synthetic Moiré data and the mesh. (b) Convergence behavior with 1% noise.

Figure 4.38(b) shows the convergence behavior when 1% noise was added. Generally, convergence was quite rapid and fairly robust as regards initial guess. For the case shown, all moduli had an initial guess of 20% of the maximum modulus.

There are many variations of this problem that can be performed. One interesting variation would be to consider a beam with continuously varying properties—a gradient material. For a certain mesh discretization, the moduli would be represented by a large number of values that would normally lead to ill-conditioning and erratic results. However, by including some regularization on the moduli the problem would be made well posed. The same approach can also be used even if the material properties have a two-dimensional distribution. This could be quite useful for determining the properties of biological systems.

II: Boundary Condition using Experimental Data

Earlier, when we considered the experimental problem in Figure 4.17, the elastic boundary condition was treated as an overhang with specified properties but with unknown applied pressure. This reduced the problem to identifying two unknown loads.

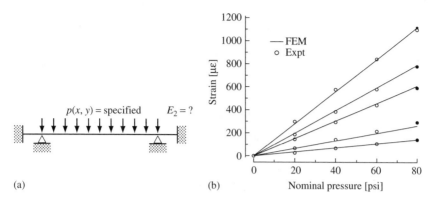

Figure 4.39: Modeling of plate with elastic boundary conditions. (a) Overhang has unknown modulus. (b) Reconstruction of strains.

Figure 4.39(a) models the problem differently; the overhang plate has specified thickness but unknown modulus. Since we will treat this problem as a parameter identification problem, the pressure load will be taken as specified.

With E_2 as the single unknown, the converged result gave $E_2 = 0.26E_1$ and the reconstructions are shown in Figure 4.39(b); the standard deviation in the strain is $18\,\mu\epsilon$ and the results are about the same as the improved results of Figure 4.19(b). The results hardly changed when both moduli are assumed unknown.

4.7 Choosing the Parameterization

The goal of the experimental work is to complete the specification of the problem. As discussed in the Introduction, there are a variety of unknowns in the problem, and, consequently, there is a choice as to how to parameterize these unknowns. This section, in the light of the examples covered so far, attempts to frame the issues involved in choosing the appropriate parameters, but first we recap the context for the discussion.

Consider the typical static FEM problem stated as

$$[K(a_j)]\{u\} = \{P\}$$

The stiffness is indicated as depending on a number of parameters a_j. From a partially specified problem point of view, the only allowable unknowns are some of the parameters a_j and some of the applied loads $\{P\}$. Thus, quantities of interest such as stresses or deformed shapes are not considered basic unknowns—once the problem is fully specified these are obtained as a postprocessing operation.

As indicated, there are two obvious cases of identification: one is where the applied load is to be identified—this is explicit, the other is when a property (modulus, say) is to be identified—this is implicit. To contrast the two approaches: the explicit formulation is accomplished in one computational pass but often requires many sensors, the implicit formulation requires many iterations but usually the number of unknowns (and hence required number of sensors) is small. But as shown in the Introduction, force identification and parameter identification are often interchangeable problems; as to which approach is preferred will depend on the specifics of any particular problem. If the Young's modulus is unknown, this is an obvious choice as an implicit parameter. If the boundary condition is unknown, then we can parameterize it as a spring or as a load as indicated in Figure 4.18(c). We will discuss further examples of both.

In other cases, the parameters are associated with a particular modeling. For example, the stress intensity factors K_1 and K_2 are associated with the LEFM (Linear Elastic Fracture Mechanics) analysis of cracks but actually do not appear in an FEM model of the same problem. Thus an interesting situation arises as to how to use FEM modeling as part of the parameter identification when the parameter does not appear, either explicitly or implicitly, in the FEM model. We will revisit the photoelastic analysis of cracks of Section 3.3 to discuss this point.

I: Parameterized Loadings

When the loading is distributed, it may require a number of parameters to characterize it, and there are a number of situations that may prevail depending on our *a priori* information and the number of sensors available.

If there are a good number of sensors, then the loading on each node can be considered the unknown. If there are fewer sensors but the distribution is expected to be somewhat smooth, then again the loading on each node can be considered the unknown, and regularization is used to make the solution well behaved; this was demonstrated in Section 4.4.

If the number of unknowns is computationally too large, then the parameterized loading scheme of Section 4.4 can be used. Actually, this is part of a more generalized approach in which the loading is assumed to be synthesized from specified forms; that is, let the nodal loadings be

$$[\,K\,]\{u\} = \{P\} = \sum_j \{\phi\}_j \tilde{P}_j$$

where $\{\phi\}_j$ are the specified distributions, $\tilde{P}_j$ their scalings, and the number of distributions is considerably less than the number of loaded nodes. The solution approach is to solve the problems

$$[\,K\,]\{\psi\}_j = \{\phi\}_j$$

and the response is then synthesized as

$$\{u\} = \sum_j \{\psi\}_j \tilde{P}_j$$

When combined with measurements, this will lead to a system of linear equations for the unknown $\tilde{P}_j$. This approach parameterizes the load distribution independent of the mesh refinement and iteration is not required.

The challenging situation is when there are not many sensors and the distribution is not simple. Consider the situation of an airfoil in an unsteady stream or a building under wind loading. Both will have a complicated loading (in space and time) but we are interested somehow in characterizing (and therefore estimating) the stresses. A direct attempt to obtain the loads would require a great number of sensors.

In these cases it is necessary to develop a model (preferably, a physics- or mechanics-based model) for the loading; we will mention two such cases. It usually happens that the distribution is in terms of parameters that appear nonlinearly in the relation. While the parameters are few in number, that they appear nonlinearly means they are effectively implicit parameters.

Reference [164] gives a discussion of the effects of water waves on submerged structures. It is a rather complicated subject but they show that the pressure distribution (at a particular frequency ω) with respect to depth (z) (see Figure 4.40(a)) is

$$p(z) = \frac{1}{2}\rho g H \frac{gh}{\omega} \frac{\cosh[k(h+z)]}{\sinh[kh]} \cos \omega t , \qquad \omega^2 = gh \tanh[kh]$$

The wave number k is related to the wavelength λ by $k = 2\pi/\lambda$. In truth, a real water wave would have a spectrum of frequencies, but for a central frequency, we could take A and k as the unknown parameters of the loading and represent the pressure distribution as

$$p(z) = A \cosh[k(h+z)] \cos \omega t$$

Reference [44, 88, 121] discusses blast loadings in both air and water. Again, the complete analysis is rather complicated, but as an illustration, the pressure loading in

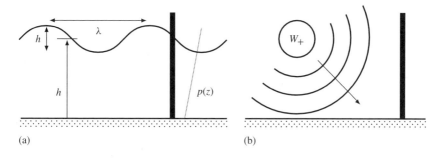

(a)   (b)

Figure 4.40: Examples of parameterized loadings based on loading models. (a) Water waves. (b) Blast loading.

water has the empirical relation

$$P(t) = P_m e^{-\alpha t}, \qquad P_m = 17570 \left[\frac{W^{1/3}}{R} \right]^{1.13}, \qquad \alpha = \frac{740}{W^{1/3}}$$

where P_m is the maximum pulse pressure in [psi], W is the equivalent charge weight of the explosive in [lbs of TNT], and R is the distance from the source in [inches]. The pressure wave is assumed to propagate spherically with the speed of sound. A possible parameterization would take P_m and α as the unknowns.

The test case we will consider here is the cantilevered beam of Figure 4.41(a) with a traction distribution given by the modeling

$$q(x) = A[1 - x/L]e^{B[1+(x/L)^4]}$$

The unknown parameters are A and B, and Figure 4.41(a) shows how the distribution varies as B is changed.

The solution procedure is similar to what we do for implicit structural parameters except that it is the load that is changed; that is, let us have a set of guesses, a_j^o, for the unknown parameters, then solve the system

$$[K]\{u\}_o = \{P\}_o$$

In turn, change each of the parameters (one at a time) by the definite amount $a_j \longrightarrow a_j + da_j$ and solve

$$[K]\{u\}_j = \{P\}_j$$

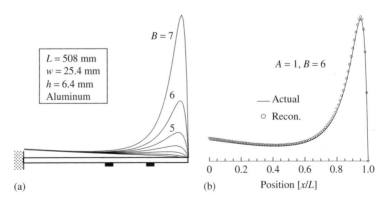

Figure 4.41: Beam with a parameterized loading. (a) Variation of distribution with parameter B. (b) Reconstructed distribution with 5% noise.

The sensitivity of the solution to the change of parameters is constructed from

$$\{\psi\}_j \equiv \left\{ \frac{u_j - u_o}{da_j} \right\} \bar{a}_j$$

where $\bar{a}_j$ are the normalizing factors, typically based on the search window for the parameter. The response solution corresponding to the true parameters is different from the guessed parameter solution according to

$$\{u\} = \{u\}_o + \sum_j \left\{ \frac{u_j - u_o}{da_j} \right\} \Delta a_j = \{u\}_o + \sum_j \{\psi\}_j \Delta \tilde{P}_j$$

Using data it is now possible to determine Δa_j and hence an improved $a_j = a_j^o + \Delta a_j$ iteratively.

Figure 4.41(b) shows the reconstruction when two strain measurements with 5% noise are used along with nominal parameters of $A = 1.0$, $B = 6.0$. The convergence rate is similar to the examples already shown and it gave the converged values of $A = 1.08$, $B = 5.92$. The achievement of the example is the determination of a complete (complicated) distribution using just two sensors.

A note of interest: static force location problems can be presented as implicit load parameter problems in which the parameters are load value and position. With sufficient sensor data (as with Moiré whole-field data), this can be made into a very robust scheme.

II: Structural Parameter ID as Force ID

What the strain gage analyses of the box beam (Figure 4.14), and plate (Figure 4.17) have in common is that they formulate the unknown structural parameter as a force or set of forces. The parameter is then obtained as a postprocessing operation; for example, for the box beam, the load position is unknown but by determining two loads, the position is then obtained as $x = WP_2/(P_1 + P_2)$. When an implicit parameter problem is recast as a force ID problem, then the solution procedure is linear, and iteration is not required.

Consider the beam shown in Figure 4.42(a) with a nonlinear spring. Let it be required to identify the spring properties. This is a nonlinear mechanics problem and will be treated in more detail in Chapter 7; for now, we illustrate how a parameter ID problem can be rephrased as a force ID problem.

The nonlinear spring is governed by the equation

$$P(u) = P_o \left[\frac{u}{\delta} \right] e^{[1-u/\delta]}$$

and is shown as the full line in Figure 4.42(c). This could be treated as an implicit parameter problem with the parameters to be identified being P_o and δ; however, we will assume we have enough loading data and will characterize the spring without stating a specific spring model.

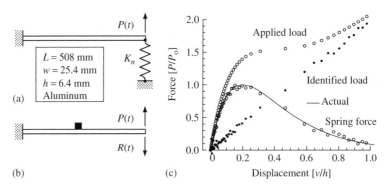

Figure 4.42: A beam attached to an unknown nonlinear spring. (a) Geometry and properties. (b) Nonlinear spring replaced with an unknown reaction force. (c) Reconstructed force displacement behavior of the spring.

Consider the problem in Figure 4.42(b), which is the same as in Figure 4.42(a) but the spring is removed with a free-body cut and replaced with a reaction force $R(t)$. Assuming load levels are low enough not to cause large deflections, this is a linear force identification with $P(t)$ known and $R(t)$ unknown. (Note that the time dependence referred to is pseudotime to indicate a changing load history but there are no inertia effects.)

The parameters for the spring were chosen as

$$P_o = 1.5h\, EI/L^3, \qquad \delta = h/5$$

and the load was chosen such that the maximum deflection reached the thickness h.

The reconstructed spring force using one sensor at the midpoint of the beam is shown in Figure 4.42(c). The correspondence with the exact forces is very good. It is worth reiterating that once the forces have been determined, the problem is fully determined and everything else about the problem (such as stress, displacements, and so on) can be determined. This includes the displacements at the nonlinear spring, which allowed making the plot for the nonlinear spring relation.

This problem has a lot in common with the implicit loading parameter problem except that because sufficient data were available, a loading model was not necessary. With the results of Figure 4.42(c), it is now possible to curve-fit the force/deflection data and extract the parameters P_o and δ.

III: Unknown Loads as Intermediate Parameters

What the holographic analysis of the plate (Figure 4.5), the interferometric analysis of the beam (Figure 4.8), the strain gage analysis of the ring (Figure 4.11), and the photoelastic analysis of the hole (Figure 4.29), all have in common is that they use the experimental data *indirectly* to infer some other information about the problem; that is, the data do not directly give the quantities of interest (the bending strains in the plate, for example), and additional processing is required to complete the stress analysis.

Loads were introduced as a convenient intermediate parameterization of the unknowns—the stresses are the unknowns of interest but from a fully specified problem point of view, stresses are not considered basic unknowns. The unknowns must be structural parameters or loads.

The loads as intermediate parameters may have physical significance as, for example, the applied pressure in the holographic case, but the traction distribution of Figure 4.30 is not intended to be the actual traction distribution (this would have required data to be collected nearer to the applied load site), and the applied load P_1 in Figure 4.10(a) is certainly fictitious. The inferred loads are understood to be a set of loads that are consistent with the given experimental data and are capable of accurately reconstructing the stress state in the vicinity of the data.

We will now revisit the photoelastic analysis of cracks of Section 3.3 to further elaborate on this point. Stress intensity factors are usually presented in graphs as $K_i = \sigma_\infty \sqrt{\pi a} f(a/W)$ where W is some geometric factor (such as the width) and $f(a/W)$ shows the influence of the geometry on the crack behavior. This formula gives us a clue as to how to approach crack problems (and singularity type problems in general): the FEM is used to characterize the geometric effects, and the role of the experiment is to characterize the effective remote loading. This is essentially what we will do in this problem.

From an analytical point of view, there are a number of ways of extracting stress intensity factors from an FEM analysis. The most obvious way is to put a very fine mesh around the crack tip and to use representations such as Equation (3.5) (or their displacement equivalent) to extrapolate the parameters. Besides being computationally expensive, this suffers all the same drawbacks of extrapolation as discussed in Section 3.3. Various special elements and methods have been proposed (see References [16, 97, 174] for a discussion) to reduce the computational cost. We discuss one such method next.

A very versatile method of crack parameter extraction is that of the virtual crack closure technique (VCCT) first introduced in Reference [151]; in this method, the work done in extending the crack an amount Δa using the very near-field crack equations is equated to the work done by the FEM discretized form. Since the crack is not actually extended (or closed), only one model is needed. The method has the additional convenience that it is applicable directly to dynamic problems and therefore will be our choice in the next few chapters.

The strain energy release rate is computed as

$$G = G_1 + G_2 + G_3 = \frac{F_x[u^+ - u^-] + F_y[v^+ - v^-] + F_z[w^+ - w^-]}{2\Delta a}$$

where the $\pm$ refers to the nodes above and below one element behind the crack, respectively, and F_i are the force components at the crack tip. The individual stress intensity factors are obtained from

$$K_1 = \sqrt{EG_1}, \qquad K_2 = \sqrt{EG_2}, \qquad K_3 = \sqrt{EG_3}$$

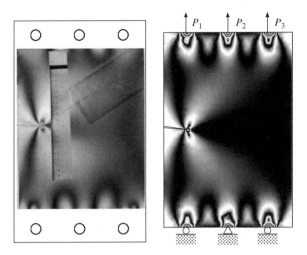

Figure 4.43: Crack under predominantly Mode I loading. (a) Experimental light-field photoelastic fringe data. (b) Modeling and reconstruction.

Tests show that this method gives quite accurate results as long as the element size is less than about 10% of the crack length. It is also used for delamination problems involving rotations.

Figure 4.43(a) shows a zoomed out image of the stress-frozen specimen described in Figure 3.11; this is a light-field image of the green separation. The way we will use the FEM modeling is to first use the photoelastic data to determine the effective loading and then use this loading with the VCCT to determine the stress intensities.

Figure 4.43(b) shows the modeling; the clamp/pins are replaced with point support boundary conditions at the bottom and three applied loads on the top. The elevated (stress-freezing) temperature properties were used; the Young's modulus was estimated as $16.75\,\text{MPa}\,(2430\,\text{psi})$. Data remote from the crack tip but not close to the load points were used. First-order regularization was used with a relatively large value of regularization—this forced the distribution to be linear. The results were

$$\{P\}^{\text{T}} = \{0.351,\ 0.330,\ 0.318\}^{\text{T}} P_{\text{o}}\,, \qquad P_{\text{o}} = 6.46\,\text{N}\,(1.45\,\text{lb})$$

The expected resultant load value is $P_{\text{o}} = 6.67\,\text{N}\,(1.50\,\text{lb})$. It is interesting that the load distribution picks up a slight bending action. Figure 4.43(b) shows a reconstruction of the fringe pattern using these load values; the comparison with Figure 4.43(a) is quite reasonable.

The loads were also used to give the stress intensity factors

$$K_1/K_{\text{o}} = 1.18\ (1.28)\,, \qquad K_2/K_{\text{o}} = 0.03\ (0.03)$$

where $K_{\text{o}} = \sigma_\infty \sqrt{\pi a}$ is the stress intensity factor for an infinite sheet. The parenthetical values are results using the expected load distributed uniformly and computed using the VCCT.

IV: Unknown Loads Parameterized through a Second Model

To help fix ideas here, we will consider thermoelastic stress analysis mentioned in the Introduction. When a body is heated, it expands, and, therefore, there is straining; however, stresses are not induced unless the expansion is restricted, that is, if the body is restrained in any way or is nonuniformly heated, significant stresses can be generated. Reference [30] is a comprehensive study of thermoelastic stresses.

Consider the thermoelastic stress analysis of a cylinder with a (slowly) moving hot fluid as depicted in Figure 4.44. Further, assume we are using integrated photoelasticity. Then the fringe distribution observed on any cross section is

$$\frac{f_\sigma}{h} N(y) = \int \sqrt{(\sigma_{xx} - \sigma_{yy})^2 + 4\sigma_{xy}^2} \, dx$$

where the integration is along the light path. There is insufficient data to determine the stress distributions even in the simple case of axisymmetric stresses [55].

The finite element treatment of temperature effects is usually by way of initial strains, that is, the temperature causes an expansion in each direction of $\epsilon_0 = \alpha \Delta T$, where α is the coefficient of thermal expansion, ΔT is the temperature change, and ϵ_0 is to be thought of as a strain preexisting in the body. The constitutive relation is written as

$$\{\sigma\} = [\, D \,]\{\epsilon - \epsilon_0\}, \qquad \{\epsilon_0\} \equiv \alpha \Delta T \{1, \, 1, \, 1, \, 0, \, 0, \, 0\}^{\mathrm{T}}$$

The strain energy for an element with initial strain is

$$\mathcal{U} = \frac{1}{2} \int \{\epsilon - \epsilon_0\}^{\mathrm{T}} [\, D \,]\{\epsilon - \epsilon_0\} \, dV$$

$$= \frac{1}{2} \int \{\epsilon\}^{\mathrm{T}} [\, D \,]\{\epsilon\} \, dV - \int \{\epsilon_0\}^{\mathrm{T}} [\, D \,]\{\epsilon\} \, dV + \frac{1}{2} \int \{\epsilon_0\}^{\mathrm{T}} [\, D \,]\{\epsilon_0\} \, dV$$

On combining with the principle of stationary potential energy, the first term yields the familiar element stiffness relation, the second yields a set of applied loads, and the third

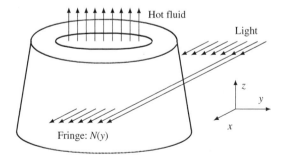

Figure 4.44: Thermoelastic stresses observed by integrated photoelasticity.

disappears since the initial strains are not varied; that is,

$$\mathcal{U} = \tfrac{1}{2}\{u\}^{\mathrm{T}}[\ k\ ]\{u\} - \{P_{\mathrm{T}}\}^{\mathrm{T}}\{u\}, \qquad \{P_{\mathrm{T}}\} = \int \{\epsilon_{\mathrm{o}}\}^{\mathrm{T}}[\ D\ ]\left\{\frac{\partial g}{\partial x}\right\} \mathrm{d}V$$

where $g(x, y, z)$ are the element interpolation functions. The assembled vectors $\{P_{\mathrm{T}}\}$ are the actual applied loads—clearly, there are too many applied loads to be determined as unknown parameters.

Many finite element books [18, 35] show that the heat conduction problem can be reduced to solve the system of equations

$$[K_{\mathrm{k}} + K_{\mathrm{c}}]\{T\} = \{Q\}$$

where $[\ K_{\mathrm{k}}\ ]$ is the conductivity matrix, $[\ K_{\mathrm{c}}\ ]$ is the convection matrix, $\{Q\}$ is the nodal point heat flow, and $\{T\}$ is the resulting nodal values of temperature. The boundary conditions are specified either in terms of the temperature or some heat flow conditions.

The solution to our stress analysis problem is to, for example, parameterize the temperature on the inner walls and determine the temperature distribution in terms of this; that is,

$$T(z) = \sum_{j}\phi(z)_{\mathrm{T}j}\tilde{P}_{j}, \qquad T(x, y, z) = \sum_{j}\psi(x, y, z)_{\mathrm{T}j}\tilde{P}_{j}$$

The temperature fields $\psi(x, y, z)_{\mathrm{T}j}$ are in turn used to compute the sensitivity responses to find the stresses. Note that the final problem is linear in the unknown parameters. This idea of using a second model or concatenating models generalizes to problems involving residual stresses, radiation, and electrostatics, to name a few.

4.8   Discussion

The power of the sensitivity response method is that it can utilize a commercial FEM program as an external processor and this greatly extends the range and type of problems that can be solved.

Irrespective of the geometrical complexity of a structure, the inverse problem gets reduced to solving an $[M_p \times M_p]$ system of equations where M_p is the number of unknown forces or parameters. Since M_p is usually substantially smaller than the total number of unknowns in the corresponding FEM problem, its solution cost is insignificant.

Using photoelastic data or identifying implicit parameters adds the complication that the system of equations must be solved iteratively. In all the problems discussed, the number of iterations required for convergence ranged from 5 to 10 with the FEM portion of the solution remaining dominant. We conclude that the computational cost of the inverse method is basically that of the forward FEM solution times the number of iterations. In comparison to the dynamic and nonlinear problems covered in the next few chapters, the cost of a forward static solution is negligible and therefore the issue is not significant.

By contrast, choosing the parameterization of the problem and selecting an appropriate number of sensors is a more important issue. In fact, the choice of parameterization emerges as the crucial issue in our inverse methods. The examples chosen illustrated an array of possibilities and the solution methods introduced are robust enough to handle each of them. In the next few chapters we show some more possibilities for parameterizations.

5

Transient Problems with Time Data

This chapter concentrates on determining force histories and structural parameters using experimentally measured point responses—a possible scenario is shown in Figure 5.1 in which accelerometers are used to identify a blast-loading history acting on a shell structure. The basic approach is an extension of the sensitivity response idea introduced for static problems in Chapter 4 but extended to time histories. The method, which can determine multiple-load histories, allows for the measured data to be supplemented with additional information based on an *a priori* bias; the addition of this regularization term provides a robust solution to these normally ill-conditioned problems. It also allows for the data collected to be of dissimilar type, for example, strain and velocity.

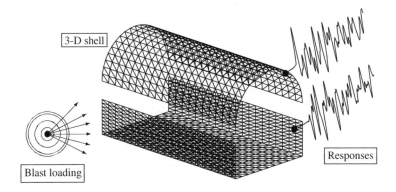

Figure 5.1: Dynamic responses are used to infer the blast loading on the shell.

Modern Experimental Stress Analysis: completing the solution of partially specified problems. James Doyle
© 2004 John Wiley & Sons, Ltd ISBN 0-470-86156-8

Wave propagation signals are good carriers of information over distances and, therefore, provide the opportunity to remotely monitor events; this is something we pay particular attention to.

5.1   The Essential Difficulty

Before developing the specifics of our actual method in the next section, we first lay out the issues of the force identification problem and then sketch the outlines of a naive method of solution. This will help explain why transient force identification problems are notoriously difficult to solve.

To put this force identification problem in context, recall that the response $u(x, t)$ of a general linear structural system to a single excitation load $P(t)$ can be written as the convolution

$$u(x, t) = \int_0^t G(x, t - \tau) P(\tau) \, d\tau$$

where $G(x, t)$ is the system response function. The force identification problem is usually posed as: Given some history measurements $u(t)$ (perhaps imperfectly), and knowledge of the system function $G(x, t)$ (perhaps imperfectly), determine the load history $P(t)$. We append to this the problem: given some space-distributed measurements $u(x)$ (perhaps imperfectly), and knowledge of the system function $G(x, t)$ (perhaps imperfectly), determine the load history $P(t)$. Both problems are quite difficult for two reasons. The first is that they are highly ill-conditioned as discussed in Chapter 3. Second, the system function $G(x, t)$ in appropriate analytical form for general structures is very difficult to determine and this has meant that most of the structural systems analyzed were relatively simple. It is the second issue that must be addressed in order for our methods to be applicable to general structures.

For convenience of discussion, the governing equations as discussed in Chapter 1 are recast as the recurrence relations

$$\{u\}_{n+1} = [\ A\ ]\{u\}_n + [\ B\ ]\{\tilde{P}\}_n \tag{5.1}$$

In this, n is the subscript over the discretized time; $\{u\}$ contains the state vectors (at least the displacement and velocity of each DoF), and is of size $M \geq 2M_u$; the vector of forcing terms, $\{\tilde{P}\}$, is of size $\{N_p \times 1\}$; and $[\ B\ ]$ is the $[M \times N_p]$ matrix that associates the forces with the DoF (Particular forms of this are given in Section 5.4).

In the forward problem, given the initial conditions as $\{u\}_1 = $ known and the applied loads' histories $\{\tilde{P}\}$, we can solve for the response recursively from Equation (5.1). In the inverse problem of interest here, the applied loads are unknown, but we know some information about the responses; we wish to use this information to determine the applied loads. In particular, assume we have a vector of measurements $\{d\}$ of size $\{M_d \times 1\}$, which is related to the structural DoF according to

$$\{d\}_n \Leftrightarrow [\ Q\ ]\{u\}_n$$

We want to find the forces $\{\tilde{P}\}$ that make the system best match the measurements. Consider the general least-squares error given by

$$E(u, \tilde{P}) = \sum_{n=1}^{N} [\{d - Qu\}_n^{\mathrm{T}} \lceil W \rfloor \{d - Qu\}_n], \qquad \{u\}_{n+1} = [\ A\ ]\{u\}_n + [\ B\ ]\{\tilde{P}\}_n$$

Our objective is to find the set of forces $\{\tilde{P}\}$ that minimize this error functional.

It is possible to establish a global system of simultaneous equations by the usual procedures for minimizing the least squares, that is, we arrange Equation (5.1) to form

$$[\ \overline{A}\ ]\{\overline{u}\} = [\ \overline{B}\ ]\{\overline{P}\}, \quad \{u\} = \{\{u\}_1, \ \ldots, \{u\}_N\}^{\mathrm{T}}, \quad \{\tilde{P}\} = \{\{\tilde{P}\}_1, \ \ldots, \{\tilde{P}\}_N\}^{\mathrm{T}}$$

where $\{\overline{u}\}$ and $\{\overline{P}\}$ are the responses and loads concatenated over time. Solve for $\{u\}$ in terms of $\{\tilde{P}\}$, that is, first decompose $[\ \overline{A}\ ]$ and then solve $[\ \Psi\ ]$ as the series of back-substitutions

$$[\ \overline{A}\ ][\ \Psi\ ] = [L^{\mathrm{T}} D U][\ \Psi\ ] = [\ \overline{B}\ ]$$

The matrix $[\ \Psi\ ]$ is the collection of forward solutions for unit loads applied for each of the unknown forces. We, therefore, can write the actual forward solution as

$$\{\overline{u}\} = [\ \Psi\ ]\{\overline{P}\}$$

Substitute this into the error equation and minimize with respect to $\{\tilde{P}\}$. The resulting system is

$$[\Psi^{\mathrm{T}} Q^{\mathrm{T}} W Q \Psi]\{\overline{P}\} = [Q\Psi]^{\mathrm{T}} \lceil W \rfloor \{\overline{d}\}$$

This, in fact, is the static solution as developed in Chapter 4 in which it was shown to be quite powerful and computationally effective.

Unfortunately, this naive scheme is not practical because of size. Consider a dynamic problem in which there are M_u degrees of freedom at N time steps; this would lead to the very large system array $[\ \overline{A}\ ]$ of size $[(M_u \times N) \times (M_u \times N)]$. A system with 10,000 DoF over 2000 time steps has nearly 100×10^6 unknowns and a system size of $[\ \overline{A}\ ]$, that is, the square of that. This would lead to a RAM memory requirement of $3 \times 10^9\, MB$, which would cost billions of euros. Furthermore, solving the system of equations would take approximately N^3 operations, which on a $1\, GFlop$ computer would take 10^{13} seconds or 300,000 years! Clearly, this approach to solving the problem does not scale very well, and the procedure is restricted to very small problems only. Even if advantage were taken of the special nature of $[\ \overline{A}\ ]$, as discussed in Section 5.4, the numbers would still be outrageously large.

The key factor of size must be addressed, and many types of solutions have been proposed. There is a sizable literature on force identification using time data. Reference [161] gives an excellent summary of the literature as well as an overview of the subject itself; more recent citations can be found in Reference [118]. A variety of methods have been used, but the most common solution scheme is some form of deconvolution. Reference [143] summarizes some of the main deconvolution methods although not directly applied to the force identification problem. References [62, 63, 68] mostly used Fourier methods because of the relative simplicity of the inversion and because it is suitably matched to the spectral element approach. Deriving analytical response functions is a formidable challenge as seen from the following references. Reference [122] and References [147, 148] looked at half-space and half-plane problems, respectively; general frame structures were solved by the introduction of spectral elements in Reference [118]; folded plate structures were solved using plate spectral elements [53]. Each of these is quite limited in that they are restricted to relatively simple geometries. The beginnings of a deconvolution method coupled with a general finite element approach were given in Reference [69], which used an FEM program to generate wavelet solutions that could be synthesized in the inverse solution. A more general approach is developed in References [3, 4] that is finite element-based and, hence, is applicable to quite general structures. In these latter papers, special attention was paid to having an algorithm whose computational cost scales as for the corresponding forward problem.

There is very little written on force deconvolution using whole-field data. Some static work that is related can be found in References [106, 112, 129]. A dynamic method is proposed in References [41, 72, 73]; this aspect of force identification has special features, so we leave its consideration until the next chapter.

With the above discussion in mind, some of the attributes the force identification method should have are the following:

- To be able to analyze structures with complexities as usually found in forward problems and handled by the finite element method.

- To be able to determine many unknown force histories forming unknown shape distributions as well as acting independently of each other.

- To be able to utilize many sensors of different types distributed throughout the body and not placed in any particularly optimized positions. Thus, be able to handle fewer significant sensors than unknown forces.

- To be scalable similar to forward problems, from a computational cost aspect, using the finite element method.

In the following, the ill-conditioning is handled by the introduction of regularization terms with respect to both time and space behaviors, and the size issue is addressed through the use of sensitivity response functions.

5.2   Deconvolution using Sensitivity Responses

We begin by recapping the developments of Reference [69] for deconvolution using time history data. This will serve as a backdrop for the refinements we introduce to improve the quality of results, and to handle multiple loads.

Single-Force History

The force history is represented as

$$P(t) = \sum_{m=1}^{M_p} \tilde{P}_m \phi_m(t) , \qquad \{P\} = [\ \Phi\ ]\{\tilde{P}\} \tag{5.2}$$

where $\phi_m(t)$ are specified functions of time; they are similar to each other but shifted an amount $m\Delta T$. We will consider specific forms later. The array $[\ \Phi\ ]$ has as columns each $\phi_m(t)$ discretized in time. The vector $\{P\}$ has the force history $P(t)$ discretized in time. This force causes the response at a particular point

$$u(x,t) = \sum_{m=1}^{M_p} \tilde{P}_m \psi_m(x,t) , \qquad \{u(x)\} = [\ \Psi\ ]\{\tilde{P}\} \tag{5.3}$$

where $\psi_m(x,t)$ is the response to force $\phi_m(t)$ at the particular point, and $[\ \Psi\ ]$ are all the discretized responses arranged as columns. To keep the connection with Chapter 4, we will refer to $\phi_m(t)$ as unit or perturbation forces and $\psi_m(t)$ as sensitivity responses. These responses may be obtained, for example, by the FEM, and, therefore, will have imbedded in them the effects of the complexity of the structure. Because of the linearity of the system, the responses are also similar to each other but shifted by an amount $m\Delta T$.

Let the measured data at a certain location be related to the displacements (DoF) as

$$\{d\} \Longleftrightarrow [\ Q\ ]\{u\}$$

where, as in Chapter 4, $[\ Q\ ]$ acts as a selector; then, the error function is

$$\mathcal{A} = \{d - Qu\}^{\mathrm{T}} \lceil\ W\ \rfloor \{d - Qu\} = \{d - Q\Psi\tilde{P}\}^{\mathrm{T}} \lceil\ W\ \rfloor \{d - Q\Psi\tilde{P}\}$$

where $\lceil\ W\ \rfloor$ is some weighting array (over time) on the data.

In most cases, least-squares minimization leads to a highly ill-conditioned system of equations. We saw in the previous chapter that this required regularization of the spatially distributed forces. The same actually occurs for the time distribution of the force history. Force reconstructions will show the manifestations of the ill-conditioning as ΔT is changed, that is, with small ΔT, the data are incapable of distinguishing between neighbor-force values $P_n + \Delta$, $P_n - \Delta$.

Time regularization connects components within $\{\tilde{P}\}$, that is,

$$\mathcal{B} = \{\tilde{P}\}^{\mathrm{T}} [\ D\ ]^{\mathrm{T}} [\ D\ ]\{\tilde{P}\} = \{\tilde{P}\}^{\mathrm{T}} [\ H\ ]\{\tilde{P}\}$$

where, for example, if $\{D\}$ is the $[(M_p - 1) \times M_p]$ first difference matrix, then

$$[\ D\ ] = \begin{bmatrix} -1 & 1 & 0 & \cdots \\ 0 & -1 & 1 & \cdots \\ 0 & 0 & -1 & \cdots \end{bmatrix}, \qquad [\ H\ ] = [\ D\ ]^{\mathrm{T}}[\ D\ ] = \begin{bmatrix} 1 & -1 & 0 & \cdots \\ -1 & 2 & -1 & \cdots \\ 0 & -1 & 2 & \cdots \end{bmatrix}$$

where $[H]$ is an $[M_p \times M_p]$ symmetric matrix. Other orders of regularization, as discussed in Chapter 3, can also be used.

The minimum principle becomes

$$E(\tilde{P}) = \mathcal{A} + \gamma \mathcal{B} = \{d - Q\Psi\tilde{P}\}^{\mathrm{T}}\lceil W \rfloor\{d - Q\Psi\tilde{P}\} + \gamma \{\tilde{P}\}^{\mathrm{T}}[H]\{\tilde{P}\} = \min$$

Minimizing with respect to $\tilde{P}_m$ then gives

$$[[Q\Psi]^{\mathrm{T}}\lceil W \rfloor[Q\Psi] + \gamma [H]]\{\tilde{P}\} = [Q\Psi]^{\mathrm{T}}\lceil W \rfloor\{d\} \qquad (5.4)$$

The core array is symmetric of size $[M_p \times M_p]$. Actually, it is of near Toeplitz- [143] type in which each row is the same but shifted one location; this was exploited in Reference [69] to give a fast algorithm for the inversion of Equation (5.4) of large size. We will revisit this issue later in Chapter 7 when systems with a large set of unknowns are considered. Also note that, unlike the naive solution presented at the beginning, the cost of the inverse solution is not affected by the system size of the FEM model. This is the primary achievement of the present method.

There are a variety of functions that could be used for $\phi_m(t)$; the main requirement is that they have compact support in both the time/space and frequency domains. Reference [69] used a Gram–Schmidt reduction scheme to generate a series of smooth orthogonal functions from the basic triangular pulse. A set of these is illustrated in Figure 5.2(b). Here, we will use the triangular pulse directly; this is equivalent to using a linear interpolation between the discretized points. The advantage of the triangle is that the scales $\tilde{P}_m$ have the direct meaning of being the discretized force history and, hence, there is no need to actually perform the summation in Equation (5.2).

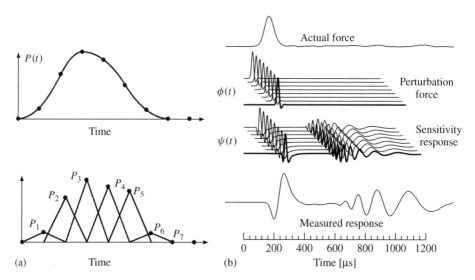

Figure 5.2: The sensitivity response concept. (a) A discretized history is viewed as a series of scaled triangular perturbations. (b) How the scaled perturbations superpose to give the actual histories.

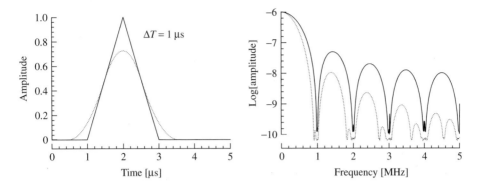

Figure 5.3: Triangle load perturbation (full line) and its 11-point moving average (dashed line). (a) Time domain. (b) Frequency spectrum.

The disadvantage of the triangle is that the frequency content of this pulse is higher than in the actual force history and, therefore, a finer mesh and time integration step must be used. A convergence study must be performed to determine an appropriate mesh and step size; typically a step size of $\Delta T/10$ is adequate where ΔT is chosen on the basis of the frequency content of the excitation force. With reference to Figure 5.3, as long as the frequency content of the actual force is inside the first lobe, the frequency content of the load perturbation is adequate.

In fact, without loosing the advantage of the straight triangular pulse, it is also possible to use a smoothed triangular pulse. These two histories, along with their amplitude spectrums, are shown in Figure 5.3. The advantage of the smoothed triangular pulse is that the side lobes in its amplitude spectrum are small, and, consequently, the requirements on the FEM mesh are less stringent than with the direct triangular pulse. Specifically, if the threshold amplitude is about one-hundredth of the peak value (a difference of -2.0 on the log plot), then the smoothed triangle has a frequency range about one-quarter of that of the unsmoothed triangle.

The force history we will use in some of the later examples is shown along with its amplitude spectrum in Figure 5.4. This shows that a discretization of $P(t)$ every $\Delta T = 10\,\mu s$ is adequate because it corresponds to a sampling frequency of $100\,KHz$. This gives a triangle of width $2\Delta T = 20\,\mu s$ and a time integration of $\Delta T/10 = 1\,\mu s$; the inset shows that this is also adequate if the small side lobes are ignored.

One final point about the amount of shifting: the choice was made to shift each perturbation force by an amount $m\Delta T$. But other choices are possible, for example, each force could be shifted $m\alpha\Delta T$ where $0 < \alpha < 1$. The motivation for doing this might be to improve the resolution of the force reconstructions without decreasing the size $(2\Delta T)$ of the triangle. If α is made small enough, then ill-conditioning becomes a problem; the use of regularization then prevents obtaining the desired force resolution. Furthermore, this approach would require performing the summation in Equation (5.2) to get the actual reconstructions. All in all, it seems that the shift of $m\Delta T$ is a good choice, and when greater resolution is required then, simply, a smaller ΔT should be used.

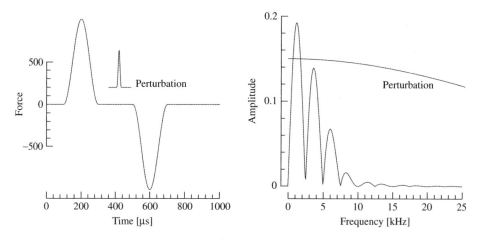

Figure 5.4: Sample input force history. (a) Time domain. (b) Frequency spectrum.

Multiple Forces

There are a number of features we now wish to add to the above developments, primary among which is the ability to determine multiple unknown force histories. There are two situations of interest here: the first is that of multiple isolated forces, and the second is that of traction distributions. As shown in Chapter 4, the main difference is whether the forces are related to each other or not; isolated forces are a finite number of forces, which are not related, but traction distributions contain very many forces, which are continuously related to their neighbors.

Let the N_p unknown force histories be organized as

$$\{\overline{P}\} = \{[\ \Phi\ ]_1\{\tilde{P}\}_1, [\ \Phi\ ]_2\{\tilde{P}\}_2, \ldots, [\ \Phi\ ]_{N_p}\{\tilde{P}\}_{N_p}\}^{\mathrm{T}}$$

where each $\{\tilde{P}\}_n$ is the history $P_n(t)$ arranged as a vector of M_p time components. This collection of force vectors causes the collection of responses

$$\begin{Bmatrix} \{u\}_1 \\ \{u\}_2 \\ \vdots \\ \{u\}_{N_d} \end{Bmatrix} = \begin{bmatrix} [\ \Psi\ ]_{11} & [\ \Psi\ ]_{12} & \cdots & [\ \Psi\ ]_{1N_p} \\ [\ \Psi\ ]_{21} & [\ \Psi\ ]_{22} & \cdots & [\ \Psi\ ]_{2N_p} \\ \vdots & \vdots & \ddots & \vdots \\ [\ \Psi\ ]_{N_d1} & [\ \Psi\ ]_{N_d1} & \cdots & [\ \Psi\ ]_{N_dN_p} \end{bmatrix} \begin{Bmatrix} \{\tilde{P}\}_1 \\ \{\tilde{P}\}_2 \\ \vdots \\ \{\tilde{P}\}_{N_p} \end{Bmatrix} \tag{5.5}$$

where $\{u\}_i$ is a history of a displacement (DoF) at the i measurement location. We also write this as

$$\{u\}_i = \sum_j [\ \Psi\ ]_{ij}\{\tilde{P}\}_j \qquad \text{or} \qquad \{\overline{u}\} = [\ \overline{\Psi}\ ]\{\overline{P}\}$$

The $[M_p \times M_p]$ array $[\ \Psi\ ]_{ij}$ relates the i location response history to the j applied perturbation force. It contains the responses $\psi_m(t)$ arranged as a vector.

Let the N_d sensor histories of data be organized as

$$\{\overline{d}\} = \{\{d\}_1, \{d\}_2, \ldots, \{d\}_{N_d}\}^{\mathrm{T}}$$

where $\{d\}_n$ is a vector of M_d components (it is simplest to have $M_d = M_p$ but this is not required). The measured data DoF are a subset of the computed responses, that is,

$$\{\overline{d}\}_i \Longleftrightarrow [\ \overline{Q}\ ]\{\overline{u}\} \qquad \text{or} \qquad \{\overline{d}\} \Longleftrightarrow [\overline{Q}\ \overline{\Psi}]\{\overline{P}\} \tag{5.6}$$

Define a data error functional as

$$E(\overline{P}) = \{\overline{d} - \overline{Q}\ \overline{\psi}\ \overline{P}\}^{\mathrm{T}}\lceil\ W\ \rfloor\{\overline{d} - \overline{Q}\ \overline{\psi}\ \overline{P}\} + \gamma_{\mathrm{t}}\{\overline{P}\}^{\mathrm{T}}[\ H_{\mathrm{t}}\ ]\{\overline{P}\} + \gamma_{\mathrm{s}}\{\overline{P}\}^{\mathrm{T}}[\ H_{\mathrm{s}}\ ]\{\overline{P}\}$$

which has incorporated regularization in both space and time. Minimizing with respect to $\overline{P}$ gives

$$[[\overline{Q}\ \overline{\Psi}]^{\mathrm{T}}\lceil\ W\ \rfloor[\overline{Q}\ \overline{\Psi}] + \gamma_{\mathrm{t}}[\ H_{\mathrm{t}}\ ] + \gamma_{\mathrm{s}}[\ H_{\mathrm{s}}\ ]]\{\overline{P}\} = [\overline{Q}\ \overline{\psi}]^{\mathrm{T}}\lceil\ W\ \rfloor\{\overline{d}\} \tag{5.7}$$

The core matrix is symmetric of size $[(M_p \times N_p) \times (M_p \times N_p)]$ and, in general, is fully populated. Only when N_p is large (there are a large number of force histories) does the computational cost of solving Equation (5.7) become significant.

Time regularization connects components within each $\{\tilde{P}\}_j$ only, that is,

$$\mathcal{B}_j = \{\tilde{P}\}_j^{\mathrm{T}}[\ D\ ]_j^{\mathrm{T}}[\ D\ ]_j\{\tilde{P}\}_j = \{\tilde{P}\}_j^{\mathrm{T}}[\ H\ ]_j\{\tilde{P}\}_j$$

We do this for each separate force to construct a total $[\ H_{\mathrm{t}}\ ]$ matrix as

$$[\ H_{\mathrm{t}}\ ] = \begin{bmatrix} H_1 & 0 & 0 & \cdots \\ 0 & H_2 & 0 & \cdots \\ 0 & 0 & \ddots & H_{N_p} \end{bmatrix}$$

Here $[\ H_{\mathrm{t}}\ ]$ is of size $[(M_p \times N_p) \times (M_p \times N_p)]$. Space regularization connects components in $\{\tilde{P}\}_j$ across near locations. For example, using first differences leads to

$$[\ D\ ] = \begin{bmatrix} -I & I & 0 & \cdots \\ 0 & -I & I & \cdots \\ 0 & 0 & -I & \cdots \end{bmatrix}, \qquad [\ H_{\mathrm{s}}\ ] = [\ D\ ]^{\mathrm{T}}[\ D\ ] = \begin{bmatrix} I & -I & 0 & \cdots \\ -I & 2I & -I & \cdots \\ 0 & -I & 2I & \cdots \end{bmatrix}$$

where $\lceil\ I\ \rfloor$ is the unit matrix of size $[M_p \times M_p]$, the difference matrix $[\ D\ ]$ is of size $[(M_p \times N_p - 1) \times (M_p \times N_p)]$, and the regularization matrix $[\ H_{\mathrm{s}}\ ]$ is of size $[(M_p \times N_p) \times (M_p \times N_p)]$. As is seen, the sparsity of the matrices $[\ H_{\mathrm{t}}\ ]$ and $[\ H_{\mathrm{s}}\ ]$ are quite different and therefore their effect on the solution is also quite different even for same order regularizations.

Isolated sensors are not especially well suited for determining distributions, so we leave the issues of space distributions and space regularization until the next chapter that uses space-distributed data.

Computer Implementation: Sensitivity Response Method

The key to the computer implementation is the series of sensitivity response solutions in the matrix $[\ \Psi\ ]$; it is these solutions that can be performed externally by any commercial FEM code. The basic relationship is shown in Figure 5.5, which is very similar to that of Figure 4.3 for static problems. Again, the role of the inverse program **StrlDent** is to prepare the scripts to run the FEM programs.

The cost of performing the least-squares reduction is

$$\text{cost} = M_p^3 N_p^2 N_d$$

It is interesting to note that, in comparison to the cost of determining one force from one sensor, the number of forces has a squaring effect. The cost for decomposing the normal system of equations is about

$$\text{cost} = \tfrac{1}{3} M_p^3 N_p^3$$

Thus, when the number of forces and number of sensors are comparable, forming the least squares and solving the system have about the same computational cost.

The sensitivity response is computed only once for a given force location and irrespective of the number of sensors, hence, the FEM cost is simply that of a forward solution. Only when N_p is large (there are a large number of separated forces) does this computational cost become significant. This is something we take up again later in the chapter and in Chapter 7.

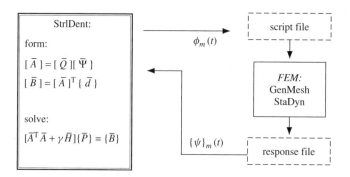

Figure 5.5: The distributed computing relationship between the inverse program and the FEM programs; a $\{\psi\}_m(t)$ history at all data locations is computed for each force location.

Preliminary Metrics

We do a study using synthetic data with the objective of understanding the nature of the ill-conditioning. A note of caution when using synthetic data: if the inverse problem is to be solved at a time step ΔT, then the synthetic data must be generated at a time step of $\Delta T/10$, otherwise there is a mismatch between the superposition solution of Equation (5.3) and the synthetic data.

Consider the very long beam of square cross section shown in Figure 5.6. An input force with the history of Figure 5.4 is applied close to the free end. The beam is modeled with 12.7 mm elements and integrated with a time step of $\Delta T = 1\,\mu s$, and the inverse problem will be solved with $\Delta T = 10\,\mu s$. The generated acceleration responses are also shown in Figure 5.6.

Note that a beam is a dispersive system, which means that a wave changes its shape as it propagates [70]. There are three stages to each trace. The first is the arrival of the initial wave, the second is the reflections from the near free end, and the third is the reflection from the far fixed end. We will be interested in determining the force history over a period of about 2500 μs, and, therefore, the only significant reflections are from the free end.

When determining force from measurements, it is not always clear whether the measurement is being integrated or differentiated. This makes a difference as regards to how errors manifest themselves in the reconstructed forces. Reference [70], working in the frequency domain, makes a nice comparison between the time differential levels for simple structural systems. This is summarized in Table 5.1 in which the symbol D stands for time differentiation.

For a rod, for example, force and strain are at the same level and, therefore, strain is a good quantity for determining force because errors are neither integrated nor differentiated. Force and acceleration have the relation

$$DP = DDu = D^2u = accn$$

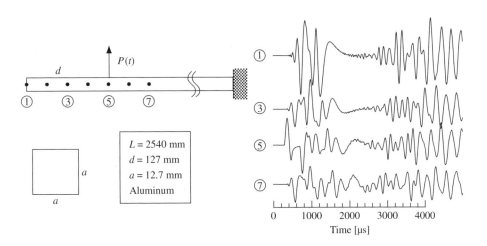

Figure 5.6: Dimensions of the beam and acceleration responses.

Table 5.1: Time differential relation among different quantities.

	Displacement	Velocity	Acceleration	Force	Stress	Strain
Rod	u	Du	D^2u	Du	Du	Du
Beam	v	Dv	D^2v	$D^{3/2}v$	Dv	Dv
Membrane	u	Dw	D^2w	Dw	Dw	Dw
Plate	w	Dw	D^2w	Dw	Dw	Dw

and, therefore, would require a time integration to get force from acceleration. The consequence of this is that any errors in the acceleration will appear as an integrated effect (increasing in time) in the force.

For a beam, the relation between force and acceleration is a half integration since

$$D^{1/2}P = D^{1/2}D^{3/2}v = D^2v = \text{accn}$$

Fractional derivatives and integrations have simple interpretations in the frequency domain; here it is sufficient to realize that force and acceleration have a time integration relation but not one as strong as a first integration.

We consider two sources of contamination of the measurements. The first is random noise and the second is a systematic error due to an offset.

The effect of 10% random noise is shown in Figure 5.7; no regularization is used. Sensor 5 is at the force location, and the results show that the force is reconstructed quite well exhibiting only the randomness of the noise. However, the ill-conditioning manifests itself dramatically as the distance from the impact site is increased; note that the error appears as an increasing function of time as expected from the integration

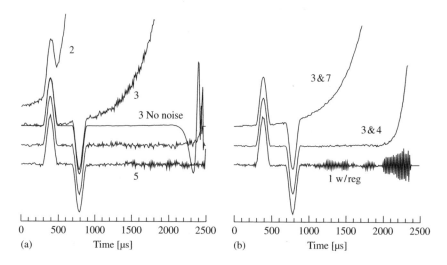

Figure 5.7: Effect of random noise on force reconstructions. (a) Effects are exacerbated as distance is increased. (b) Special combinations of sensors.

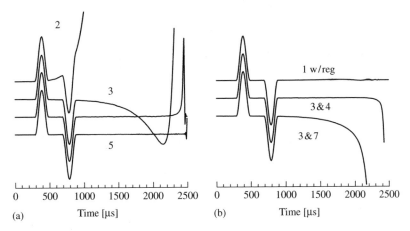

Figure 5.8: Effect of offset error on force reconstructions. (a) Effects are exacerbated as distance from force increases. (b) Special combinations.

relation between force and acceleration. It is curious that the random noise (with zero mean) causes this definite trend in the reconstruction. Also shown are the reconstructions using Sensor 3 without any noise. Here, the results are almost exact up to the time it takes the signal to propagate to the sensor (note that this is different for different frequency components).

Intuitively, it might be thought that the ill-conditioning can be reduced by adding more data. Figure 5.7(b) shows that combining data from Sensors 3 and 7 does not reduce the problem (all it achieves is smoothing of the noise) because Sensor 7 suffers the same ill-conditioning. On the other hand, combining data from Sensors 3 and 4 does reduce the problem; indeed the result is not just the average of the two, but the proper solution. Finally, the figure also shows what can be achieved with regularization; the result is for the sensor at the free end. We will return to regularization later.

The effect of offset contamination is shown in Figure 5.8. The amount used is 1/500th of the maximum acceleration; for a 10-bit system, this is about one least-significant bit (1 LSB). The symptoms are similar to that for the random noise with the manifestations of ill-conditioning increasing away from the force location. Again, combining data from Sensors 3 and 7 does not improve things, but regularization and mixing with good sensor locations does improve the results.

To summarize, ill-conditioning is a system problem and not affected by errors in the data. However, errors in the data can exaggerate the manifestations of the ill-conditioning. Regularization works directly on reducing the ill-conditioning.

As a practical matter, most algorithms work better using velocity data. Therefore, it is useful to first do the process of converting accelerations to velocity. The accelerometer outputs can be integrated according to the trapezoidal rule

$$v_{n+1} = v_n + \tfrac{1}{2}(a_n + a_{n+1})\Delta t$$

to determine the velocities. Because of slight offsets in the acceleration voltages, a trend is sometimes observed in the velocity. This trend is estimated and subtracted; we will refer to this detrended data as the raw data.

5.3   Experimental Studies

We do two experimental studies, both of which are on dispersive systems. The first is the impact of a 3-D cylindrical shell, and the second is of a beam similar to the test case of the previous section, but it highlights the importance of the proper selection of structural model.

Experimental Setup

The setup for both experiments is similar to that shown in Figure 5.9.

Data for the experiments were collected by means of an Omega Instruments DAS-58 data-acquisition card; this is a 12-bit card capable of a 1-MHz sampling rate and storing up to 1 M data points in the on-board memory. A total of eight channels can be sampled; however, increasing the number of channels decreases the fastest possible sampling rate. For this experiment, four channels were sampled at 250 kHz, or a 4 μs time step.

Accelerometers (PCB 309A) and a modified force transducer (PCB 200A05 as discussed in Chapter 2) were used to collect the data. The connections between the accelerometers and the computer were through an Omega Instruments BNC-58 multiplexing unit.

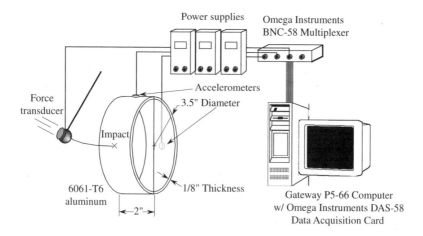

Figure 5.9: Typical experimental setup using accelerometers to identify an impact force history.

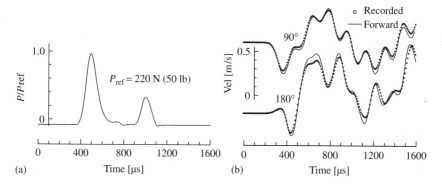

Figure 5.10: Recorded data for the shell. (a) Force history. (b) Velocity history and its comparison to the forward problem results.

Application to a Cylindrical Shell

As a first experimental application of our method, we consider a study of a cylindrical shell.

The relatively short aluminum cylinder is 51 mm (2 in.) long with a diameter of 89 mm (3.5 in.); the wall thickness is 3 mm (1/8 in.). Accelerometers were placed on the top (90°) and 180° from the impact site, as shown in Figure 5.9, and attached using beeswax. The specimen was suspended by strings and impacted midway 25 mm (1 in.) along the length.

The recorded force and (processed) velocity data are shown in Figure 5.10. Note that the impact is a double impact on a very small time scale.

A good inverse analysis is predicated on a high fidelity finite element model of the structure being analyzed, that is, this model must correspond accurately to the real experimental situation in terms of material properties (Young's modulus, mass density), dimensions (diameter, thickness), and sensor locations. Therefore, before any solution to an inverse problem is attempted, the modeling must be verified by doing a forward problem.

A convergence study was performed to determine the proper element size for the finite element model. The data from the force transducer provided the input force history for this study. From the results, it was concluded that 3 mm (1/8 in.) modules would be an acceptable size for the inverse analysis. The resulting mesh is shown in Figure 5.11(a) and has 2816 folded plate elements. This creates a problem with a system size of 8976 degrees of freedom.

The results for the forward problem are shown in Figure 5.10. The two sets of results are almost in phase with each other. To achieve these results, the effective radius of the cylinder was taken as the average of the inside and outside radii.

The force reconstructions from the experimental velocity inputs are shown in Figure 5.11. It is pleasing to note that the double impact is detected. Accelerometer

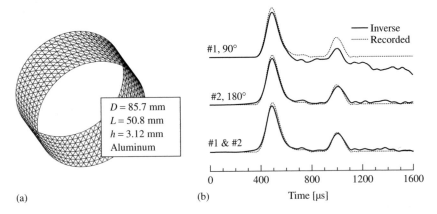

(a)   (b)

Figure 5.11: Shell experiment results. (a) Mesh and properties. (b) Comparison of force reconstructions with measured force history.

#2 on its own does a good job but the same is not true for Accelerometer #1. At first sight, this might appear to indicate that somehow this data is corrupted. However, the fact that using both accelerometers together gives somewhat improved results does not bear this out: if Accelerometer #1 were indeed contaminated, then mixing both acceleration data would also show the contamination. A more reasonable conjecture is that position #1 is very sensitive, in an ill-conditioning sense, to slight errors in the data.

This conjecture has important implications for the inverse methods so it was decided to further investigate the nature of the discrepancy by manipulating synthetic data in a couple of ways and comparing the performance of Positions #1 and #2.

First the sensitivity due to the positioning of the accelerometer was examined. These results are shown in Figure 5.12(a) and demonstrate that, for a given misalignment, the

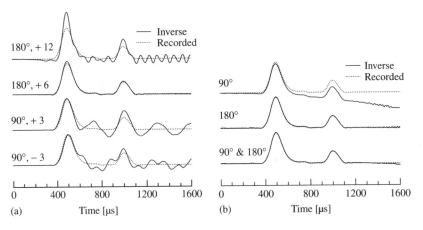

(a)   Time [μs]   (b)   Time [μs]

Figure 5.12: Effect of sensor error on force reconstructions. (a) Sensor mispositioning in the hoop direction (numbers are in mm). (b) Drift added to the velocity inputs.

90° position is more sensitive than the 180° position. However, the symptom of the error (a constant frequency superposition) does not reflect the symptom shown in the experimental results.

As a second study, approximately the same amount of drift seen in the experimental velocities was added to the velocities generated synthetically. This amount was determined by examining the raw data from the accelerometers and determining the digitizing level of the voltages recorded. This value was found to be around $15\,\mathrm{m/s^2}$ ($600\,\mathrm{in./s^2}$). Again, these adjusted velocities were used to reconstruct the force from the response at the two nominal positions. These results are shown in Figure 5.12(b). Again, the 90° position is more sensitive to slight deviations in the input data. This time the symptom is similar to that of the experimental results. It is noted that using both accelerometers does not give a simple average but gives a result that is better than the individual results.

These studies indicate a concern that is quite crucial in inverse studies and is different from the experimental issues of position accuracy and signal fidelity. When doing an inverse analysis, the location of the sensor(s) and not just its accuracy is very important. This is not something that can always be decided in advance; for example, if the impact were at the 270° location then the roles of the accelerometers would be reversed. The implication is that multiple sensors must be used, and that they be combined through the use of regularization. Both of these aspects are part of the program StrIDent.

Impact of a Beam

The beam dimensions and accelerometer locations are shown in Figure 5.13. The impact and accelerometers are close to the free end and, therefore, the only significant reflections are from that end. The measured accelerations are shown in Figure 5.14 and the measured force in Figure 5.15. The force was recorded with the transducer calibrated in Figure 2.30.

A comparison of the forward solution is shown in Figure 5.14(a) using elementary beam theory (element length of 6.35 mm) for the FEM modeling. The acceleration comparisons are quite good at the early times but deviate significantly in the reflections. This is typical of incorrect modeling parameters. The three significant possibilities are Young's modulus, density, and the FEM mesh.

It is generally possible to get accurate material properties, and a demonstration of doing this will be given in Section 5.7. Consequently, the reason for the acceleration mismatch must be that the FEM beam model is not correct or that the measured locations are not sufficiently accurate. In the present case, the main cause of the mismatch is the beam

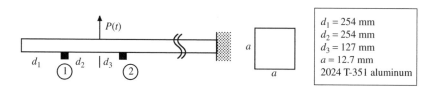

Figure 5.13: Dimensions of the beam showing the accelerometer locations.

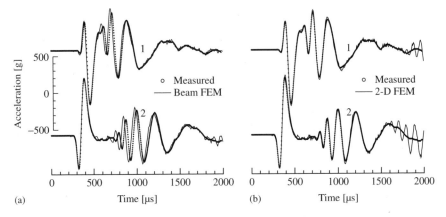

Figure 5.14: Comparison of measured accelerations with those computed using different beam theories. (a) Bernoulli–Euler elementary beam theory. (b) 2-D FEM modeling.

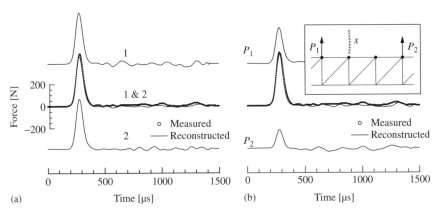

Figure 5.15: Force reconstruction for the impact of a beam using 2-D modeling. (a) Determining forces at the nominal impact location. (b) Simultaneously determining two forces surrounding the nominal impact location.

modeling. It happens that for thicker beams and/or for higher frequency content inputs, the elementary Bernoulli–Euler beam model is no longer adequate and a higher order model (such as the Timoshenko beam theory) must be used [70]. In the present case, we will use a 2-D FEM model. The comparisons are shown in Figure 5.14(b); the agreement is a significant improvement over Figure 5.14(a). The 2-D mesh used square modules with two elements per module and eight modules through the thickness of the beam.

The force reconstructions are shown in Figure 5.15(a) in which they are compared with the measured value. Zero-order regularization was used with the value of γ chosen so as to detrend the results. The agreement is quite good.

It might be thought that a disadvantage of using an FEM-based inverse method is that the resolution is restricted by the element size, that is, customized meshes must be used to accurately place a node at a measurement site or a load point. This is not necessarily so. As a simple illustration, we will use both sensors to reconstruct two forces in the vicinity of the applied load. Since these forces are equipollent to the actual load, with reference to the inset of Figure 5.15(b), taking moments about the first load gives

$$x P = x[P_1 + P_2] = L P_2 \qquad \text{or} \qquad x = \frac{P_2}{P_1 + P_2} L$$

where L is the distance between the two forces. Figure 5.15(b) shows the reconstruction of the individual forces and a comparison of the resultant with the measured history. The comparison is quite good. Also shown is the computed location using only force values within 80% of the peak value. The nominal measured position was at the node point.

5.4 Scalability Issues: Recursive Formulation

A disadvantage of the least-squares-based sensitivity response method (SRM) just discussed is that it solves for all time components simultaneously. Therefore, as the number of time steps increase, the computational cost increases as a cubic. This puts a practical limit to the number of forces and/or the number of time steps that can be computed. Intuitively, it would seem feasible to do the inverse solution incrementally in time—using the data at the current and earlier time to determine the current forces. A direct implementation of this idea will not work because the current responses do not necessarily depend on the current forces. A simple illustration is that of the impact of a long rod: if the sensor is at a large distance away from the impact site, the applied force will have gone to zero long before the sensor shows any response.

The main idea to be developed in the following text is to perform the minimization recursively (in time), rather than globally, working from the last time step. The original form of the algorithm developed here was given in References [21, 113] and has much in common with the Kalman filter [95, 103, 104]. Further developments were given in References [33, 170, 168] for heat conduction problems and References [34, 91] for force identification problems. However, as pointed out in Reference [34] "one of the disadvantages of the method is that the amount of computations increases dramatically as the order of the model increases". This necessitated the authors to introduce a modal reduction scheme. The final size of the problem studied in References [34, 91] was just nine degrees of freedom. More details and applications can be found in the Monograph [169], and more details on the current formulation can be found in References [4, 5].

State Vector Forms of the Governing Equations

For discussion of some of the algorithms, it is preferable to have the discretized equations in the state vector form of Equation (5.1). Some specific forms for [A] and [B] will now be given.

Using the central difference finite difference scheme, our system of equations is

$$
\left\{ \begin{array}{c} u \\ \dot{u} \\ g \end{array} \right\}_{n+1} = \left[\begin{array}{ccc} (I - M_1^{-1} K \Delta t^2) & (I - M_1^{-1} C \Delta t) \Delta t & M_1^{-1} B_g \Delta t^2 \\ -M_1^{-1} K \Delta t & (I - M_1^{-1} C \Delta t) & M_1^{-1} B_g \Delta t \\ 0 & 0 & I \end{array} \right] \left\{ \begin{array}{c} u \\ \dot{u} \\ g \end{array} \right\}_n + \left\{ \begin{array}{c} 0 \\ 0 \\ \dot{g} \end{array} \right\}_n \Delta t
$$

with $[M_1] \equiv [M + \frac{1}{2} C \Delta t]$. Note that we have made the rate of force, $\{\dot{g}\}_n$, the actual applied load; this will give the equivalent of first-order regularization in the time direction.

Using the Newmark implicit scheme, the governing system of equations can be written as

$$
\left\{ \begin{array}{c} u \\ \dot{u} \\ \ddot{u} \\ g \end{array} \right\}_{n+1} = \left[\begin{array}{cccc} Z_1 & \Delta t \, Z_2 & \Delta t^2 Z_3 & Z_4 \\ \dfrac{2}{\Delta t}(Z_1 - I) & (2Z_2 - I) & 2\Delta t \, Z_3 & \dfrac{2}{\Delta t} Z_4 \\ \dfrac{4}{\Delta t^2}(Z_1 - I) & \dfrac{4}{\Delta t}(Z_2 - I) & (4Z_3 - I) & \dfrac{4}{\Delta t^2} Z_4 \\ 0 & 0 & 0 & I \end{array} \right] \left\{ \begin{array}{c} u \\ \dot{u} \\ \ddot{u} \\ g \end{array} \right\}_n + \left\{ \begin{array}{c} Z_4 g \\ \dfrac{2}{\Delta t} Z_4 g \\ \dfrac{4}{\Delta t^2} Z_4 g \\ \dot{g} \end{array} \right\}_n \Delta t
$$

with

$$
Z_1 = \beta \lceil 2C \Delta t + 4M \rfloor / \Delta t^2, \quad Z_2 = \beta \lceil C \Delta t + 4M \rfloor / \Delta t^2,
$$
$$
Z_3 = \beta \lceil M \rfloor / \Delta t^2, \quad Z_4 = \beta \lceil B_g \rceil
$$

and $\beta = [K + 2C/\Delta t + 4M/\Delta t^2]^{-1}$. Note that the DoF vector, $\{u\}$, is of size ($m = 3m_u + m_g) \times 1$.

Ricatti Equation

We will not give the detailed derivation but just hint at its essentials; a full derivation is given in Reference [169].

The global minimum principle is

$$
E(u, \tilde{P}) = \sum_{j=1}^{N} [\{d - Qu\}_j^T \lceil W \rfloor \{d - Qu\}_j + \gamma \{\tilde{P}\}_j^T [H]\{\tilde{P}\}_j] = \min
$$

where N is the total number of time steps. Consider the arbitrary time step n, the optimized partial error sum is

$$
E_n(u, \tilde{P}) \equiv \sum_{j=n}^{N} [\{d - Qu\}_j^T \lceil W \rfloor \{d - Qu\}_j + \gamma \{\tilde{P}\}_j^T [H]\{\tilde{P}\}_j]
$$

where E_n means the summation begins at $j = n$. The partial error sum at the earlier time is

$$E_{n-1}(u, \tilde{P}) \equiv \left[\{Qu - d\}_{n-1}^{\mathrm{T}} \lceil W \rfloor \{Qu - d\}_{n-1} \right.$$

$$\left. + \gamma \{\tilde{P}\}_{n-1}^{\mathrm{T}} [H] \{\tilde{P}\}_{n-1} E_n(u_n = Au_{n-1} + B\tilde{P}_{n-1}) \right]$$

where the parenthesis of E_n means "function of". The key to the algorithm is to replace the second partial sum with the solution to the previously optimized problem. It can be shown [169] that E_n is quadratic in $\{u\}_n$ for any n, hence leading to the recursion relation

$$E_n(u) = \{u\}_n^{\mathrm{T}} [R_n] \{u\}_n + \{u\}_n^{\mathrm{T}} \{S_n\} + C_n$$

where $[R_n]$, $\{S_n\}$, and C_n are coefficients. Recurrence relations can be established for these coefficients. Define the inverse term

$$[D_n] \equiv [2[B]^{\mathrm{T}} [R_n][B] + 2\gamma [H]]^{-1} \qquad (5.8)$$

then the recurrence relations are

$$[R_{n-1}] = [Q]^{\mathrm{T}} \lceil W \rfloor [Q] + [A]^{\mathrm{T}} [R_n - 2R_n B D_n B^{\mathrm{T}} R_n][A]$$

$$\{S_{n-1}\} = -2[Q]^{\mathrm{T}} \lceil W \rfloor \{d\}_{n-1} + [A]^{\mathrm{T}} [I - 2R_n B D_n B^{\mathrm{T}}]\{S_n\}$$

This is a form of the Ricatti equation [94]. Starter values for the recursion are obtained by looking at the optimized error function at the end point $n = N$. This gives

$$[R_N] = [Q]^{\mathrm{T}} \lceil W \rfloor [Q], \qquad \{S_N\} = -2[Q]^{\mathrm{T}} \lceil W \rfloor \{d\}_N$$

In this way, starting with the end values, we recursively determine and store (either internally or to disk) the quantities $[D_n][2R_n B]^{\mathrm{T}}[A]$ and $[D_n][B]^{\mathrm{T}}\{S_n\}$. This is called the backward sweep. During the forward sweep, we compute

$$\{\tilde{P}\}_n = -[D_{n+1}][2RB]_{n+1}^{\mathrm{T}}[A]\{u\}_n - [D_{n+1}][B]^{\mathrm{T}}\{S_{n+1}\}$$

$$\{u\}_{n+1} = [A]\{u\}_n + [B]\{\tilde{P}\}_n$$

These are the form of the equations used in References [34, 91] for force identification and References [33, 168, 170] for heat conduction problems. As discussed next, these equations are still unsuited for scaling to large-sized finite element problems.

Efficient Formulation

The significant arrays introduced as part of the Ricatti formulation are of size

$$\{d\} = [M_d \times 1]$$

$$[\,Q\,] = [M_d \times M]$$

$$[\,R_n\,] = [M \times M] \qquad \text{symmetric}$$

$$\{S_n\} = \{M \times 1\}$$

$$[\,D\,],\ [D_n^{-1}] = [M_p \times M_p] \qquad \text{symmetric}$$

These are not in a good form for scaling to large size problems; in particular, the square, fully populated array $[\,R_n\,]$ is of size $[M \times M]$, and recall that $M > 2M_u$. For our target system of 10,000 DoF, this is on the order of 1000 Mbytes (when stored in double precision). Hence, any manipulations involving it will be computationally intensive as well as requiring large storage. Furthermore, the system requirements are then orders of magnitude greater than that needed for the corresponding forward analysis via FEM.

The equations will now be modified so as to avoid having to form the matrix $[R_n]$ and thus avoid any computations involving it. In addition, we discuss the manipulation of the matrix $[\,A\,]$, since it too is large.

I: Time-Invariant Formulation

In many structural dynamics problems, the system parameters do not change over time. As a result, many of the arrays in the recursive relations also do not change over the increments in time; this leads to the possibility of formulating the equations in terms of changes of quantities rather than the quantities themselves. The following equations for an efficient formulation are due to Kailath [102, 124] who discusses it in good detail. Derivations can also be found in References [5, 169].

To motivate the approach, consider the symmetric array $[\,R_n\,]$ at two times in the recursion and given by

$$[R_{n-1}] = [Q^{\mathrm{T}} W Q] + [\,A^{\mathrm{T}}\,][\,R_n\,][I - 2BD_n B^{\mathrm{T}}][\,R_n\,][\,A\,]$$

$$[R_{n-2}] = [Q^{\mathrm{T}} W Q] + [\,A^{\mathrm{T}}\,][R_{n-1}][I - 2BD_{n-1} B^{\mathrm{T}}][R_{n-1}][\,A\,]$$

The only recursive term needed at the second level is $[\,D_n\,]$, which is also obtained from $[\,R_n\,]$ according to Equation (5.8). Thus, the key to the recursion is the matrix $[\,R_n\,]$. In looking at the two levels for $[\,R_n\,]$ (which must be archetypes for all two neighboring levels), it was noticed that the structure of all the arrays are of the form

$$[\,T\,]^{\mathrm{T}}[\,S\,][\,T\,]$$

with the center array, and thus also the resulting product, being symmetric. Furthermore, there is a great deal of duplication in going from one level to the next and this motivates us to express the quantities in terms of differences, that is, form the difference

$$[\Delta R_{n-1}] \equiv [R_{n-1}] - [R_{n-2}] = [A^T][[\Delta R_n] - 2[KD^{-1}K^T]_n + 2[KD^{-1}K^T]_{n-1}][A^T]$$

where we have used $[\ I\ ] = [D_n^{-1}][\ D_n\ ]$ and introduced a new variable defined as

$$[\ K_n\ ] \equiv [\ R_n\ ][\ B\ ][\ D_n\ ] = \text{size}[M \times M][M \times M_p][M_p \times M_p] = [M \times M_p]$$

This will be the key variable in the new formulation.

To put the expression for $[\Delta R_{n-1}]$ into proper recursive form, we need to replace $[K_{n-1}]$ with quantities at the n level. First look at $[D_n^{-1}]$, we get the sequence

$$[D_n^{-1}] = 2\gamma\ [\ H\ ] + 2[B^T R_n B] = 2\gamma\ [\ H\ ] + 2[\ B\ ]^T[R_{n-1} + \Delta R_n][\ B\ ]$$

$$= 2\gamma\ [\ H\ ] + 2[B^T R_{n-1} B] + 2[B^T \Delta R_n B] = [D_{n-1}^{-1}] + 2[B^T \Delta R_n B]$$

Now utilize this in the computation for $[K_{n-1}]$

$$[K_{n-1}] = [R_n - \Delta R_n][\ B\ ][D_{n-1}] = [\ K_n\ ] + [2K_n B^T - I][\Delta R_n B D_{n-1}]$$

Substitute this in the expression for $[\Delta R_{n-1}]$ then after some manipulation get

$$[\Delta R_{n-1}] = [\ A^T\ ][I - 2K_n B^T][\Delta R_n + 2\Delta R_n B D_{n-1} B^T \Delta R_n][I - 2K_n B^T]^T[\ A\ ]$$

This is of the form $[\ Y\ ][\ L\ ][\ Y\ ]^T$, hence let

$$[\Delta R_{n-1}] = [Y_{n-1}][L_{n-1}][Y_{n-1}]^T, \qquad [\Delta R_n] = [\ Y_n\ ][\ L_n\ ][\ Y_n\ ]^T$$

Substituting the first on the left and the second on the right, and equating like terms leads to the following recursion relations

$$[D_{n-1}^{-1}] = [D_n^{-1}] - 2[B^T Y_n][\ L_n\ ][Y_n^T B]$$

$$[D_{n-1}] = [D_{n-1}^{-1}]^{-1}$$

$$[L_{n-1}] = [\ L_n\ ] + 2[L_n Y_n^T B][D_{n-1}][B^T Y_n L_n]$$

$$[Y_{n-1}] = [\ A^T\ ][Y_n - 2K_n B^T Y_n]$$

$$[K_{n-1}] = [\ K_n\ ] - [Y_n - 2K_n B^T Y_n][L_n Y_n^T B][D_{n-1}]$$

$$\{S_{n-1}\} = -2[Q^T W]\{d\}_{n-1} + [\ A^T\ ]\{\{S_n\} - 2[K_n B^T]\{S_n\}\} \tag{5.9}$$

It is clear that these computations can be sequenced so that many of the partial products can be reused.

This new form of the recursive relations deals with arrays of size

$$[\, L_n \,] = [M_d \times M_d] \qquad \text{symmetric}$$
$$[\, Y_n \,] = [M \times M_d]$$
$$[\, K_n \,] = [M \times M_p]$$

The largest arrays are of size $[M \times M_d]$ and $[M \times M_p]$, and since both M_d and M_p (number of sensors and number of forces, respectively) are significantly less than M (the system size), there is a huge reduction in the storage requirements.

There are many contributions to the computational cost but they can be divided as matrix products and inversion to get $[\, D \,]$. The most significant matrix product is $[\, A^T \,][\, Y \,]$. Since all computations occur at a given time step, then the cost is linear in the number of time steps. An estimate of the cost is

$$\text{cost:} \qquad O(NM^2 M_d) + O(NMM_p M_d) + O(NM_p^3)$$

Since $M_p \approx M_d \ll M$, then it is seen that the system size dominates the cost. Furthermore, because the cost varies as $O(M^2)$, it is higher than the corresponding cost for a forward FEM problem, which can take advantage of the banded symmetric nature of its system array. We will try to add that feature next.

II: Efficient Storage and Computation

The structure of the matrix $[\, A \,]$ is shown in Figure 5.16. Because of the nature of the structural systems, the stiffness matrix $[\, K \,]$ is banded and symmetric of size $[M_u \times B]$; consequently, the two upper left portions can be made banded (although not symmetric). As is usual with explicit discretization schemes, both the mass $[\, M \,]$ and damping $[\, C \,]$ matrices are taken as diagonal and, hence, the two upper middle portions are diagonal only of size $\{M_u \times 1\}$. The first two bottom matrices are zero and the third is diagonal with

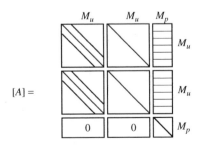

Figure 5.16: Structure of the system matrix for the explicit formulation.

unity. The remaining far left matrices are mostly zeros but because usually $M_p \ll M_u$, we do not give it any special treatment.

We never do actually assemble the [A] matrix but instead retain the individual component matrices. The storage used for the array is then

$$[(2M_u + M_p) \times (2M_u + M_g)] \quad \Longleftrightarrow \quad [M_u \times B + 2M_u]$$

This affords a substantial reduction for large systems. Reference [4] shows products involving [A] can also be achieved somewhat efficiently, in particular, taking advantage of the banded symmetric nature of [K]. Consequently, the overall cost scales as for the corresponding forward problem; of course, the absolute cost is higher.

Computer Implementation

Both the Ricatti and time-invariant equations were implemented in a program called Inverse [4, 5]. Implicit and explicit versions of the time integrations are selectable. The existing finite element code, to produce the stiffness, mass, and damping arrays of 3-D thin-walled shell structures with reinforcements, was modified and incorporated into the program. The model of the structure is created using the mesh-generating program associated with the finite element program.

In order to exercise the program and determine some of the metrics, we applied it to a shell problem using synthetic data. The chosen example is similar to that shown in Figure 5.11, and presents a complex problem in that it is three dimensional with a large number of degrees of freedom. The elements used are such that there are six degrees of freedom at each node—three displacements and three rotations. The total number of DoF for the mesh shown is 3456 and this translates into total system size (state vector plus unknown forces)

implicit:	$Size = 10375$,	$\Delta t = 5.0\,\mu s$,	$N = 400$,	time: 53 mins
explicit:	$Size = 6375$,	$\Delta t = 0.4\,\mu s$,	$N = 5000$,	time: 636 mins

The timings shown are for the solution of seven forces distributed along the west edge of the shell. The relative cost between the backward and forward portions of the solution is approximately 8:1 for both integration schemes. The difference between the two schemes is that the implicit method can use a time step related to the frequency content of the excitations, whereas the step size for the explicit method is dictated by the element size. In this instance, the advantage lies with the implicit method.

Discussion

At the core of the recursion method is the imbedding of the finite element modeling. As an advantage, because special effort was made to insure that the computational costs scale in a manner similar to the forward problem using the FEM, the computational costs scale linearly with the number of time steps. This is a notable advantage especially when long time traces are to be modeled.

On the other hand, the imbedding of the finite element modeling can be perceived as a disadvantage because it requires internal programming of the finite element method. This restricts the application of the method; it would make application to nonlinear problems especially difficult to implement. Indeed, from a programming point of view, the least-squares-based SRM is much simpler to implement, and this is a significant advantage when tackling unfamiliar problems. The SRM has two additional advantages that will only become apparent later. The first is that it can handle space-distributed data at coarse time intervals, and, therefore, is suitable for use with the whole-field optical methods as we will see in Chapter 6. Second, it has greater flexibility in handling data of different types, in particular, of simultaneously handling space-distributed and time traces.

In the following, therefore, the SRM will be the method of choice in spite of its potential computational bottlenecks.

5.5   The One-Sided Hopkinson Bar

It often occurs that we need to apply a known force history to a structure. In some instances, where the frequency of excitation is low, we can use an instrumented hammer as is common in modal analysis [76] and as was done with the cylinder problem. In other instances, we need a high-energy, high-frequency input, and here the common force transducers do not have the required specifications. The one-sided Hopkinson bar was designed for this purpose.

Basic Concept of Hopkinson Bars

A widely used method for dynamic material analyses is by means of the split Hopkinson bar [110], sometimes called the Kolsky Bar. The standard configuration has evolved into one which has two long bars of identical material, instrumented with strain gages, with the small specimen placed between them as shown in Figure 5.17(a). The first rod, or input bar, is subjected to a high-energy, high-frequency pulse usually achieved by impacting it with a high-velocity projectile, although electrically generated pulses have been also used [87]. A stress wave is generated, which, over time, causes the specimen in between the two bars to be dynamically loaded.

Typically, because the specimen is small, the wave propagation effects within the specimen are negligible, and it quickly reaches a uniform stress state. Under this circumstance, both the reflected and transmitted signals in the long bars have a simple relationship to the specimen stress (and strain) state. Using the strain gages to measure the waves propagated in both the input and output bars, the dynamic stress/strain behavior of the specimen can be determined. Since the bars are such that the cross-sectional radius

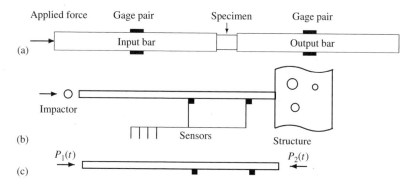

(a)

(b)

(c)

Figure 5.17: Hopkinson bars. (a) Standard split Hopkinson pressure bar configuration. (b) The one-sided Hopkinson bar. (c) Free-body diagram for the input bar.

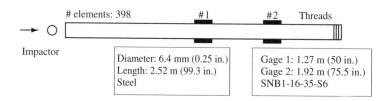

Figure 5.18: Dimensions and gage positions for the one-sided Hopkinson bar.

is not very large, dispersion effects are minimal and, therefore, the analysis can be done using 1-D rod theory.

In the one-sided Hopkinson bar configuration, shown in Figure 5.17(b), only the input bar is the same as above—there is no output bar and the specimen is not small or uniform. Since the bar is attached to a general structure, the transmitted force does not have a simple relation to the incident pulse. As indicated by the free-body diagram in Figure 5.17(c), the one-sided Hopkinson bar, therefore, requires the simultaneous determination of two force histories.

The construction of the one-sided Hopkinson bar is shown in Figure 5.18. The semi-conductor gages are arranged so as to cancel any bending effects. The rather long length of the rod allows for pulses of up to 200 μs in duration to be generated.

The basic test to be performed on the bar is when it is not attached to a specimen or structure. Here, $P_1(t)$ will just be the impacting force while $P_2(t)$ should be zero. Results are shown in Figure 5.19 for when a 25.4 mm (1.0 in.) steel ball was used as the impactor. Both gages, separately, do a good job when determining the single impact force. Note that the time scale corresponds to many reflections of the initial pulse, and there is not a noticeable deterioration of the results.

There is a slight deterioration of results when both gages are used simultaneously to determine forces at each end. The force reconstruction at the free end then gives an idea of the basic quality achievable from the Hopkinson bar.

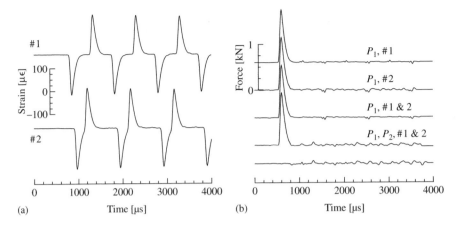

Figure 5.19: Impacted free–free rod. (a) Measured strains. (b) Reconstructed forces.

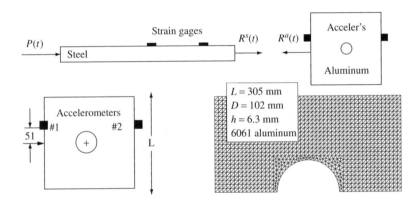

Figure 5.20: Square plate with central hole. The Hopkinson bar is attached to the left side.

Two-Dimensional In-Plane Problem

Figure 5.20 shows the schematic of a problem used to test the ability of accelerometers to reconstruct forces on a two-dimensional plate. This force will be compared with that determined by the strain gage responses from the one-sided Hopkinson bar.

Figure 5.20 shows the specimen dimensions and placement of the accelerometers. The specimen was machined from a 6.3 mm (1/4 in.) thick sheet of aluminum. It was cut to be a 305 mm (12 in.) square with a 102 mm (4 in.) diameter hole bored in its center. A small hole was drilled and tapped at the midpoint of the left edge for the connection of the one-sided Hopkinson bar.

Accelerometers were placed on the edge of the force input and on the opposite edge of the plate at 51 mm (2 in.) off the force input line. The data were collected at 4 μs rate over 4000 μs.

A convergence study was first performed to determine the proper element size for the finite element model. A symmetric model was used in the convergence study to allow for more elements of a smaller size. The force history of Figure 5.19(b) was used for this purpose. This study indicated that a maximum element size of 6.3 mm (0.25 in.) modules could be used in the analysis. The model and mesh are shown in Figure 5.20; the number of elements for the resulting model was 2159 plate elements, which creates a problem with a system size of 3407 degrees of freedom.

The acceleration data were converted to velocities and detrended, the strain gage data were modified to account for the slight nonlinearity of the Wheatstone bridge (since the resistance change of the semiconductor gages is relatively large). These data are shown in Figure 5.21.

The inverse problem was divided into two separate problems corresponding to the bar and the plate. The data from the two strain gages on the bar were used simultaneously to reconstruct the impact force and the force input to the plate. These force reconstructions are shown labeled as P and R^s, respectively, in Figure 5.22. The impact force, as expected, is essentially that of a single pulse input. The reconstructions of the force at the connection point, on the other hand, are quite complicated persisting with significant amplitude for the full duration of the recording. Note that the connection force is larger than the impact force because of the high impedance of the plate.

The R^s force from the bar was used as input to the forward problem for the plate, and the computed velocities are compared with those measured in Figure 5.21. The comparison is quite good considering the extended time period of the comparison. The data from the accelerometers were then used as separate single sensor inputs to reconstruct the force at the connection point. These force reconstructions are shown labeled as R_1^a and R_2^a in Figure 5.22. The comparison of the reconstructed forces at the connection point agrees quite closely both in character and magnitude. This is significant because the force transmitted across the boundary is quite complex and has been reconstructed from quite different sensor types placed on quite different structural types.

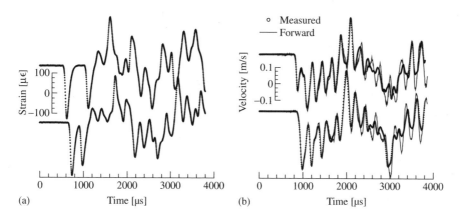

Figure 5.21: Experimentally recorded data. (a) Strains from the bar. (b) Velocities of the plate computed from accelerations.

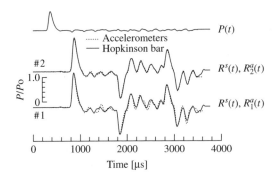

Figure 5.22: Force reconstructions for the plate with a hole. The reference load is $P_0 =$ 2224 N (500 lb).

Possible Variations of the One-sided Hopkinson Bar

Because a general finite element modeling underlies our force identification method, and because we can determine multiple unknown forces simultaneously, we are given flexibility in the design of the rod and its attachment to the structure. Here, we discuss some possible future applications of the basic idea of the one-sided Hopkinson bar.

Keep in mind that in verification studies, especially of finite element codes, we need to know the precise history of force, but the history itself need not be of any particular waveform. Thus, the goal here is to produce a measurable force history rather than one of a particular shape.

The first case is that of a dynamic three- or four-point bend as shown in Figure 5.23(a). For the four-point case, on the left side there are three unknown forces. If necessary, the "fixed" boundary conditions could also be instrumented.

The second case is that of drop weight loading as shown in Figure 5.23(b). It is worth noting that it is not necessary to include in the modeling that part of the rod below the bottom gage, that is, the drop weight and impact need not be modeled.

The final example in Figure 5.23(c) is taken from the standard split Hopkinson bar. In the high strain rate testing of composites, for example, a shear component of force can develop at the interface with the bars. By monitoring the top and bottom gages separately, this shear component of force (as well as the axial components) can be determined.

Another point worth making is that the bar need not be fixed to the specimen. That there could be multiple contacts between the rod and specimen does not affect the analysis. Finally, should loading histories of a particular waveform be required, there is the flexibility of changing the striker and/or bar without complicating the analysis.

5.6   Identifying Localized Stiffness and Mass

When a structure experiences some change, its effect on the responses is that of a change of stiffness and/or mass. In the present discussion, we will take these changes as localized

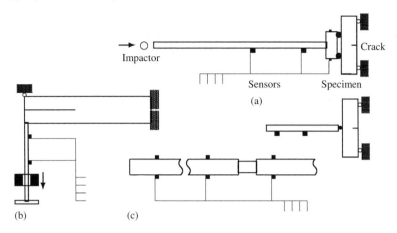

Figure 5.23: Possible Hopkinson bar arrangements. (a) Three- and four-point bending. (b) Drop weight test. (c) Split Hopkinson bar.

in the sense of being easily isolated from the structure and replaced by a series of forces, that is, the identification occurs in two steps: first the equivalent loads are determined, then the identification is performed on the reduced structure knowing (completely) the loads and responses. The next section covers the treatment of the implicit parameter case.

Beam Attached to a Spring and Mass

Consider the beam with a concentrated mass and stiffness (in the form of an attached spring) as shown in Figure 5.24(a). This identification problem is first converted into a two-force identification problem as shown in Figure 5.24(b); the spring and mass are replaced with an unknown reaction history $P_2(t)$. In an actual identification experiment, the applied load would probably be known, but we will take it as an additional unknown here because reconstructing it allows it to be used as a monitor for the quality

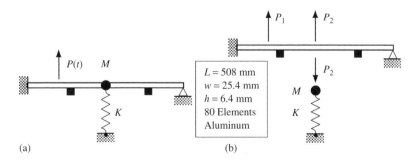

Figure 5.24: A beam with unknown localized mass and stiffness. (a) Geometry and properties. (b) Two-force identification problem and free-body diagram of mass and spring system.

of the force identification problem. Two sensors are used because there are two unknown forces. It is also worth pointing out that the spring stiffness could be identified by purely static methods.

The concentrated stiffness was taken as equivalent to 25% of the stiffness of a clamped–clamped beam, and the concentrated mass was taken as equivalent to 20% of the beam mass, that is,

$$K = 0.25 \times 192EI/L^3, \qquad M = 0.2 \times \rho AL$$

There is nothing particular about these values other than that they put the resonance frequency of the separated systems close to each other. The beam was modeled with 40 elements.

The synthetic $v(x, t)$ velocity data were generated for a sine-squared input pulse of duration 2000 μs as shown in Figure 5.25(a); this load history would be similar to that obtained from an impact hammer. An integration time step of 50 μs was used.

The two identified forces are shown in Figure 5.25(a) when the input data are contaminated with 1% noise. Note that although $P_1(t)$ is compact in time, P_2 persists over the full time window. Indeed, if there were no damping in the system, it would persist for all time. The consequence of this is that a large time window was used so that the trailing portion could be discarded.

Because the forces are known, we can also reconstruct (as a postprocessing operation) any other quantity of interest. The displacement and acceleration at the mass point are also shown in Figure 5.25(a). The acceleration shows the presence of the noise.

While the problem of the beam is fully specified, the substructure of the spring–mass system is only partially specified. We will use the force and response information to make it fully specified. The equation of motion for the separated spring–mass system is

$$P = Kv + M\ddot{v}, \qquad P = -P_2$$

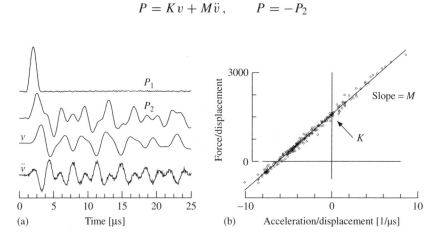

Figure 5.25: Beam with spring and mass identification results. (a) Identified forces and responses. (b) Data plotted as linear curve-fit.

Because $P(t)$, $v(t)$, and $\ddot{v}(t)$ are known, this is a two-parameter identification problem. It is possible to work directly with this relation and use the least-squares approach to determine K and M, that is, arrange the data as

$$\{u\}K + \{\ddot{u}\}M = \{P\}$$

where the braces indicates the collection of time values. Premultiply by $\{u\}^T$ and $\{\ddot{u}\}^T$ to get the system of equations

$$\begin{bmatrix} \{u\}^T\{u\} & \{u\}^T\{\ddot{u}\} \\ \{\ddot{u}\}^T\{u\} & \{\ddot{u}\}^T\{\ddot{u}\} \end{bmatrix} \begin{Bmatrix} K \\ M \end{Bmatrix} = \begin{Bmatrix} \{u\}^T\{P\} \\ \{\ddot{u}\}^T\{P\} \end{Bmatrix}$$

It is not likely that this system is ill-conditioned since there are significantly more data than unknown parameters, but just in case, add regularization. Let

$$[\,G\,] = \begin{bmatrix} \{u\}^T\{u\} & \{u\}^T\{\ddot{u}\} \\ \{\ddot{u}\}^T\{u\} & \{\ddot{u}\}^T\{\ddot{u}\} \end{bmatrix} + \gamma\,[\,H\,]$$

then, the unknown parameters are obtained from

$$\begin{Bmatrix} K \\ M \end{Bmatrix} = [\,G\,]^{-1} \begin{Bmatrix} \{u\}^T\{P\} \\ \{\ddot{u}\}^T\{P\} \end{Bmatrix} \tag{5.10}$$

Note that $[\,G\,]$ and its inverse are symmetric.

Solely, so that we can plot results, we will also work with the equation in the form

$$\frac{P}{v} = K + M\frac{\ddot{v}}{v}$$

This is now of the form $u = a + bx$ with $u \equiv P/v$, $x \equiv \ddot{v}/v$ as discussed in Chapter 3. A drawback is that $v(t)$ alternately goes through zero making division by it problematic. This is circumvented by dividing only when the displacement is above a certain threshold magnitude; the data shown in Figure 5.25(b) are for a threshold of 50% of the average magnitude. Additionally, a seven-point moving-average smoothing was performed on the acceleration. The identified K and M from Figure 5.25(b) are essentially the same as obtained from Equation (5.10).

Beam Attached to an Oscillator

In the previous example, the motion of the mass was identical to the motion at the point of attachment. We now look at a situation in which the elastic attachment between the mass and beam renders the motion of the mass as an additional unknown.

Consider the beam with an attached oscillator as shown in Figure 5.26(a). This identification problem is first converted into a two-force identification problem as shown in

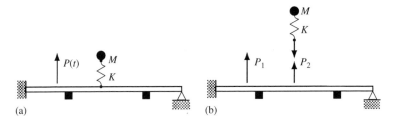

Figure 5.26: A beam with an unknown attached oscillator. (a) Geometry. (b) Two-force identification problem and free-body diagram of mass and spring system.

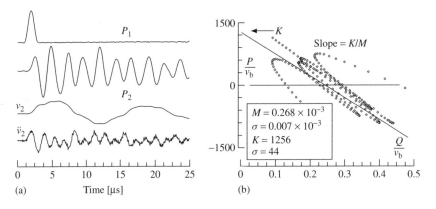

Figure 5.27: Beam with oscillator identification results. (a) Identified forces and responses. (b) Data plotted as linear curve-fit.

Figure 5.26(b), which actually is identical to that of Figure 5.24(b). However, the analysis of the separated spring–mass system will be different.

The procedure for the force identification problem follows identically to that of the previous example, and the forces, displacement, and acceleration are shown in Figure 5.27(a). The character of the attachment force and the displacement is different in comparison to Figure 5.25(a).

The equation of motion for the separated spring–mass system is written in two parts

$$P = K[v_b - v_m], \qquad P = M\ddot{v}_m, \qquad P = -P_2$$

where v_b and v_m are the displacements of the beam (base) and mass, respectively. Because v_m is not known, it must be eliminated from the system of equations. There are a number of ways of doing this, we will choose to obtain v_m by integrating the force, that is,

$$v_b(t) = \frac{1}{M} \iint P(t)\, dt\, dt \equiv \frac{1}{M} Q(t)$$

This integration is accomplished numerically giving $Q(t)$ as a history that is like $P_2(t)$ with an undulation similar to $v_2(t)$.

The equation used for identification is now

$$P(t) = K v_b(t) - \frac{K}{M} Q(t) \qquad \text{or} \qquad \frac{P}{v_b} = K - \frac{K}{M} \frac{Q}{v_b}$$

The second form, which is purely for illustration, is shown plotted in Figure 5.27(b) using a 50% threshold on v_b. Note that the spread of the data is significantly larger than in the previous example.

Frequency Domain Analysis

Frequency domain methods play an important role in the identification of linear systems, and we take this opportunity to introduce some concepts. Some discussion of spectral analysis was given in Chapter 2 and more details can be found in Reference [38, 70].

The time signals shown in Figures 5.25(a) and 5.27(a) were transformed using FFT parameters of $N = 4096$, $\Delta t = 50\,\mu s$ after a seven-point moving-average smoothing was performed. The rearranged transformed equation for the first problem can be put in the form

$$\frac{\hat{P}}{\hat{v}} = K + M \frac{\hat{\ddot{v}}}{\hat{v}} \qquad \text{or} \qquad u = a + bx$$

where the hat " $\hat{}$ " means a transform quantity that is complex. A plot against frequency of the new u and x are shown in Figure 5.28(a). The relation between displacement and acceleration should be $\hat{\ddot{v}} = -\omega^2 \hat{v}$, and this is closely substantiated in the bottom plot of Figure 5.28(a).

Separating the real and imaginary parts gives

$$\text{Re}\left\{\frac{\hat{P}}{\hat{v}}\right\} = K + M\,\text{Re}\left\{\frac{\hat{\ddot{v}}}{\hat{v}}\right\}, \qquad \text{Im}\left\{\frac{\hat{P}}{\hat{v}}\right\} = 0 + M\,\text{Im}\left\{\frac{\hat{\ddot{v}}}{\hat{v}}\right\}$$

It is interesting to note in Figure 5.28(a) that the two imaginary components seem directly proportional to each other as indicated by this equation; however, we will not use this imaginary component. Figure 5.28(b) shows the curve-fit using only the real components. The results are comparable to those shown in Figure 5.25(b).

We will rearrange the equations for the second problem by differentiating the spring relation to eliminate v_m to get

$$\ddot{v}_b = \frac{1}{M} P - \frac{1}{K} \ddot{P}$$

Note that the displacement v_b does not appear in the equation; this is because we are using a higher (differential) order model in comparison to the previous example.

Double differentiating the reconstructed load history would, in general, pose a difficulty; however, using frequency domain methods allows us to write $\hat{\ddot{P}} = -\omega^2 \hat{P}$. This is

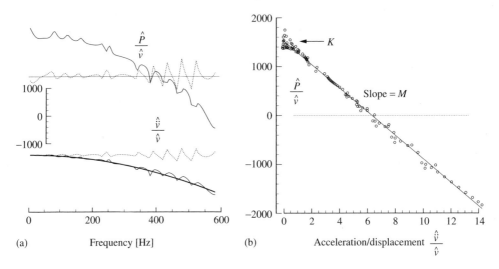

Figure 5.28: Frequency domain formulation of beam with attached spring and mass. (a) Fourier transforms of signals arranged as converted data. (dashed lines are imaginary components). (b) Linear curve-fit with frequency domain data.

more acceptable than double differentiating the displacement because the force is nearly stationary. Rearranging as before, we get

$$\frac{\hat{\ddot{v}}_b}{\hat{P}} = \frac{1}{M} + \frac{1}{K}\omega^2$$

Again, this is a two-parameter identification problem.

The Fourier spectrum of the force shows a relatively large spectral peak at $\omega = \sqrt{K/M}/2\pi = 400\,\text{KHz}$, which is not reflected in the response. The converted data are shown in Figure 5.29(a). The final set of data and the curve-fit are shown in Figure 5.29(b). Although the identified parameters are noticeably different from the time domain values, it is interesting to note that the corresponding statistics are about the same.

Discussion

A significant advantage of the frequency domain methods (within the present context) is gained when the parameters themselves are frequency dependent. For example, the modulus is frequency dependent in viscoelastic materials, a loose joint if represented by a stiffness would also be frequency dependent. While both situations are linear, the frequency domain identification can give a much richer description.

The oscillator example highlights a difficulty with this approach to parameter identification: while it is straightforward to identify the load, the parameter identification itself

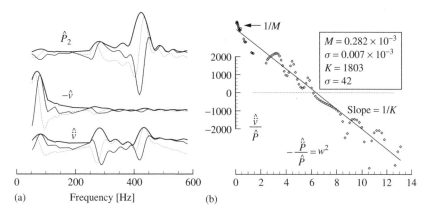

Figure 5.29: Identification results for beam with attached oscillator. (a) Fourier transforms of force and responses. (b) Data plotted as linear curve-fit.

could be a significant task. In fact, it could be comparable to obtaining the parameters directly from the complete structure. This is what we take up next.

5.7 Implicit Parameter Identification

An unknown parameter such as density is generally not associated with a localized region of the structure and therefore cannot be isolated as done in the previous section. These situations must be treated as implicit parameter identification. We will consider two methods, both of which are extensions of methods discussed in Chapter 4. Our preference will be for the SRM but it is instructive to consider the extensions of the direct differentiation method (DDM).

Direct Differentiation Method (DDM)

The assembled stiffness and mass matrices are implicit functions of its parameters a_j. To make the parameters "visible", do a Taylor series expansion of the system matrices about a known (guessed) configuration a_j^o,

$$[K] \approx [K^o] + \sum_j \left[\frac{\partial K}{\partial a_j} \bar{a}_j \right] \frac{\Delta a_j}{\bar{a}_j} = [K^o] + \sum_j \left[\frac{\partial K}{\partial a_j} \bar{a}_j \right] \Delta \tilde{P}_j$$

where $\bar{a}_j$ are normalizing values of the parameters and $\Delta \tilde{P}_j$ are nondimensional scale factors. Similarly, for the mass and damping

$$[M] \approx [M^o] + \sum_j \left[\frac{\partial M}{\partial a_j} \bar{a}_j \right] \Delta \tilde{P}_j, \qquad [C] \approx [C^o] + \sum_j \left[\frac{\partial C}{\partial a_j} \bar{a}_j \right] \Delta \tilde{P}_j$$

Substitute these into the equations of motion to get

$$[K^o]\{u\} + [C^o]\{\dot{u}\} + [M^o]\{\ddot{u}\}$$

$$+ \sum_j \left[\left[\frac{\partial K}{\partial a_j}\bar{a}_j\right]\{u\} + \left[\frac{\partial C}{\partial a_j}\bar{a}_j\right]\{\dot{u}\} + \left[\frac{\partial M}{\partial a_j}\bar{a}_j\right]\{\ddot{u}\} \right] \Delta\tilde{P}_j = \{P\}$$

Assume we have a guess, $\{u^o\}(t)$, of the displacement histories, then move the parameter terms to the right-hand side to get

$$[K^o]\{u\} + [C^o]\{\dot{u}\} + [M^o]\{\ddot{u}\} = \{P\} - \sum_j \{\phi\}_j \Delta\tilde{P}_j$$

where

$$\{\phi\}_j \equiv \left[\frac{\partial K}{\partial a_j}\bar{a}_j\right]\{u^o\} + \left[\frac{\partial C}{\partial a_j}\bar{a}_j\right]\{\dot{u}^o\} + \left[\frac{\partial M}{\partial a_j}\bar{a}_j\right]\{\ddot{u}^o\}$$

We recognize $\{\phi\}_j$ as a known (since $\{u^o\}$ is known through iteration) load distribution and its contribution is scaled by $\Delta\tilde{P}$. Solve the series of forward problems

$$[K^o]\{\psi_P\} + [C^o]\{\dot{\psi}_P\} + [M^o]\{\ddot{\psi}_P\} = \{P\}$$

$$[K^o]\{\psi_K\}_j + [C^o]\{\dot{\psi}_K\}_j + [M^o]\{\ddot{\psi}_K\}_j = -\{\phi\}_j$$

Note that the left hand structural system is the same for each problem. An approximation for the displacement is then given by

$$\{u\} = \{\psi_P\} + \sum_j \{\psi_K\}_j \Delta\tilde{P}_j$$

Replacing $\Delta\tilde{P}_j$ with $\tilde{P}_j - \tilde{P}_j^o$, and using subscript i to indicate measurements, we can write the error function as

$$\mathcal{A} = \sum_i W_i \left\{ d_i - \psi_{Pi} + \sum_j \tilde{P}_j^o \psi_{Kij} - \sum_j \tilde{P}_j \psi_{Kij} \right\}^2$$

This is reduced to standard form by adding regularization and minimizing with respect to $\tilde{P}_j$.

The responses $\{\psi_K\}_j$ are called *sensitivities* because they indicate the change in response for a unit change of parameter. The key computational issue is determining these sensitivities. For each parameter, finite difference is used on the stiffness and mass matrices to compute

$$\left[\frac{\partial K}{\partial a_j}\right] \approx \frac{[K(a_j^o + \Delta a_j)] - [K(a_j^o)]}{\Delta a_j} , \qquad \left[\frac{\partial M}{\partial a_j}\right] \approx \frac{[M(a_j^o + \Delta a_j)] - [M(a_j^o)]}{\Delta a_j}$$

We, therefore, see that this method is essentially the dynamic equivalent of the DDM as discussed in Section 4.6. As pointed out there, its major disadvantage is that it is necessary to have access to the system matrices. It has the further practical disadvantage that the loads $\{\phi\}_j$ comprise individual load histories on every degree of freedom. Storage of these can require a good deal of computer resources, but perhaps more importantly, most FEM packages would not be capable of solving a problem with so many loads without resorting to some elaborate superposition scheme. The interpretation of $\{\phi\}_j$ being associated with a set of element nodal forces $\{F\}$, as in the static case, is not available here in the dynamic case.

We conclude that the DDM is not especially suited for our purpose, and therefore will pursue the simpler SRM discussed in Sections 3.4 and 4.6.

Sensitivity Response Method

Let us have a set of guesses for the unknown parameters, then solve the system

$$[\, K\,]_o\{u\}_o + [\, C\,]_o\{\dot{u}\}_o + [\, M\,]_o\{\ddot{u}\}_o = \{P\}$$

In turn, change each of the parameters (one at a time) by the definite amount $a_j \longrightarrow a_j + da_j$ and solve

$$[\, K\,]_j\{u\}_j + [\, C\,]_j\{\dot{u}\}_j + [\, M\,]_j\{\ddot{u}\}_j = \{P\}$$

The sensitivity of the solution to the change of parameters is constructed from

$$\{\psi\}_j \equiv \left\{\frac{u_j - u_o}{da_j}\right\}\bar{a}_j$$

where $\bar{a}_j$ is a normalizing factor, typically based on the search window for the parameter. The small change da_j should also be based on the allowable search window; typically it is taken as 5% but in some circumstances could be chosen adaptively.

The response solution corresponding to the true parameters is different from the guessed parameter solution according to

$$\{u\} = \{u\}_o + \sum_j \left\{\frac{u_j - u_o}{da_j}\right\}\Delta a_j = \{u\}_o + \sum_j \{\psi\}_j \Delta\tilde{P}_j$$

It remains to determine Δa_j or the equivalent scales $\Delta\tilde{P}_j$.

Replacing $\Delta\tilde{P}_j$ with $\tilde{P}_j - \tilde{P}_j^o$, and using subscript i to indicate measurements d_i, we can write the error function as

$$\mathcal{A} = \sum_i W_i \left\{d_i - Qu_{oi} + Q\sum_j \psi_{ij}\tilde{P}_j^o - Q\sum_j \psi_{ij}\tilde{P}_j\right\}^2, \qquad u_{oi} = u(x_i, a_j^o)$$

where, as usual, Q acts as a selector. This is reduced to standard form by minimizing with respect to $\tilde{P}_j$ to give

$$[[Q\Psi]^{\mathrm{T}}\lceil W \rfloor[Q\Psi] + \gamma [H]]\{\tilde{P}\} = [Q\Psi]^{\mathrm{T}}\lceil W \rfloor\{d - Qu_{\mathrm{o}} + Q\psi\tilde{P}^o\}$$

The system of equations is solved iteratively with the new parameters used to compute updated responses.

Although the situation here is dynamic, the schematic of the algorithm as implemented is identical to the static schematic of Figure 4.36, and we achieve a nice separation of the inverse problem and the FEM analysis. Indeed, a significant advantage of the approach is that the underlying problem could even involve nonlinear dynamics.

By way of example, consider the plate shown in Figure 5.30, which has an inset plug whose mass and stiffness must be identified. The applied load history $P(t)$ is taken as known.

There are two unknown parameters in this problem; however, if the removed material of Figure 5.30 is replaced by a series of tractions, then there are too many tractions that need to be identified. This problem is therefore more suitable for obtaining the parameters directly.

Figure 5.30 shows the sensor placement and Figure 5.31(a) shows the recorded data. The first issue to be addressed is the number of sensors required. The previous examples show that, approximately, one sensor history is required for each unknown force history. Here, however, there are just two unknowns and a single sensor has the equivalent of a large amount of data.

Figure 5.31(b) shows the convergence results using just one sensor. In spite of there being 10% noise, the single sensor correctly identified the parameters. These results were quite robust as regards initial guess and choice of sensor.

A significantly more complicated version of this problem would take the position and size of the plug as additional unknowns. Besides the practical challenge of remeshing for

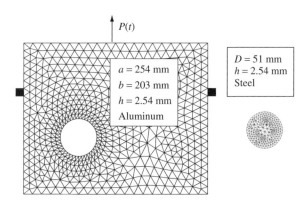

Figure 5.30: An aluminum plate with an unknown plug. The plug is displaced for easier viewing of its location and size.

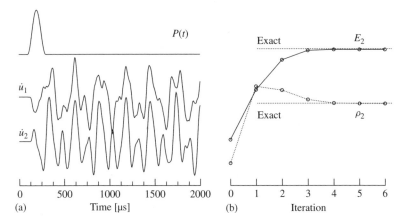

Figure 5.31: Plug identification. (a) Applied load and velocity responses. (b) Convergence results using a single sensor with 10% noise.

each guess, it is not clear that size and density or size and modulus can be adequately distinguished in the dynamics. That is, in broad terms, the dynamics as recorded at the sensor locations are dominated by the total mass and stiffness of the plug as if it were concentrated, and mass is density times size. Therefore, it would be first necessary to decide an appropriate set of parameters that could be determined.

Experimental Study

As discussed in the Introduction, a very important stage of the experimental analysis is when the nominal structural parameters and properties are "tweaked" for their actual values. Typical properties that could be determined are Young's modulus and mass density, but it also includes damping. Although there are often alternate ways of determining the modulus and density, damping is a property, which has a value only for a particular setup and, therefore, must be determined *in situ*. Parameters of a particular experimental setup must also be determined *in situ*; for example, accelerometers mounted with beeswax need precise position measurement after mounting.

This example shows how the dynamic properties of an impacted beam were obtained. Note that wave propagation characteristics are dominated by the ratio $c_0 = \sqrt{E/\rho}$, and therefore, wave responses are not the most suitable for obtaining separated values of E and ρ. Since mass density can be easily and accurately determined by weighing, we will illustrate the determination of Young's modulus and damping.

Figure 5.32 shows the geometry and the acceleration data recorded for the impact of the beam. The force transducer of Figure 2.29 was used to record the force history, and PCB 303A accelerometers used for the response. The DASH-18 system recorded the data. The beam was hit on the wide side, which made it suitable for modeling with the Bernoulli–Euler beam theory; however, this arrangement increases the amount of damping due to movement through the air.

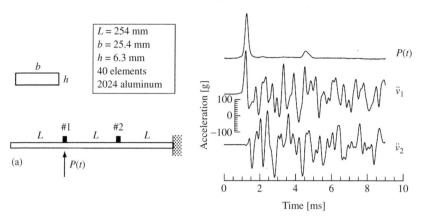

Figure 5.32: An impacted aluminum beam. (a) Geometry and properties. (b) Recorded accelerations.

The damping is implemented in the FEM code as

$$[\,C\,] = \alpha[\,M\,] = \frac{\eta}{\rho}[\,M\,]$$

which is a form of proportional damping. The parameter α has the units of s^{-1} and this will be identified along with the modulus. The search window for this type of problem can be set quite narrow since we generally have good nominal values. Consequently, convergence was quite rapid (three to four iterations) and the values obtained were

$$\text{nominal:} \quad c_0 = 5080.00\,\text{m/s}\ (200000\,\text{in./s})\,, \qquad \alpha = 0.0\,\text{s}^{-1}$$
$$\text{iterated:} \quad c_0 = 4996.18\,\text{m/s}\ (196700\,\text{in./s})\,, \qquad \alpha = 113.37\,\text{s}^{-1}$$

Figure 5.33 shows the response reconstructions using the nominal values and the iterated values compared with the measured responses. It is apparent that taking the damping into account can improve the quality of the comparisons considerably.

To improve the correspondence over a longer period of time would require iterating on the effective properties of the clamped condition. Physically, the beam was clamped in a steel vice and not fixed as in the model. For wave propagation problems, some energy is transmitted into the vice.

Discussion

Parameter identification problems are different from force identification problems in that the inputs are usually under the control of the experimentalist. For example, the applied load can be adjusted (in terms of frequency content) to be most appropriate for the range of parameters.

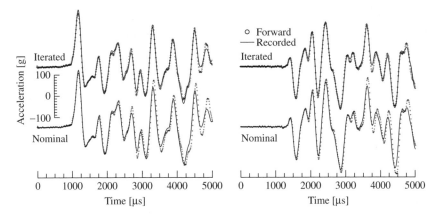

Figure 5.33: Comparison of the recorded responses and the forward solution responses.

One can envision more complex identification problems involving very many parameters. It is notoriously difficult to get good, consistent results, and some very sophisticated methods must be used. It may be possible to arrange the identification problem as a series of separate identification problems and thus reduce the ill-conditioning. For example, in the beam attached to a spring and mass problem, it would be possible to do a separate static identification of the spring, and the dynamic problem would involve only one parameter, the mass. Note that this option is not available in the beam attached to an oscillator problem.

For some problems, it makes sense to segment the parameter identification problem into two stages: first the force identification, then the parameter extraction. These are problems in which the unknowns can be easily isolated from the rest of the structure. The example of the steel plug is a case in which it is simpler to tackle the unknown parameters directly. Consider a situation in which the properties of the plug are time dependent (as could occur if the plug actually represented ablation of material), then it would not be so obvious as to which approach would be the most effective. One thing is clear, however, and that is that many more sensors would be required. These situations are revisited over the next few chapters.

5.8 Force Location Problems

Force location is one of the most difficult problems that can be tackled by the inverse methods. On the face of it, it would seem to be just a matter of adding three more unknown parameters (the (x, y, z) coordinates) to the problem. The complication comes because on the one hand the force history contains very many unknowns but are visible, and on the other hand the three unknown positions are implicit parameters requiring iteration for solution. Treating all force history values as implicit parameters (as was suggested in Section 4.6 for static loads) is not an option because of the great number of unknowns. Some hybrid scheme involving force identification and implicit parameter identification must be developed, and therefore, must involve iteration.

In this section, we will try to lay out some of the key issues involved. The problem of a finite plate, transversely impacted, will be used as the first example problem. Then, we consider the somewhat simpler problem of a frame, but one with multibays. One of the important issues to be addressed is the number of sensors needed.

Plated Structures

To make the problems definite, we will consider a single localized impact. Note that the traction could be actually distributed and we are looking at the resultant behavior.

Consider a plate whose mesh and properties are shown in Figure 5.34. The plate is clamped on the south and west sides and free on the east and north sides, and has four sensors. The impact force of Figure 5.35(a) is applied at the intersection of the dashed lines, and Figure 5.35(b) shows the four velocity responses. The object here is to use these responses to determine the history of the impact and where it occurred.

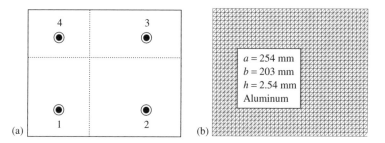

Figure 5.34: A plate impacted at the intersection of the dotted lines. (a) Sensor locations. (b) Mesh and properties.

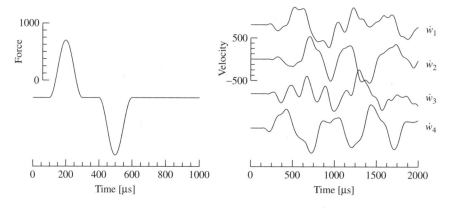

Figure 5.35: Impact of a plate. (a) Impact force history. (b) Four velocity responses.

There are two sets of unknowns here: the first is the location, which comprises two unknowns, and the second is the force history, which comprises a great number of unknowns. The previous developments have shown that if we know the force location we can determine the history.

References [53, 119] developed a location scheme that stochastically searched the allowable space. The score (or cost function) was based on the difference between force reconstructions from separate sets of sensors; that is, when the guessed location is correct, the score is a minimum. The significant drawback of the scheme is that for each guessed location a complete force identification problem needs to be solved. Indeed, for the folded plate structures of Reference [53], this was computationally feasible only by the combination of spectral elements on a massively parallel computer. The use of the SRM in such a scheme would be computationally prohibitive.

We will take a two-stage approach to the problem. The essential idea is that the early time part of the responses must have come directly from the impact and, therefore, are not affected by boundaries. Hence, if we assume there are no boundaries, then the deconvolution problem becomes significantly simpler. Once a reasonable approximation for the location is obtained, then a more refined search scheme can be used to zoom in on the precise location.

It must be emphasized that the force reconstruction history depends on precisely incorporating the actual boundary conditions, and for that we will use the SRM, but the force location problem can be more lax about these conditions.

I: Approximate Deconvolution Scheme

Reference [70] shows that the spectral response of an impacted infinite plate is given by

$$\hat{w}(r, \omega) = i\omega \mathbf{A} \left[J_0 - iY_0 - i\frac{2}{\pi}K_0 \right] = \mathbf{A}\hat{G}(r, \omega)$$

where r is position, J_0, Y_0, K_0 are Bessel functions of argument

$$z = \beta r, \qquad \beta = \sqrt{\omega}\left[\rho h/D\right]^{1/4}, \qquad D = Eh^3/12(1 - \nu^2)$$

and $\mathbf{A}$ is the amplitude at a particular frequency. Sensors other than accelerometers can be used; the transfer function for strains, for example, is [70]

$$\hat{\epsilon}_{rr} = \mathbf{A}\frac{ih}{32D}\left[(J_0 - J_2) - i(Y_0 - Y_2) + \frac{i2}{\pi}(K_0 + K_2) \right]$$

$$\hat{\epsilon}_{\theta\theta} = \mathbf{A}\frac{ih}{32D}\left[(J_0 + J_2) - i(Y_0 + Y_2) + \frac{i2}{\pi}(K_0 - K_2) \right]$$

Acceptable approximate transfer functions for the impact of composite plates is given in References [64, 66, 70].

Consider an impact location guess (x^o, y^o), then the amplitude spectrum is deconvolved simply as

$$\mathbf{A} = \frac{\hat{w}}{\hat{G}(r^o, \omega)}$$

Because $\hat{G}$ is for an infinite plate there is no worry about the zeros that occur in finite structures [70, 118]. If the guess were correct, then this amplitude would be the same from each sensor. Consequently, a measure of the goodness of the guess is the difference between the $\mathbf{A}$'s computed from the different sensors

$$\text{error} = |\mathbf{A}_1 - \mathbf{A}_2| + |\mathbf{A}_2 - \mathbf{A}_3| + \cdots$$

This can be used at any frequency, so a robust error can be obtained by summing over many frequency components

$$\text{error} = \sum \left[|\mathbf{A}_1 - \mathbf{A}_2| + |\mathbf{A}_2 - \mathbf{A}_3| + \cdots \right]$$

It is noted that this deconvolution procedure is not just based on the propagation characteristics of flexural waves; it also has incorporated the evanescent effects (through K_o), and therefore, there is little restriction on the location and/or frequency content of the signal used. More specifically, a propagation-waves approach would have required the wave to have propagated a significant distance such that the evanescent effects had spatially attenuated. This would be achieved through a combination of large distance and/or high frequency. Such a requirement is not necessary here.

II: Effect of Number of Sensors and Nearness of Boundaries

Figure 5.36 shows contour plots for a number of sensor combinations. Parenthetically, the computational cost of the present deconvolution scheme is so insignificant that the complete domain can be scanned in a brute force manner and the results presented as contours.

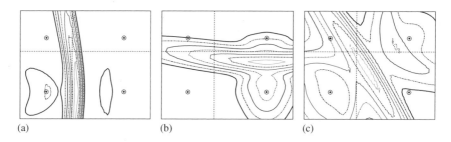

(a)   (b)   (c)

Figure 5.36: Error contours for different sensor pair combinations. The impact site is at the intersection of the dashed lines. (a) Sensors 1 & 2. (b) Sensors 2 & 3. (c) Sensors 1 & 3.

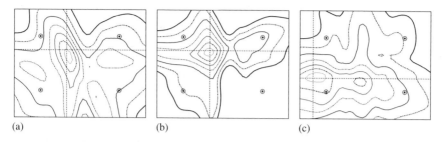

(a)   (b)   (c)

Figure 5.37: Effect of nearness of boundaries to the impact location. (a) Boundaries are somewhat remote, three sensors. (b) Boundaries are somewhat remote, four sensors. (c) Boundaries are somewhat close, four sensors.

The first issue to be addressed is the number of sensors needed to give a good location estimate. Figures 5.36(a, b, c) show results for two sensor combinations. It appears that in each case the minimum error correctly identifies the bisector of the sensors along which the impact occurred but cannot accurately tell its lateral position. Figure 5.37(a) uses all three sensors simultaneously to get a more definite 2-D location of the impact. In performing the FFT to get the amplitude spectrums $\hat{w}(r, \omega)$, the only preconditioning done to the time signals was to use an exponential window filter to bring the late time response to zero in a smooth fashion. We conclude that a minimum of three sensors is required.

Figures 5.37(b) shows results using four sensors; the definition of the location is sharpened in comparison to Figures 5.37(a).

The second issue to be addressed is that of the nearness of the boundaries because the deconvolution scheme does not include the effect of reflections from boundaries even though they are included in the response.

Figure 5.37(c) shows results using four sensors where the impact is somewhat close to the left boundary. The scheme is surprisingly successful in identifying the region of the impact although we begin to see the emergence of a competing minimum, and the identified minimum is shallower than in Figure 5.37(b).

As discussed next, we will refine the location by taking the explicit boundaries into account. What the present results show is that the boundaries do not have an overriding effect. The implications of this are that the plate as analyzed could have come from, say, a folded plate structure with many faces, the boundaries being those of plate attachments, and still the identification scheme would work. Aspects of this are considered in the multibay frame example.

III: Refinement of Location

The primary objective of the scheme just discussed is to identify high-probability regions for the location of the force. What we will now do is refine the location information by taking the full structural information into account.

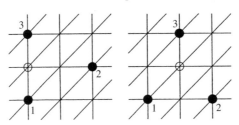

Figure 5.38: Local meshes and the locations where three forces are obtained simultaneously. (a) Impact is somewhat close to boundary. (b) Impact is somewhat remote from boundary.

One possibility is to evaluate the force at every node in the suspected area. Since this is a small area, the computational cost will not be too high. We will look at an alternative scheme that takes advantage of the ability to determine multiple forces.

Figure 5.38 shows the local meshes for the two cases considered. On the basis of the contour plots, a region spanned by a few elements is chosen, and the large black dots in the figure indicate where the forces are obtained simultaneously; the empty circle indicates the exact impact location, that is, in each case three forces are identified simultaneously.

The three forces (if not separated too much) must be equipollent to the actual applied load. Consequently, the resultant applied load must be

$$P = P_1 + P_2 + P_3$$

and using the first node as the reference node for the moments leads to the position of the resultant as

$$x = \frac{P_2 x_2 + P_3 x_3}{P_1 + P_2 + P_3}, \qquad y = \frac{P_2 y_2 + P_3 y_3}{P_1 + P_2 + P_3}$$

Since the resultant can go through zero, this is not applied at all times but only when the magnitude of P is above a threshold value; in the present examples we take it as half the mean magnitude.

Figure 5.39(a) shows the three-force reconstructions for the first case. Keep in mind that it is not expected that these forces actually behave like the resultant; what is most important is that they exhibit the appropriate moments about the first node. The calculated positions at each time step are shown in Figure 5.39(b). When the positions are averaged, we get

$$x = 0.248, \qquad y = 0.246$$

which is very close to the actual value. This value was then used for a single-force identification and this is shown as the circles in Figure 5.39(b).

The corresponding results for the somewhat near boundaries are shown in Figure 5.40. The average estimate of the location is

$$x = 0.035, \qquad y = 0.248$$

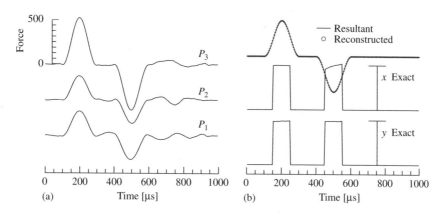

Figure 5.39: Refined results for somewhat remote boundaries. (a) The three-force reconstructions. (b) Resultants and estimated locations.

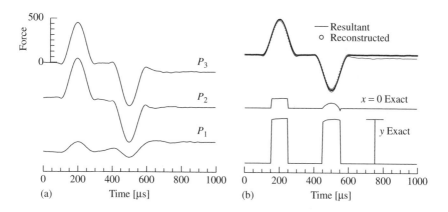

Figure 5.40: Refined results for somewhat near boundary. (a) Three-force reconstructions. (b) Resultants and estimated locations.

Again, the single-force reconstruction based on this refined estimate of the force location is quite close to the actual.

In performing the three-force identifications, some spatial regularization was used. The results are not sensitive to the actual value used. Also, the synthetic data were corrupted with 0.1% noise.

Frame Structures

We now consider the location problem for frame structures. In some ways, it is simpler than the plate/shell problem but it allows us to explore some global aspects of the problem without making it computationally expensive.

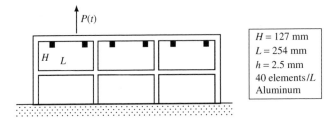

Figure 5.41: Force location on frame structures.

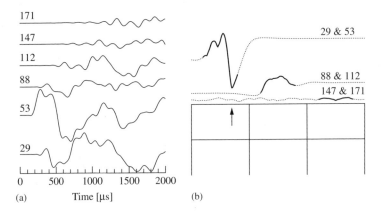

Figure 5.42: Frequency domain estimation of force location. (a) Velocity responses. (b) Error function using different sensor combinations.

Consider the multibay frame shown in Figure 5.41. An array of sensors is placed along the top and Figure 5.42(a) shows the velocity responses when the force history of Figure 5.35 was applied. It is the objective here to use these responses to determine the impact location and the load history.

We begin by using the frequency domain idea of force location as just used for plates, that is, the top horizontal members will be considered as a single continuous, infinitely long beam. The appropriate transfer function relation (assuming Bernoulli–Euler beam theory [70]) is

$$\hat{v}(r, \omega) = i\omega A[e^{-i\beta x} - i e^{-\beta x}] = A\hat{G}(x, \omega), \qquad \beta = \sqrt{\omega}\left[\rho A / EI\right]^{1/4}$$

The corresponding expression for strain is very similar, the essential difference is that the bracketed term becomes $[e^{-i\beta x} + i e^{-\beta x}]$.

Consider an impact location guess (x^o), then the amplitude spectrum is deconvolved as

$$A = c\hat{v} / \hat{G}(x^o, \omega)$$

A measure of the goodness of the guess is the difference between the **A**'s computed from the different sensors over many frequency components

$$\text{error} = \sum \left[|\mathbf{A}_1 - \mathbf{A}_2| + |\mathbf{A}_2 - \mathbf{A}_3| + \cdots \right]$$

The frequency range can be appropriately delimited as done for the plate.

On the basis of the discussion of the plate, the minimum number of sensors needed per member is two. Figure 5.42(b) shows the error function using different sensor combinations (the solid part of the line is inside the sensor locations and they all have the same zero reference at the top of the frame). It is clear that Sensors 29 and 53 correctly identify the impact site. The problem, however, is that the other sensors give a global minimum that is smaller. This can be understood by looking at Figure 5.42(a) and noting that the more remote the sensor, the smaller the response. As a consequence, the reconstructed amplitudes $\mathbf{A}_i$ are also smaller.

If all six sensors are used simultaneously, the correct global minimum is obtained; however, if only the first, third, and fifth sensors (one sensor per member) are used, an erroneous location is obtained. We conclude that sensor pairs within a member can correctly identify the location, but they cannot globally identify which member is impacted because of the joints. We need a method that will first identify the member, and then, use a refinement method to zoom in on the actual location.

Looking at the head of the responses of Figure 5.42(a), it is visually clear as to which sensors are near and which are remote to the impact site; all we need do is quantify this. Wave propagation in beams and frames is dispersive, that is, the waveform shape changes because different frequency components travel at different speeds. We will focus on the first arrival of a disturbance that exceeds a certain threshold as a way to delimit a group behavior; Figure 5.43(a) shows the arrival times at each sensor. The data are pointing to the correct source of the disturbance, but if only one sensor per member were available, then the origin would be ambiguous.

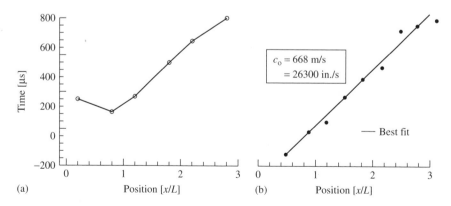

Figure 5.43: Estimating the arrival time. (a) Arrival times for actual pulse. (b) Calibration curve for sine-squared pulse.

The head of each response travels nearly nondispersively (as noted by the dominant linear relation of the arrival times for the right five sensors), and therefore, the space/time relation can be written as

$$x - x_0 = \pm c_0[t - t_0]$$

In this, we do not know the space origin (x_0) or time origin (t_0) of the pulse, nor do we know the speed (c_0) or direction ($\pm$) it has. Since we have arrival data, we can do a parameter identification but because of the ambiguity of wave direction (left moving waves must be treated as having a negative wave speed), the identification is actually a nonlinear one. A possible error function is

$$E = [[x - x_0]^2 - c_0^2[t - t_0]^2]^2$$

The parameters were correctly identified when all six sensor data were used. However, the results were quite unstable when only one sensor per member was used.

The speed c_0 is a structural property, which can be estimated independently since we know the actual structure. A simple way of doing this is to take an FFT of a typical sensor response to estimate the highest frequency content and to construct a sine-squared pulse that has the same frequency range. As an approximation, the pulse duration is related to the highest frequency by $T = 2/f$, which in the present case gives a duration of $150\,\mu s$ for a highest frequency of 13 KHz. This pulse was then applied to the actual structure, and Figure 5.43(b) shows the arrival times and best-fit line. As shown next, we do not need a very accurate value of the speed, just a reasonable approximation.

Figures 5.44(a, b) show the contours of the error function for two quite different values of speed. Only one sensor per member is used and the search is only for x_0 and t_0 ranging over the scales of Figure 5.43. Even with these inaccurate values of speed, the contours have delimited the search space considerably. But as shown in Figure 5.44(c), we can get a precise value of location by monitoring the minimum value of the error at each speed.

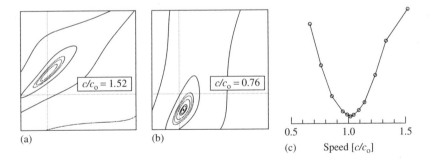

Figure 5.44: Search for the impact location when the speed is specified. The impact site and initiation time are at the intersection of the dashed lines. (a) Error contours with a high value of speed. (b) Error contours with a low value of speed. (c) Minimum error against speed.

It is interesting to note that the group speed for waves in a beam is given by

$$c_g = 2\sqrt{\omega} \left[EI/\rho A \right]^{1/4} = 1092 \, \text{m/s} \, (43{,}000 \, \text{in./s})$$

for the 13 KHz component. This is larger than what is observed above. Reference [70] has an in-depth discussion of the effect of coupling of elastic systems on their wave propagation behavior; in short, it shows that there is a cutoff frequency (low frequencies do not propagate) and that the propagating speeds are slower. The vertical members of the frame seem to have this effect.

It must be pointed out, however, that agreement or otherwise with Bernoulli–Euler beam theory does not matter since we use the actual structure to calibrate the speed. In this way, the structure could be quite complex and our approach would still be valid since all that we want is to delimit the search area so that the more refined SRM technique can be used.

Force Location on Arbitrary Structures

The main achievement of the presented force location algorithm is that it can identify regions where the force is most likely applied. Indeed, on a multifaceted structure, it can be applied to each face separately—of course, a sufficient number of sensors must be available. Once these regions have been identified, then a number of schemes are available to refine the actual location. These are made computationally feasible because the search space is small.

Consider now an arbitrary structure that is not amenable to any particular simplification: what is the computational cost of determining the force location?

We can set up a cost function similar to before, that is, use four sensors and compute four force identifications at the guessed location. For the arbitrary structure, the force deconvolution is inseparable from the location problem and the full SRM must be used for these force identifications. Four force identifications are needed for one cost function evaluation.

Suppose we use a variation of the steepest descent method to move in the downhill direction, then we need two additional cost function evaluations. Therefore, at each step, a total of 12 force identification problems must be performed. This is a significant computational cost.

Actually, the computational cost would be acceptable if it were to give the correct answer. The problem, as judged from any of the contour plots of Figures 5.36 and 5.37, is that there can be many minima that are not the global minimum required. It seems that one of the characteristics of the force location problem is that any search method adopted must be of the global type. To incorporate that feature will substantially increase the computational cost.

This global nature to the force location problem requires the use of many sensors and the application of algorithms that can utilize many sensors. We will add more to this topic in the next chapter when we develop whole-field methods, which seem inherently very suited to the force location problem. But, at present, it can be said that the force location problem does not yet have a satisfactory solution, and a good deal more effort must be expended to find such a solution.

5.9   Discussion

This chapter brought together a number of technologies to solve the force and parameter identification problem on complex structures.

The primary advantage of the dynamic SRM is that, like its static counterpart, it can use a commercial FEM program as an external processor and thus greatly extend the range and type of problems that can be solved. Its major disadvantage is that it solves for all time components simultaneously. Because the computational cost accrues as the cube of the number of time steps and the square of the number of forces, there is, therefore, an upper limit on the number of forces it can determine. The recursive formulation was introduced as an attempt to overcome this drawback. While the objective was achieved, the price to be paid is that the FEM code must be imbedded in the inverse code, thereby limiting the versatility of the method. Because versatility is an important attribute, this method was not pursued any further.

From a programming point of view, the SRM is relatively simple to implement and this is a significant advantage when tackling unfamiliar problems. It has two additional advantages that will only become apparent later. The first is that it can handle space-distributed data at coarse time intervals, and, therefore, is suitable for use with the whole-field optical methods as we will see in the next chapter. Second, it has greater flexibility in handling data of different types, in particular, of simultaneously handling space-distributed data and time trace data.

6

Transient Problems with Space Data

A sensor such as a strain gage or an accelerometer is at a discrete space location but gives information continuous in time. On the other hand, a photographic frame such as one of those shown in Figure 6.1, is at a discrete time but gives information continuous in space. Ideally, we would like many frames at a fine time discretization; however, when the data recorded is in the form of high-speed photographs, the total number of frames will be severely limited. Hence, if we wish to cover an event of significant duration, we need to supplement the recorded data so as to effectively supply the missing frames.

The purpose of this chapter is to develop such a method. The sensitivity response method of Chapter 5 is modified for using this space data and recovering complete time histories.

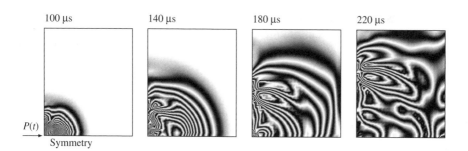

Figure 6.1: A sequence of simulated photographic frames depicting photoelastic light-field fringes in an impacted epoxy plate.

Modern Experimental Stress Analysis: completing the solution of partially specified problems. James Doyle
© 2004 John Wiley & Sons, Ltd ISBN 0-470-86156-8

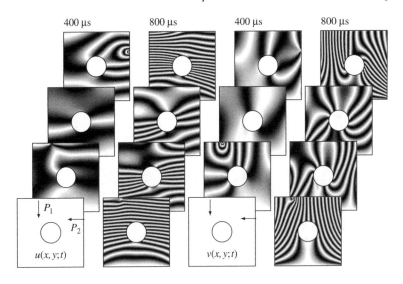

Figure 6.2: A sequence of simulated frames depicting Moiré fringes in a doubly impacted plate with a hole. The two force histories are uncorrelated and the interframe timing is 100 μs.

6.1 Space–Time Deconvolution

Figure 6.2 shows the simulated results of an impact experiment that uses high-speed photography to record Moiré fringes. It is our objective to be able to reconstruct the complete dynamic event from this limited data. The scenario we envision here is that the force history is unknown (it is due to blast loading or ballistic impact, say) and we are especially interested in local (such as at the notch root) stress behavior.

One thing that is clear from the figure is that we cannot use the Moiré fringes directly to get the stresses since the fringes are too sparse to allow for spatial differentiation. Second, since the total number of frames is severely limited, the frames cannot be used directly to obtain history behavior. Conceivably, we can overcome these limitations by increasing the sensitivity of the Moiré (e.g., using Moiré interferometry [142]) and simply paying the price of having multiple cameras [49]. Here, we propose to tackle the problem directly; that is, given a limited number of frames of (sometimes) sparse data, determine the complete history of the event including stresses and strains.

Relationship between Perturbation Forces and Frames

Figure 6.3 tries to explain how (within a wave propagation context) a single frame of data will contain both early time and late time information about the input; areas close to the input contain the latest information, remote areas contain the earliest information. When reflections are present, this simple association does not hold, and this is considered

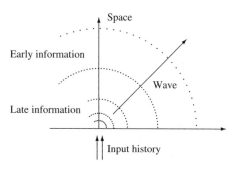

Figure 6.3: How time information from a space-localized event is spread out over space.

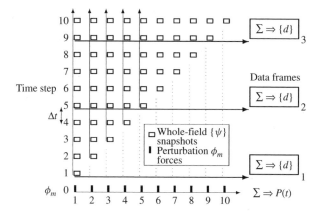

Figure 6.4: Relationship between frames due to the perturbation forces and the measured frames.

later. Clearly, the frame has no information concerning the event after the time at which the frame was taken.

Figure 6.4 shows a little more explicitly the relationship between the force and the observed frames. As in the previous chapter, let the single unknown force history be represented by

$$P(t) = \sum_{m=1}^{M_p} \phi_m(t) \tilde{P}_m$$

where each perturbation force history $\phi_m(t)$ is similar but shifted one time step $m\Delta t$, and M_p is the total number of time steps (or unknown number of force components). Each perturbation force causes a snapshot at each time step; the sequence is the same for each force just shifted one time step. Thus, it is sufficient to consider just the first perturbation force and its snapshots:

$$\phi_1(t) \Longrightarrow \{\psi\}_1, \{\psi\}_2, \{\psi\}_3, \ldots, \{\psi\}_{M_p}$$

A frame of data has contributions only up to that time but the sequence of snapshots are reversed in time. For example, let the time step be $\Delta t = 10\,\mu s$ and consider data frame $\{d\}_2$, the frame of all the displacements (DoF) at $50\,\mu s$ caused by $P(t)$. In this, $\{\psi\}_5$ is a snapshot of all the displacements at $50\,\mu s$ caused by ϕ_1. Since ϕ_m are similar to each other but shifted an amount $m\Delta T$, $\{\psi\}_4$ is a snapshot of all the displacements at $50\,\mu s$ caused by ϕ_2. Similarly, $\{\psi\}_3$, $\{\psi\}_2$, and $\{\psi\}_1$ are caused by ϕ_3, ϕ_4, and ϕ_5 at $50\,\mu s$, respectively. Therefore, for data frame $\{d\}_2$, $\tilde{P}_5$ contributes to $\{u\}_1$, $\tilde{P}_4$ contributes to $\{u\}_2$, all the way to $\tilde{P}_1$ contributing to $\{u\}_5$; any component greater than 5 does not contribute. It is this reversal in time that allows the scheme to be computationally efficient.

Thus, a snapshot of the displacements at time step i is

$$\{u\}_i = \{\psi\}_i\,\tilde{P}_1 + \{\psi\}_{i-1}\tilde{P}_2 + \{\psi\}_{i-2}\tilde{P}_3 + \cdots + \{\psi\}_1\tilde{P}_i + 0\tilde{P}_{i+1} + \cdots + 0\tilde{P}_{M_p}$$

where $\{u\}$ is a snapshot of all the displacements (DoF) at this instant of time. Note that force components occurring at time steps greater than i do not contribute.

We can write this relation as

$$\{u\}_i = \sum_{m=1}^{i}\{\psi\}_{(i+1-m)}\tilde{P}_m + \sum_{m=i+1}^{M_p} 0\tilde{P}_m \qquad \text{or} \qquad \{u\} = [\ \Psi\ ]\{\tilde{P}\} \qquad (6.1)$$

The $[M_u \times M_p]$ array $[\ \Psi\ ]$ (M_u is the total number of DoF) relates the snapshot of displacements to the applied force.

Let the N_d frames (photographs) of data be organized as

$$\{\overline{d}\} = \{\{d\}_1, \{d\}_2, \ldots, \{d\}_{N_d}\}^{\mathrm{T}}$$

where $\{d\}_n$ is a vector of M_d components. The measured data DoF are a subset of the total number of DoF, that is,

$$\{d\}_n \Longleftrightarrow [\ Q\ ]_n\{u\}_n \qquad \text{or} \qquad \{\overline{d}\} \Longleftrightarrow [\overline{Q}\ \overline{\Psi}]\{\tilde{P}\} \qquad (6.2)$$

As discussed in the previous chapters, $[\ Q\ ]$ acts as a selector of the DoF participating in the data, and products such as $[\ Q\ ][\ \Psi\ ]$ are actually never performed.

The error function in the present case is

$$E(\tilde{P}) = \{\overline{d} - \overline{Q}\,\overline{\psi}\tilde{P}\}^{\mathrm{T}}\lceil\ W\ \rfloor\{\overline{d} - \overline{Q}\,\overline{\psi}\tilde{P}\} + \gamma_{\mathrm{t}}\{\tilde{P}\}^{\mathrm{T}}[\ H_{\mathrm{t}}\ ]\{\tilde{P}\}$$

Minimizing with respect to $\tilde{P}$ gives

$$[[\overline{Q}\ \overline{\Psi}]^{\mathrm{T}}\lceil\ W\ \rfloor[\overline{Q}\ \overline{\Psi}] + \gamma_{\mathrm{t}}[\ H_{\mathrm{t}}\ ] + \gamma_{\mathrm{s}}[\ H_{\mathrm{s}}\ ]]\{\tilde{P}\} = [\overline{Q}\ \overline{\psi}]^{\mathrm{T}}\lceil\ W\ \rfloor\{\overline{d}\} \qquad (6.3)$$

The core matrix is symmetric, of size $[M_p \times M_p]$, and, in general, is fully populated.

Multiple Forces

There are two situations of multiple forces of interest here. The first is that of multiple isolated forces and the second is that of traction distributions. The main difference is whether the forces are related to each other or not; isolated forces are a finite number of forces that are not related, but a traction distribution contains an infinite number of forces that are continuously related to each other through spatial proximity.

Let the N_p unknown force histories be organized as

$$\{\overline{P}\} = \{\{\tilde{P}\}_1, \{\tilde{P}\}_2, \ldots, \{\tilde{P}\}_{N_p}\}^{\mathrm{T}}$$

where $\{\tilde{P}\}_n$ is a vector of M_p components. This collection of force vectors causes the collection of responses

$$\begin{Bmatrix} \{u\}_1 \\ \{u\}_2 \\ \vdots \\ \{u\}_{N_d} \end{Bmatrix} = \begin{bmatrix} [\Psi]_{11} & [\Psi]_{12} & \cdots & [\Psi]_{1N_p} \\ [\Psi]_{21} & [\Psi]_{22} & \cdots & [\Psi]_{2N_p} \\ \vdots & \vdots & \ddots & \vdots \\ [\Psi]_{N_d1} & [\Psi]_{N_d1} & \cdots & [\Psi]_{N_dN_p} \end{bmatrix} \begin{Bmatrix} \{\tilde{P}\}_1 \\ \{\tilde{P}\}_2 \\ \vdots \\ \{\tilde{P}\}_{N_p} \end{Bmatrix} \tag{6.4}$$

which we abbreviate as

$$\{\overline{u}\} = [\overline{\Psi}]\{\overline{P}\}$$

where $\{u\}_i$ is a snapshot of all the displacements (DoF) at the i instant of time. The $[M_u \times M_p]$ array $[\Psi]_{ij}$ relates the i snapshot of displacements to the j applied force.

As we did for the single force case, let the N_d frames (photographs) of data be organized as

$$\{\overline{d}\} = \{\{d\}_1, \{d\}_2, \ldots, \{d\}_{N_d}\}^{\mathrm{T}}$$

and the computed sensitivity responses be related to the data by

$$\{d\}_n \Longleftrightarrow [Q]_n\{u\}_n \qquad \text{or} \qquad \{\overline{d}\} \Longleftrightarrow [\overline{Q}\,\overline{\Psi}]\{\overline{P}\} \tag{6.5}$$

The error function is

$$E(\overline{P}) = \{\overline{d} - \overline{Q}\,\overline{\psi}\,\overline{P}\}^{\mathrm{T}} \lceil W \rfloor \{\overline{d} - \overline{Q}\,\overline{\psi}\,\overline{P}\} + \gamma_{\mathrm{t}}\{\overline{P}\}^{\mathrm{T}}[H_{\mathrm{t}}]\{\overline{P}\} + \gamma_{\mathrm{s}}\{\overline{P}\}^{\mathrm{T}}[H_{\mathrm{s}}]\{\overline{P}\}$$

Minimizing with respect to $\overline{P}$ then gives

$$[[\overline{Q}\,\overline{\Psi}]^{\mathrm{T}} \lceil W \rfloor [\overline{Q}\,\overline{\Psi}] + \gamma_{\mathrm{t}}[H_{\mathrm{t}}] + \gamma_{\mathrm{s}}[H_{\mathrm{s}}]]\{\overline{P}\} = [\overline{Q}\,\overline{\psi}]^{\mathrm{T}} \lceil W \rfloor \{\overline{d}\} \tag{6.6}$$

The core matrix is symmetric of size $[(M_p \times N_p) \times (M_p \times N_p)]$ and, in general, is fully populated.

The above has incorporated both time and space regularization. Time regularization connects components within $\{\tilde{P}\}_j$ only, while space regularization connects components in $\{\tilde{P}\}_j$ across spatially near locations. The matrices have the structure

$$[\,H_t\,] = \begin{bmatrix} H_1 & 0 & 0 & \cdots \\ 0 & H_2 & 0 & \cdots \\ 0 & 0 & \ddots & H_{N_g} \end{bmatrix}, \qquad [\,H_s\,] = \begin{bmatrix} I & -I & 0 & \cdots \\ -I & 2I & -I & \cdots \\ 0 & -I & 2I & \cdots \end{bmatrix}$$

where $[\,H\,]_j$ is an $[M_p \times M_p]$ symmetric matrix and $\lceil\, I \,\rfloor$ is the unit matrix of size $[M_p \times M_p]$. The program StrIDent has implemented zero-, first-, and second-order regularization.

Computational Issues

The sensitivity responses $\{\psi\}$ are obtained by the finite element method as illustrated in Figure 5.5 for point sensors. The only significant difference with the point sensor algorithm is that complete snapshots (as opposed to discrete location responses) are stored at each time step and these are then assembled into the $[\,A\,] = [Q\Psi]$ matrix as indicated in Figure 6.4; that is, although the data frames may be quite sparse in time, each data frame is composed of every snapshot up to that instant in time. Depending on the problem, these could require a considerable amount of computer storage apart from being slow to access; for this reason StaDyn stores the snapshots in binary format.

As the number of time steps (M_p) or the number of forces (N_p) increases, the computational cost of solving Equation (6.6) can become significant. The cost for the solution stage is about

$$\text{cost} = \tfrac{1}{3}(N_p \times M_p)^3$$

To put this into perspective, consider solving a medium-sized problem ($M_p = 1000$) with a single unknown force on a benchmark machine of 20 MFlops, then

$$\text{cost} \approx \tfrac{1}{3} \times 1000^3 \div (20 \times 10^6) = 17\,\text{seconds}$$

On the other hand, a large problem with 10,000 time steps or with 10 unknown forces would take about 4 hours. This is an issue to be addressed later.

6.2 Preliminary Metrics

We will concentrate, in the following examples, on determining uncorrelated force histories and leave the discussion of multiple forces forming a traction distribution (which

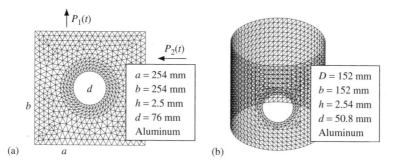

Figure 6.5: Mesh and properties of models tested. (a) In-plane loading of a plate with two forces. (b) Backside loading of a cylinder.

requires the use of space regularization) until later. In each case, we will determine the forces using synthetically generated data. An important consideration to emerge is the number of frames needed for a correct solution.

The specimen geometries are shown in Figure 6.5 with the location of the forces indicated. The flat plate has two uncorrelated forces, whereas the cylinder has a single force obscured from the camera. The force histories are shown (along with their amplitude spectrums) in Figure 6.6. Note that the spectrum of the perturbation force encompasses that of both forces.

A variety of factors contribute to the manifestations of ill-conditioning; noise and inadequacy of the data are the two of significance here. As discussed in Chapter 3, noise does not cause ill-conditioning, but it can be used to manifest the presence of ill-conditioning. All the synthetic data were corrupted with Gaussian noise having a standard deviation of up to 10% of the average displacement. Additionally, only a 70% subset of the image data is used. This subset is taken from interior points away from

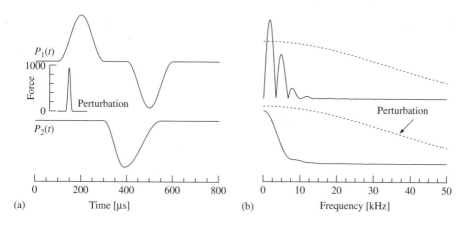

Figure 6.6: Two uncorrelated forces and the perturbation force. (a) Histories. (b) Amplitude spectrums.

any boundaries; as a consequence, no data come from close proximity to the load points. Finally, since these two forces are uncorrelated, space regularization is not needed.

For various reasons, the data collected may come from just a limited region of the structure, and further, this region may not necessarily include the location of the applied load. Consider, for example, a cylinder structure: this problem is such that there is naturally at most 50% data collected by a single image. Thus, frames taken from the front of the cylinder are remote data if used to determine inputs on the backside. The following problems were designed to investigate the effect of remote data on the force reconstruction.

Flat Plate with a Hole

Consider the in-plane loading problem of Figure 6.5. The synthetic data were generated with a time step of $\Delta T = 1\,\mu s$ and are shown in Figure 6.2.

Additionally, only a 70% subset of the image data is used. This subset is taken from interior points away from any boundaries; as a consequence, no data come from close proximity to the load points.

Effect of Frame Rate

Consider the in-plane loading problem of Figure 6.5. The synthetic data were generated with a time step of $\Delta T = 1\,\mu s$ and are shown in Figure 6.2. The inverse analysis will be done with a time step of $\Delta T = 10\,\mu s$.

Figure 6.7(a) shows the reconstructed force histories using 8 and 4 frames distributed over the total time period of $800\,\mu s$ (These correspond to framing rates of $100\,\mu s$ and $200\,\mu s$, respectively). The time step was $\Delta T = 10\,\mu s$ and no regularization was used. Clearly, it is necessary to have a sufficient number of frames to get good results—in this case 8 frames seem sufficient. The center plot in each set uses 8 frames but the u and v data are at the $200\,\mu s$ framing rate. Adding extra data like this does not make a significant difference. We conclude that it is not the number of frames *per se* that is important, but rather the frame rate.

Figure 6.7(b) shows the reconstructed force histories for different amounts of noise and no regularization. All eight frames of v displacement data were used. The 10% noise is relatively large but, nonetheless, a good reconstruction is achieved even though the regularization is zero. This means that ill-conditioning is not a problem and the data are adequate for a good reconstruction.

Multiple Reflections and Frame Rate

It can be shown that a single photograph for a semi-infinite body is capable of determining a force history up to the instant of the photograph. It seems that the presence of multiple reflections requires the simultaneous use of multiple photographs. This is an important issue, so we explain it with the following example.

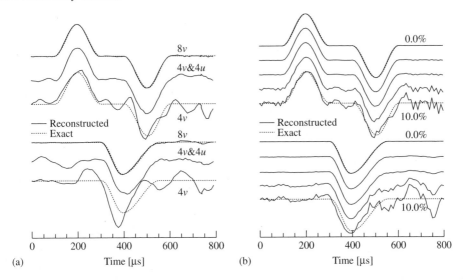

Figure 6.7: Performance. (a) Effect of framing rate for 1% noise, 70% data and no regularization. (b) Effect of noise on the force reconstructions for 70% data and no regularization; noise values are 0.0, 0.5, 1.0, 5.0, and 10.0%.

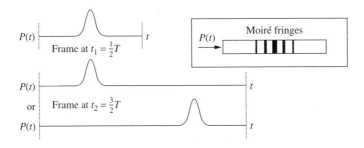

Figure 6.8: The worst-case scenario showing the ambiguity caused by multiple reflections. All three pulses give the same image.

Consider a worst-case scenario in the sense that the reflected waves look identical to the incident waves. With reference to Figure 6.8, a rod of length L is impacted at one end with a relatively short duration pulse. The nondispersive longitudinal wave takes a time $T = L/c_g$ (where c_g is the group speed) to travel from one end to the other. A photograph at time $t = \frac{1}{2}T$ (when the pulse is midway down the rod) is capable of determining the force history. A photograph at time $t = \frac{3}{2}T$ (again when the pulse is midway down the rod) is incapable of knowing if it is the first transit of a pulse initiated at time $t = T$ or the first reflection of a pulse initiated at the earlier time $t = 0$. This ambiguity increases as the time of the photograph (i.e., the total duration of the unknown force) is increased.

Suppose now we record a different type of data taken at the same framing rate, then it too will suffer from the same ambiguity. Thus, adding a second photograph of different data but also taken at time $t = \frac{3}{2}T$ will not remove the ambiguity.

Combining photographs taken at times $\frac{1}{2}T$ and $\frac{3}{2}T$ will unambiguously determine the force history up to time $t = \frac{3}{2}T$. Thus, we conclude that the rate at which the photographs need to be recorded depends on

$$\Delta T = L/c_g \qquad (6.7)$$

where L is the smallest significant wave transit length. For in-plane problems, this is given more specifically as

$$\Delta T = L/c_P \approx L/\sqrt{E/\rho}$$

where E is the Young's modulus and ρ is the mass density. In complex structures, the appropriate wave speed needs to be used.

The simple formula predicts a framing rate of about $25\,\mu s$ for the model of Figure 6.5 because the distance from the boundary to the hole is about $125\,mm$. This is a conservative estimate because it does not take into account the redundancy of data or the fact that only some of the wave is reflected. Note that a frame rate based on the overall dimension of the plate is $50\,\nu s$. Other test results show that identifying a single force for this geometry, or two forces for the plate without the hole, requires fewer frames.

Again, regularization can improve the results when the number of frames are marginally adequate. However, it seems that all effort should be made to have an adequate set of frames to begin with.

Remoteness of Data

On the basis of the study of the previous section, we conclude that, if the hole is taken as the significant boundary, then a framing rate of approximately $25\,\mu s$ should be used. But a framing rate of approximately $100\,\mu s$ seemed adequate. We now consider this in more detail and the factor in the effect of remoteness of the data.

Figure 6.9(a) shows a problem similar to the hole problem except that it has a relatively sharp notch. This problem is highly overdetermined in the sense that there is a great deal of nodal information (up to $M_d = 896$) for each frame versus the number of force unknowns ($M_p = 800/10 = 80$). We therefore discuss the selection of subgroups of this data; Figure 6.9(a) shows the two types of data selections that we will use: sparse points distributed over the plate and dense points remote from the load. To enhance any effects of ill-conditioning, we use 1% noise.

Figure 6.9(b) shows force reconstructions using data from selected vertical sets of nodes. The bottom is from a single line of data (almost along the line of symmetry), the second is from two vertical lines, and the top plot is from four vertical lines. Although the number of data has increased in this sequence, the true reason the top plot is converged is that its extra data corresponds with different propagation paths and therefore helps remove the ambiguity due to reflections. We conclude that for 2-D problems, we can

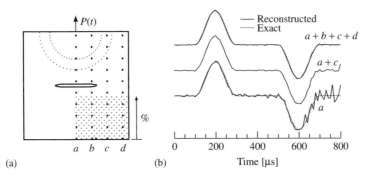

Figure 6.9: Effect of locations of data: (a) Vertical strips and remote region. (b) Force reconstructions for data taken along vertical strips.

utilize the spatial distribution of data to relax the framing rate requirement. It must be pointed out, however, that this conclusion is based on the case of a single unknown force; a spatial distribution of unknown forces would revert to the worst-case scenario of the previous section.

Figure 6.10(a) shows force reconstructions using data that is increasingly remote from the force location. The data set begins at the bottom of the specimen as shown in Figure 6.9(a); thus the 40% result is for data taken exclusively from below the notch in the shadow region. Surprisingly, the main effect of the remoteness of the data is on the trailing edge of the reconstructions. The amount of contamination is approximately equal to the propagation time from the force to the edge of the data.

If the trailing edge of the reconstruction can be sacrificed, then we conclude that the method can robustly determine the force history from remote data. Alternatively, by adjusting the amount of regularization, we can remove the erratic behavior of the trailing edge. Figure 6.10(b) shows force reconstructions using 40% of the data with different

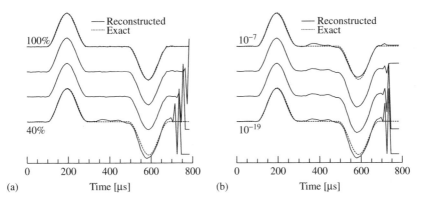

Figure 6.10: Effect of remoteness of data on the force reconstruction: (a) Different % data with $\gamma = 10^{-19}$. (b) Different γ with 40% data.

values of γ_t. It may appear from this figure that regularization helps in giving an estimate of the force history even during the trailing edge; this is not true—all that is happening is that first-order regularization will replace the unknown values with the nearest good values (which in this case is close to zero).

As a different type of problem, consider the cylinder shown in Figure 6.5(b). A single camera, at most, can record only 50% of the data; this is the amount that will be used here. Figure 6.11 shows Moiré fringes for the cylinder when impacted with a single force on the backside. To enhance any effects of ill-conditioning, 1% noise is added to the w component of displacement—this is the component in the direction of the camera.

Figure 6.12(a) shows the force reconstruction for different single frames. In each case, the reconstruction is done over $600\,\mu s$ but clearly a frame at $300\,\mu s$, say, will not be expected to have information beyond this time and this is manifested as a horizontal reconstruction in the figure. However, because of the remoteness of the data, the frame does not even have force information up to its current time indicated by the horizontal reconstruction beginning at a time earlier than the frame. The figure shows that as the frame time is delayed, a more accurate early time behavior of the force is obtained.

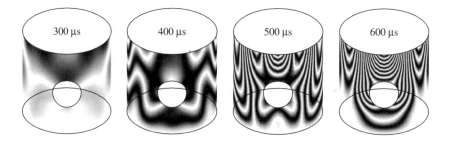

Figure 6.11: Cylinder with hole. Front side $w(x, y, z)$ Moiré fringe patterns.

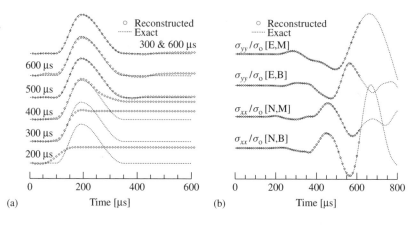

Figure 6.12: Reconstructions for the cylinder. (a) Effect of frame. (b) Stress histories at the edge of the hole ($\sigma_0 = E/10{,}000$).

However, the 600-μs frame result begins to show deterioration; this is due to the framing rate problem. The top plot in Figure 6.12(a) uses two frames and the reconstruction is very good.

Figure 6.12(b) shows reconstructions of the membrane and bending stresses at the east and north edges of the hole using the top force of Figure 6.12(a). The comparison with the exact stresses is very good, which is not surprising because of the good quality of the force reconstruction. What this comparison highlights (among other things) is that because the parameterization of the unknown was done in terms of the force, the effective differentiation of the Moiré data has already been accomplished as a by-product of using the FEM-based formulation.

Discussion

This ability of converting space-based data into time-histories information is a unique feature of the method. What the examples show is that it is robust enough to determine these histories using remote and somewhat incomplete data. In addition, it can also determine local histories of interest such as the stress at the edge of the hole. This sort of information is very difficult to extract using conventional processing techniques.

The above developments imply that the data are frames-of-displacement information; however, as seen in Chapter 4, through the ability to postprocess the FEM sensitivity responses, different and mixed types of information can be used. For example, it can be slope information as obtained from shearography [93], transverse displacements from holography [13, 12], first invariant of stress also as obtained from holography [142], direct displacement using electron beam Moiré [47], or thermography [132, 146]. Photoelasticity [15, 51] can also be used but it requires further developments because the relation corresponding to Equation (6.2) is nonlinear, and hence a nonlinear solver must be used. We cover this case later.

A further point of potential significance is that since the algorithm of Figure 5.5 is the same for both time-continuous data and space-distributed data, both types of data can be mixed together in a problem. Thus it is possible, say, to combine a few strain gage responses with a few frames of data to effect a more robust solution than what would be available if only one type of data is used. It would be interesting to pursue this idea.

6.3 Traction Distributions

Many problems of structural engineering include not only a point force history but also a space distribution. For example, a structure with wind loading has a continuous distribution of tractions. If the structure is modeled using coarse elements, then there is a coarse distribution of single point forces. However, when modeling the structure with fine elements, there is a collection of very many point forces. Clearly, high-resolution data would be needed to resolve the difference between these closely spaced single forces. This would be impossible to achieve. As was done in Chapter 4, space regularization will be introduced to overcome this difficulty.

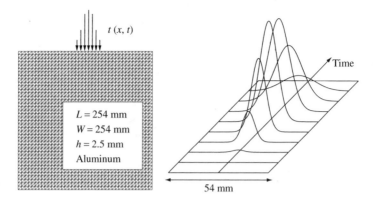

Figure 6.13: Time and space varying traction distribution on a plate.

Consider a plate with a blast impact on the top side as shown in Figure 6.13. The tractions generated would vary in time as expected, but the distribution shape would also change. Furthermore, if a blast impact actually occurred, it would be difficult to measure the data close to the impact side. The objective of this study is to determine the changing traction distribution using somewhat remote data.

Although the shape of the traction distribution remains symmetric, it spreads out over time and therefore constitutes a changing shape. Synthetic data in the form of v displacement Moiré frames are used, this data is contaminated with 1% noise. Only data from the upper half of the model is used; the amount and closeness to the load line was varied.

Resultant Behavior

When only remote data is used, it is very difficult to resolve the actual distribution shape but robust estimates of the resultant can be obtained. We therefore begin by illustrating this aspect of the problem.

Rather than determining a single force, we determine two forces, a distance of 25.4 mm (1.0 in.) or four element lengths apart, centered at different locations along the load line. The resultant, moment, and position are given by

$$P = P_1 + P_2, \qquad T = P_1 - P_2, \qquad x = \frac{P_2 L}{P_1 + P_2}$$

Space regularization was not used here because the two forces are uncorrelated with each other. The results are shown in Figure 6.14. A framing rate of 50 μs for a total of three frames was judged adequate. No data closer than 25.4 mm to the load line was used.

The resultant history of Figure 6.14(a) shows very good comparison with the actual resultant. Figure 6.14(b) shows how the peak resultant changes with the location of the

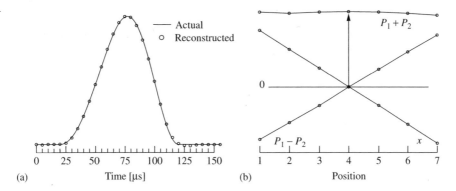

(a) Time [μs]

(b) Position

Figure 6.14: Two-force reconstructions with 1% noise. (a) Resultant history. (b) Peak values and position.

two forces; the peak hardly changes. Where the moment and location changes sign is the estimated center of the distribution. (Note that, since x is relative to P_1, it changes as the location of P_1 is changed.) These position estimates and the actual location coincide quite well. We conclude that even when somewhat remote data are used, we can get good estimates of the resultant as well as its location.

Time-Varying Traction Distribution

Figure 6.15(a) shows the reconstructed traction histories for 13 points. Space regularization is used here because the 13 forces are related to each other. The comparisons are quite good; however, to get this quality of result, it was necessary to use a frame rate of 25 μs (giving six frames of data) and to take data within 12 mm (a distance of two elements) of the load line.

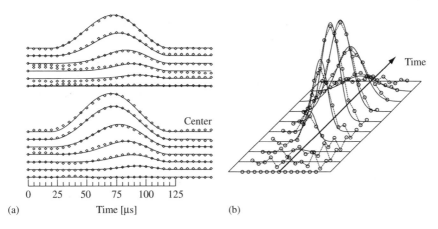

(a) Time [μs]

(b)

Figure 6.15: Traction reconstructions. (a) Time histories. (b) Distribution shapes.

It is also possible to use the space sensitivity method (as discussed in Chapter 4) to solve this problem. The advantage is that the number of unknowns could be reduced or a more complicated distribution could be tackled. This will not be done here, but a more elaborate example of determining a time-varying traction is given in the next chapter where a case of dynamic crack propagation is discussed.

6.4   Dynamic Photoelasticity

Figure 6.16 shows a set of dynamic photoelastic fringe photographs. The purpose of this section is to be able to use data like this to determine the applied load history.

As was discussed in Chapter 4, the use of photoelasticity requires an iterative approach to the use of the data. When the dynamic problem is considered, the iterative algorithm of Chapter 4 must be combined with the time reversing algorithm introduced earlier; while no new essential difficulty is introduced, the implementation details become more complex.

Dynamic Algorithm

The derivation of the formulation follows that of the static photoelastic problem in Chapter 4; the difference that time makes is accounted for by using the space–time deconvolution algorithm already introduced.

I: Iterative Formulation

As discussed in Chapters 2 and 4, the measured fringe data at any time are related to the stresses by [48]

$$\left(\frac{f_\sigma}{h}N\right)^2 = (\sigma_{xx} - \sigma_{yy})^2 + 4\sigma_{xy}^2$$

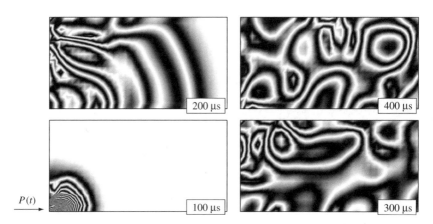

Figure 6.16: Synthetic light-field photoelastic fringe pattern for dynamic loading.

The data are a nonlinear relation of the stresses and hence the applied loads. As is usual in nonlinear problems, we begin with a linearization about a known state; that is, suppose we have a good estimate of the applied load histories $\{P\}_0(t)$ from which we can obtain estimates of the stress histories as $[\sigma_{xx}, \sigma_{yy}, \sigma_{xy}]_0$. Do a Taylor series expansion about this known state to get

$$\left(\frac{f_\sigma}{h}N\right)^2 \approx [(\sigma_{xx} - \sigma_{yy})^2 + 4\sigma_{xy}^2]_0 + 2[(\sigma_{xx} - \sigma_{yy})_0(\Delta\sigma_{xx} - \Delta\sigma_{yy}) + (4\sigma_{xy})_0\Delta\sigma_{xy}]$$

The key step is to relate the stress increments $(\Delta\sigma_{xx}, \Delta\sigma_{yy}, \Delta\sigma_{xy})$ to the scaled sensitivity responses so that for all components of force we have

$$[\Delta\sigma_{xx}, \Delta\sigma_{yy}, \Delta\sigma_{xy}] = \sum_j \{\psi_{xx}, \psi_{yy}, \psi_{xy}\}_j \Delta\tilde{P}_j$$

(The responses $[\psi]$ must be computed for each component of force.) In anticipation of using regularization where we wish to deal directly with the force distributions, let $\Delta\tilde{P} = \tilde{P} - \tilde{P}_0$ so that the approximate relation can be written as

$$\left(\frac{f_\sigma}{h}N\right)^2 \approx \sigma_0^2 - \sum_j G\tilde{P}_{oj} + \sum_j G_j\tilde{P}_j$$

with

$$\sigma_0^2 \equiv [(\sigma_{xx} - \sigma_{yy})^2 + 4\sigma_{xy}^2]_0, \qquad G_j \equiv [2(\sigma_{xx} - \sigma_{yy})_0(\psi_{xx} - \psi_{yy})_j + 8(\sigma_{xy})_0\psi_{xyj}]$$

The data are only from a subset of the total number of possible points and we indicate this by $\{d\} \Leftrightarrow Q\{N\}$ as we have done before. Let the data be expressed in the form $\overline{N} \equiv Nf_\sigma/h$ so that we can express the data relation as

$$\overline{N}^2 = Q\sigma_0^2 + \sum_j G\tilde{P}_0 - \sum_j G\tilde{P} \qquad \text{or} \qquad [QG]\{\tilde{P}\} = \{\overline{N}^2 - Q\sigma_0^2 + QG\tilde{P}_0\}$$

Establish the overdetermined system and reduce to get

$$[(QG)^T W QG + \gamma H]\{\tilde{P}\} = [QG]^T[W]\{\overline{N}^2 - Q\{\sigma_0^2 + G\tilde{P}_0\}\}$$

This looks identical to the relation obtained in Chapter 4, the significant difference is that all quantities with subscript 'o' are complete functions of time. Since the equations are solved repeatedly using the newly computed $\tilde{P}$ to replace $\tilde{P}_0$, a way must be found to keep track of the stress histories at all data points.

II: Initial Guesses and Regularization

In all iterative schemes, it is important to have a method of obtaining good initial guesses. In the present case, we must have an initial guess for all of the unknown applied load histories. A number of possibilities may suggest themselves depending on the particular problem; the scheme discussed here has the virtue of always being available.

Since the solution scheme for a linear data relation does not require an initial guess, this suggests that a linearized version of the photoelastic data relation may be a useful source for an initial guess; that is, assume

$$N f_\sigma / h \approx \sigma_{xx} - \sigma_{yy}, \qquad \sigma_{xy} \approx 0$$

Or one might even try

$$N f_\sigma / h \approx \alpha(\sigma_{xx} - \sigma_{yy}) + \beta 2 \sigma_{xy}$$

where α and β can be chosen as appropriate. In many circumstances, these relations will give poor histories but, at least, they will always give correct orders of magnitude. Thus, the initial step in the iterative process will entail a complete force identification as described earlier in this chapter. The rate of convergence depends on the number of unknown forces, this suggests (if determining traction distributions) using fewer forces to obtain an idea of the unknowns to use as starter values.

Also, as not in Chapter 4, if only a single force history is unknown, then the problem can be phrased in a linear fashion with no iteration. For multiple isolated loads or traction distributions, this can be used to get an initial estimate of the forces.

The next step is to determine the amount of regularization. The explicit form for the regularization is

$$\gamma [\, H \,] = \gamma_t [\, H_t \,] + \gamma_s [\, H_s \,]$$

where both time and space regularization is incorporated. The same discussion as given earlier also applies here. The new aspect of regularization worth discussing arises because the algorithm to be proposed will combine both the linear algorithm (with the linear data relation) and the nonlinear algorithm (with the nonlinear data relation). Because the respective core matrices could be of significantly different magnitudes (since the former is order stress and the latter is order stress squared), the amount of regularization must be adjusted appropriately.

Rather than input separate amounts of regularization, a more consistent approach is to normalize the equations. A simple normalization is to use the average diagonal value of the normal equations, that is,

$$\beta^2 = \mathrm{Tr}[[\, A \,]^T [\, A \,]]/N_p$$

where N_p is the total number of unknowns. The regularized form of the system of equations becomes

$$\left[\frac{1}{\beta^2} [\, A \,]^T [\, A \,] + \gamma [\, H \,] \right] \{\tilde{P}\} = \left\{ \frac{1}{\beta^2} [\, A \,]^T \{d\} \right\}$$

Actually, in all the implementations of the algorithms presented in this book, this normalizing of the least squares equations is made available irrespective of the need. In this way, irrespective of the particular problem, similar orders of magnitude of regularization can be used.

III: Computer Implementation

The computer implementation is essentially that of static photoelasticity as shown in Figure 4.23. The sensitivity solutions $\{\psi\}$ are determined only once (for each load), but they must be stored. These are complete snapshots of the stress fields for each force component at each time increment ΔT. The analyzer produces the snapshots of displacements $\{u\}$, the postprocessor produces the stress field snapshots $[\ \psi\ ]$, which are stored to disk. Depending on the problem, this could require a large amount of disk space.

The updating of the stress fields $[\ \sigma_o\ ]$ requires a FEM call on each iteration. This stress information needs to be stored only at the time values corresponding to the data frames. Both the left- and right-hand side of the system of equations must be re-formed on each iteration and therefore the UDU decomposition to solve the system of equations must be done on each iteration.

The particular makeup of the matrix $[\ A\ ]$ corresponds to that developed earlier in the chapter.

Test Cases using Synthetic Data

We look at two test cases. The first considers the frame rate and how it affects the initial guess. The second deals with the fact that photoelastic fringe orders are positive only and cannot distinguish between tensile and compressive stresses.

I: Single Force—Effect of Frame Rate

As a first test case, consider the problem shown in Figure 6.17. This is an epoxy plate with a single concentrated load, and because of symmetry, only one half of the plate is modeled. The load history is shown as the bottom plots in Figure 6.18 and the fringe

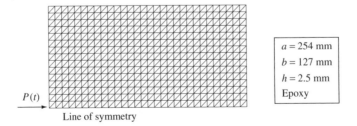

Figure 6.17: Mesh and properties.

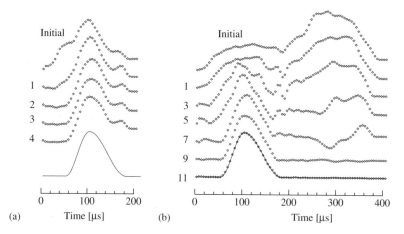

Figure 6.18: Convergence behavior of the force reconstructions with 1% noise. (a) One frame at 200 μs. (b) Two frames, at 200 and 400 μs.

patterns are those shown in Figure 6.16. Because the inverse problem is solved with $\Delta T = 5\,\mu\text{s}$, these synthetic data were generated with $\Delta T = 0.5\,\mu\text{s}$. Only 50% of the interior data will be used.

The iteration results over two time windows are shown in Figure 6.18. Note that the second window encompasses a great number of reflections. Only data from nodes two elements or more from the left-hand side were used.

The two issues that need to be considered are the robustness of the convergence relative to the initial guess and the number of frames required.

The top plots of Figure 6.18 are the initial guesses based on the linear approximation of the data relation; the bottom full line plots are the exact force history. The initial guess of Figure 6.18(a) is somewhat like the actual force and therefore convergence is rather rapid. But the trailing edge is rather poorly reconstructed. The initial guess of Figure 6.18(b) is only like the actual force in overall magnitude and therefore convergence is rather slow. However, it does converge to the exact value. All in all, the convergence of Figure 6.18(b) is quite remarkable considering the initial guess and the fact that there is noise in the data.

The reason that a long time window is required to give good results is not entirely clear but based on the earlier studies of remoteness of data, it seems to be related to propagation effects. Consider the force component $\tilde{P}_n\phi(t)$ close to the end of the time window; data for this component is spatially located close to the impact site. Therefore, the recorded data (being predominantly remote from this point) is somewhat deficient in information about the component. Regularization will estimate this component based on the estimates of the earlier time forces. For the larger time window, these earlier time forces have already gone to zero and therefore the estimate of the force is zero. The conclusion seems to be that for an arbitrary force history that persists for a long time, a sacrificial amount of reconstructed history must be included at the trailing edge.

II: Two Uncorrelated Forces—Ambiguity of Sign

A deficiency of photoelasticity is that the sign of the applied load cannot be inferred solely from the fringe pattern. For static problems, a small rotation of the analyzer relative to the polarizer causes a slight shift in the fringe pattern and the direction of this shift gives the sign of $\sigma_1 - \sigma_2$. This facility is not available in dynamic problems unless the dynamic event is repeated with the changed polariscope setting.

For single applied loads, this ambiguity of load sign does not cause a serious problem since the convergence behavior is not affected and the sign can be *a posteriori* inferred from the particulars of the problem (e.g., if it is an impact, the normal traction must be compressive).

The ambiguity does cause a serious problem when there are multiple loads. Figure 6.19 shows the early time fringe pattern for the block of Figure 6.17 when two (similar but not identical) load histories are applied. At the earlier time there is little difference between the left and right patterns within the model and no difference between the (a) and (b) patterns. This is so even though they are different in sign. At the later time, there is only a slight difference between the (a) and (b) patterns. What this means is that if the initial guess has the wrong sign, then it will be very difficult for the iterative process to correct for this. There will be convergence but it will be to the wrong answer.

In the absence of outside information, a way to cope with this ambiguity problem is to run the inverse analysis multiple times having forced a difference in the sign of the loading. The case that has the minimum χ^2 is most likely the true situation. It may seem that this is computationally expensive especially if many forces representing a traction distribution are to be determined. Actually, this ambiguity difficulty does not arise in traction distribution problems because, through the regularization mechanism, the sign of a given force component is constrained to be similar to its neighbors. The

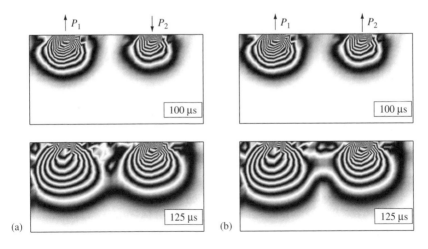

Figure 6.19: Early time fringe patterns for two uncorrelated loads. (a) Force histories are opposite sign. (b) Force histories are same sign.

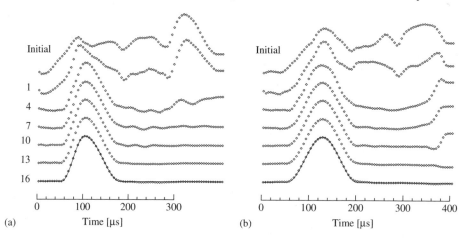

Figure 6.20: Convergence behavior of the force reconstructions with 1% noise. (a) Left force $P_1(t)$. (b) Right force $P_2(t)$.

ambiguity would arise only if there were multiple uncorrelated traction distributions to be determined.

Figure 6.20 shows the convergence behavior for the case of Figure 6.19 where the loads are of opposite sign and the initial guess histories are forced to be different. Again, the scheme seems to be quite robust as regards the actual initial history. For this case, four frames of data were needed (a frame every 100 µs). Additionally, only data from the lower 85% of the model was used, thus no data came from the immediate vicinity of the loads.

Test Case using Experimental Data

We conclude this section with an experimental example. The practical issues of fringe counting and discretizing the data are discussed.

I: Experimental Setup

The experimental setup is shown in Figure 6.21; the Cordin high-speed camera can record a sequence of 27 photographs with a framing rate up to every 8 µs.

The impact is generated by a steel ball. If the ball contacted the epoxy directly, then the contact time would be relatively long and wave propagation effects would be hardly noticeable. The epoxy plate is attached to the one-sided Hopkinson bar so that the impact is actually steel on steel and the strain gages on the bar allow the force at the plate to be computed as discussed in Chapter 5. The epoxy properties are given in Chapter 2.

A frame of data is shown in Figure 6.22(a). Reference [150] has complete sets of such fringe patterns for different specimen arrangements and different layered materials.

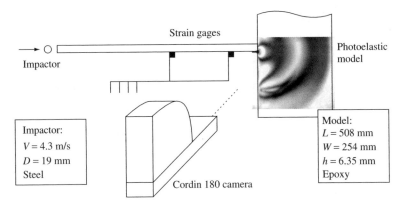

Figure 6.21: Experimental setup for recording dynamic photoelastic photographs.

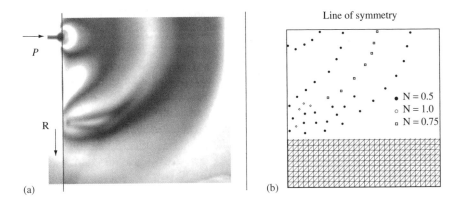

Figure 6.22: Photoelastic experiment. (a) Recorded image. (b) Discretized data superposed on (partial) mesh.

The photograph shows the Rayleigh wave (indicated by R) propagating downward along the surface. It is the objective here to determine the contact force history from this single frame of data.

One of the difficulties in using photoelasticity for dynamic problems is in identifying the fringe orders. In static problems, a variety of techniques can be used including the use of white light where the zero fringe order can be identified as the black fringe. For dynamic crack propagation problems [50], the pattern close to the crack tip has a definite monotonic ordering easily counted from the remote state. For general dynamic problems with multiple reflections and data similar to the 400-µs frame in Figure 6.16, there is just the grey scale image with no obvious ordering.

Furthermore, because of sharp stress gradients, there is a good deal of ambiguity in the fringe pattern. For example, in Figure 6.22(a) it would be natural to assume that the fringe pattern near the Rayleigh wave is like a contact point such as in Figure 4.24.

However, as shown in the discretized data points of Figure 6.22(b), all the black fringes have the same order of 0.5. This was determined primarily through experience with such problems, but unambiguously identifying the fringe orders can be a challenging problem for general cases in dynamic photoelasticity.

It is worth mentioning that all fringe orders need not be identified since only a subset of the data needs to be used for the inverse method to succeed. Thus, in Figure 6.22(b), only those fringes that are obvious or clearly defined were identified and discretized. However, the data used must be distributed over the entire model; otherwise the time information would be distorted.

The discretized data of Figure 6.22(b) were relocated onto the mesh nodes using the scheme introduced in Chapter 3 and used in Chapter 4. One half of the epoxy plate was modeled because of the line of symmetry.

Figure 6.23(a) shows the convergence behavior. As previously demonstrated, in spite of the very poor initial guess, the solution eventually converges to the right solution. The frame of data was taken at time step $t = 300\,\mu s$ but the solution was determined over a time period of $400\,\mu s$; the horizontal trailing edge is indicative of the first-order time regularization used when no real information is in the data.

The final converged history is reminiscent of Figure 6.18(a) indicating a need for more frames of data. Unfortunately, the available frames did not span a sufficient time window. Parenthetically, a disadvantage of the Cordin rotating mirror camera for this type of problem is that it cannot be synchronized with the event. As a consequence, it is a hit-and-miss situation having the mirror correctly aligned to coincide with the position of the stress wave.

A detailed plot of the force history is shown in Figure 6.23(b); the comparison is with the force determined from the Hopkinson bar. All in all, the comparison is quite good. It is expected that the results could be improved if the parameters of the model (modulus, density, fringe constant, etc.) were "tweaked" as described with the experiment

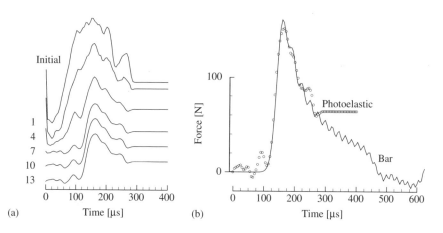

Figure 6.23: Convergence behavior of the force reconstructions. (a) Force reconstructions during iterations. (b) Comparison with force measured from the one-sided Hopkinson bar.

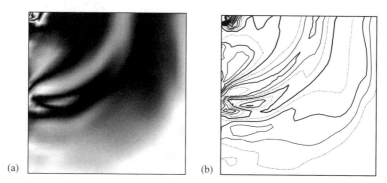

(a)   (b)

Figure 6.24: Data reconstructions. (a) Isochromatic fringe pattern. (b) Contours of fringe pattern at every 0.5 fringe order.

in Section 5.7. This was not done here because the limited number of frames did not warrant it.

As additional processing, the determined force history was used to reconstruct the fringe pattern and the results are shown in Figure 6.24(a); it has all the characteristics of Figure 6.22(a). What is very interesting is the actual contours of intensity as shown in Figure 6.22(b). These contours are done at every 0.1 of a fringe with the whole and half orders being the thick lines and they show the very complex nature of the fringe pattern especially in the vicinity of the Rayleigh wave. In particular, they show that the center of the white and black fringes cannot be relied upon to be whole and half-order fringes, respectively. This reinforces the point that fringe identification can be a significant challenge in dynamic photoelasticity.

Discussion of Dynamic Photoelasticity

The discussion of photoelasticity given in Chapter 4 also applies here to the dynamic case. But the dynamic problem has highlighted a few more difficulties with photoelasticity that are worth discussing.

When there are multiple loads, distinguishing between their signs based solely on the fringe data is nearly impossible and this complicates considerably the solution of general problems. But even if ways were devised to overcome the above difficulty, identifying the fringe order for such problems would be a formidable task in itself. Mixing of strain gage or accelerometer point data with the image data might help here. In retrospect, it would have been beneficial to have recorded accelerometer data at the time of doing the experiment.

We conclude that photoelasticity is not especially suitable for doing inverse dynamic problems. However, as pointed out in Chapter 4, its main reason for inclusion here is as a test case of the problems that may arise when the new whole-field methods (such as fluorescent paints) are used.

6.5   Identification Problems

We now consider some identification problems. In both examples, we consider dynamically recorded whole-field Moiré data, but in the first case it is applied to frames, while
in the second it is applied to a plate.

Moiré Analysis of Frames

The examples in this chapter have dealt with plates and shells since it is natural to associate whole-field distributed data with them. This example will look at a Moiré analysis
of a frame structure. We also consider the situation of unknown boundaries that can be
parameterized as a set of unknown forces.

I: One Unknown Force

The geometry is given in Figure 6.25, and the input force is that shown as $P_1(t)$ in
Figure 6.6. Each member was modeled with 40 frame elements.

   Figure 6.26 shows some Moiré fringes at two times. The frame is essentially inexten-
sible; that is, there is very little relative displacement along the axis of each member as
recognized by the almost uniform shade of grey. As expected, the dominant displacement
is transverse to the member.

   The force reconstructions using a single frame of data at $1000\,\mu s$ are shown in
Figure 6.27. A time step of $\Delta t = 5\,\mu s$ was used. It is quite surprising that the single frame is adequate for the reconstruction. On the basis of the worst-case scenario
introduced earlier, it would have been thought that because of the multiple reflections,
a faster framing rate would be needed. It seems that there are two factors at play here
that mitigate against the need for using more frames. First, the wave propagation is dispersive [70] (the flexural waves themselves and the fact that the joints act as partial

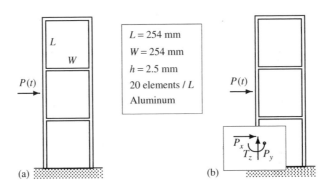

Figure 6.25: Impacted frame. (a) Geometry. (b) Left boundary replaced with unknown reactions.

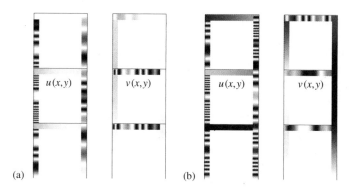

Figure 6.26: Moiré fringe patterns for an impacted frame. The fringe area was doubled for easier viewing; white corresponds to zero displacement. (a) Time 500 μs. (b) Time 1000 μs.

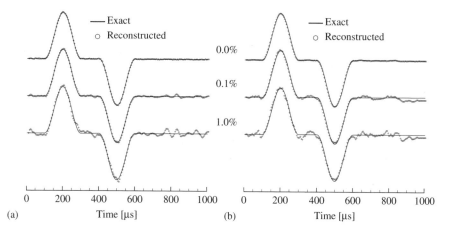

Figure 6.27: The effect of noise on the force reconstructions using a single frame of data. (a) Using $u(x, y)$ displacements. (b) Using $v(x, y)$ displacements.

reflectors) and therefore there is less likelihood of ambiguity after superposition. Second, the effective path length for a complete transit of the wave to the remotest area is a minimum of two and a half member lengths; and that seems to be the characteristic length.

The plots of Figure 6.27 are for different amounts of noise. First note that even if there is no noise (the data are perfect), (time) regularization was required. In each of the cases shown, the value of regularization used was that just at the point where the solution becomes oscillatory. There is not a great deal of difference between the $u(x, y)$ and $v(x, y)$ data results.

By and large, frame structures seem a little more sensitive to noise than the corresponding in-plane plate problem. This could be due to the fact that there is less redundancy in the data.

II: Unknown Boundary Condition Parameterized as Unknown Loads

For a real structure, the boundary condition as indicated in Figure 6.25(a) would not be completely fixed. Indeed, if it was an actual frame with a foundation in soil, the true boundary condition would be an unknown.

We will look at solving the problem with an unknown boundary condition.

As an illustration, make a free-body cut as shown in Figure 6.25(b) and replace the unknown boundary with a complete set of tractions; because we are dealing with frame structures, we can replace the tractions in terms of resultants. Thus the complete set of unknowns for the inverse problem is the applied load plus the three histories $P_x(t)$, $P_y(t)$, and $T_z(t)$. Note that the free-body cut need not be at the boundary but can be anywhere convenient along the member.

The force reconstructions are shown in Figure 6.28. The main criterion for the goodness of the reconstruction is the quality of $P(t)$—at present, it is not our intention to determine the actual boundary reactions. In this case, one frame is no longer adequate and two frames must be used.

If the forward problem is now solved using this set of four applied loads, then the displacement fields as indicated in Figure 6.26 would be obtained. In particular, if the free-body cut is taken at the base and the displacement reconstructed, it is found that the maximum displacement at the base while not zero is four orders of magnitude smaller than the displacements at the $P(t)$ load point.

Implicit Parameter Identification

The algorithm for distributed parameter identification is essentially the same as given in Section 5.7; the fact that the data are in the form of whole-field images does not make a significant difference.

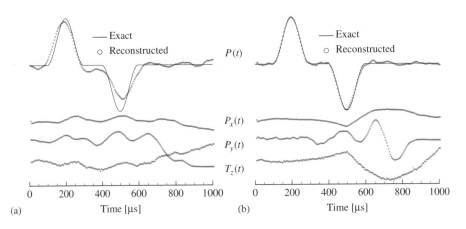

(a)   Time [μs]   (b)   Time [μs]

Figure 6.28: Applied force and boundary reaction reconstructions for 0.1% noise. (a) Using one frame at 1000 μs. (b) Using two frames at 500 μs and 1000 μs.

The only practical point to note is that the measured data are a subset of the total responses, that is, they are taken at a limited number of times and that they are taken from a limited region of the structure. Consequently, it is important that the images contain data relevant to the parameters being identified.

Figure 6.29 shows convergence results where the modulus of the horizontal members of the frame of Figure 6.25 are identified. Relative to Figure 6.25, the values were reduced to 60, 90, and 70%, bottom to top. The early frame of data has difficulty identifying the top member because not enough of the wave has reached it. The later frame gives almost the exact result even though there is 10% noise. In each case, the initial guess was the original nominal value. We can infer that if the change in modulus was due to damage, then we have just demonstrated the ability to determine multiple damages in a structure.

As a second example, we reconsider the problem posed in Figure 5.30 where a somewhat localized change of mass and stiffness in the form of a plug is to be identified.

Figure 6.30(a) shows the synthetic Moiré data; the steel inset is just about discernible. Figure 6.30(b) shows the convergence results using just the single frame of data. Convergence is rapid but it is interesting to note that the density is slightly more sensitive to noise than the modulus. Furthermore, the converged result using the Moiré displacement data is about the same as that shown in Figure 5.31(b) using velocity data. Looking at Figure 5.31(a), note that the 600-μs frame involves multiple reflections; identification using a 300-μs frame of data gave poorer results that were more sensitive to noise.

This example, and the examples for the frame, show that we have a parameter identification scheme that can interchangeably use either point sensor data or whole-field data. As shown in the next section, this versatility can be put to good use when the unknown parameters have an unknown spatial component.

A final point worth mentioning is that extra independent data can be generated by repeating the experiment with the load applied at a different point.

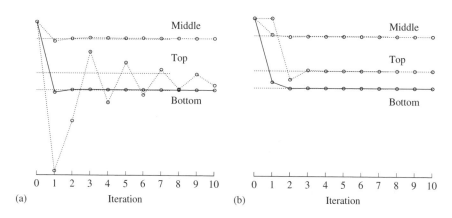

Figure 6.29: Identification of horizontal member stiffnesses. (a) Convergence results for single data frame at 500 μs and 10% noise. (b) Convergence results for single data frame at 1000 μs and 10% noise.

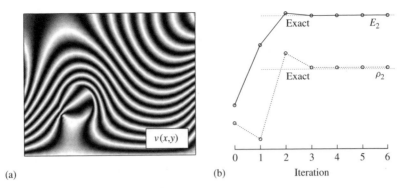

(a)   (b)   Iteration

Figure 6.30: Plug identification. (a) Synthetic Moiré fringe data at time 600 μs. (b) Convergence results for 10% noise.

6.6   Force Location for a Shell Segment

As a second example, consider the continuous region shown in Figure 6.31. This is approximately the same problem covered in Section 5.8 except that the plate has rib reinforces making it a 3-D shell problem. The impact will be transverse using the same force history.

The objective here is to find the approximate global location of the impact and then use one of the methods of Chapters 5 or 6 for refinement. The scenario we envision is that the structure has both point sensors and whole-field capabilities. The sensors detect the event (when the signal exceeds a preset threshold) and then triggers the optical recording. One of the points we are interested in is the allowable delay time between the initial event and the trigger that will still give good location estimates.

Figures 6.32 and 6.33 show a sequence of out-of-plane fringes for two impact locations on the panel; an obvious point about these figures is that if an image can be captured at an early enough time, then the impact location is easily detected. The impact event is actually initiated at 100 μs.

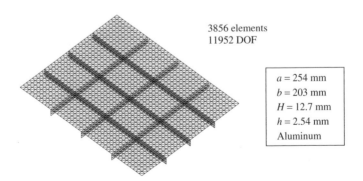

3856 elements
11952 DOF

| $a = 254$ mm |
| $b = 203$ mm |
| $H = 12.7$ mm |
| $h = 2.54$ mm |
| Aluminum |

Figure 6.31: Mesh and properties for a shell segment.

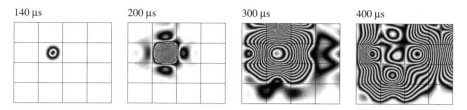

Figure 6.32: Sequence of out-of-plane fringes when impacted on the panel.

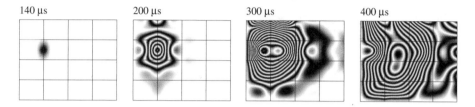

Figure 6.33: Sequence of out-of-plane fringes when impacted on the reinforcing rib.

Our challenge is that we may not actually have caught the event early enough and so the question arises as to the adequacy of the later images for detecting the impact site; that is, as time evolves, there are more superpositions of reflections and the global maximum deflection does not necessarily coincide with the impact location.

The basic strategy we will adopt is based on the impact response of an infinite sheet [56]. An infinite sheet impacted with a pulse of finite duration has a maximum deflection at the impact site and a slowly expanding radius. At the edge of the radius, the deflections are undulating. Thus predominantly, the deflected shape looks like a mound.

Our algorithm comprises three steps. First, any points with negative deflection are set to zero. Second, thinking of the deflections w_i as the height of the mound, the center of gravity of the mound is found from

$$x_c = \frac{\sum w_i x_i}{\sum w_i}, \qquad y_c = \frac{\sum w_i y_i}{\sum w_i}$$

The summations are over all the nonzero data points in the image. Third, the deflected shape is modified according to the scale

$$\text{scale:} \qquad \alpha = e^{-\xi^2} e^{-\eta^2}, \qquad \xi = (x - x_c)/L, \qquad \eta = (y - y_c)/L$$

where L is a characteristic dimension—the diagonal of a rectangular plate, for example. This has the effect of accentuating the displacements in the vicinity of (x_c, y_c) relative to remote regions. These last two steps are then repeated until convergence.

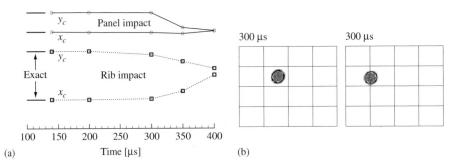

Figure 6.34: Estimated impact sites. (a) Position using different frames of data. (b) Converged modified fringe patterns for the 300-μs frames.

Figure 6.34 shows the estimated impact sites using the different frames. There are two points to be made about this figure. First, that it is possible, for a large time range, to get quite accurate estimates of the impact site. Second, even if the estimate is not precise, it has at least delimited the search space sufficiently so that the refined search schemes now become computationally feasible.

6.7 Discussion

The examples show that the developed inverse method is robust enough to solve general problems with a limited number of frames of data. Furthermore, the data can even be remote from the region of load application. It was also successfully adapted to handle multiple forces. In the case of continuous distributions of tractions, this required the use of space regularization.

A significant attribute of the method is illustrated by the cylinder problem: the normally difficult task of differentiating Moiré displacement to get strains is automatically taken care of by having an FEM formulation underlying the method. Consequently, data need not come from the region of interests (such as at a crack tip) in order to get good results for that region.

One of the significant side benefits of the success of the method is that we now have a consistent underlying framework for handling different experimental data. Thus, essentially the same method is used for static and dynamic data, and for space-discrete/time-continuous and time-discrete/space-continuous data. This opens up the possibility of combining both into a single inverse solution; this idea is taken up in the next chapter.

7

Nonlinear Problems

The situations of interest in this chapter are when the underlying mechanics of the problem is nonlinear. This nonlinearity arises primarily in one of two ways: either the material behavior is nonlinear, or the deformations/deflections are large. An extreme example is given in Figure 7.1, which shows the FEM simulation of the very high velocity penetration of an aluminum rod into a steel block. For this modeling to be reasonably accurate, it is required that reasonable experimental values for the elastic/plastic/erosion properties of both materials are available. Furthermore, an experimental ability to determine and verify the loads would help validate the modeling.

Nonlinear problems are very difficult to solve; even the formulation for the forward problem is challenging as seen in the brief discussion given in Chapter 1. The purpose of this chapter is to lay out some of the basic issues involved and some of the strategies that can be adopted. The algorithm developed is quite elaborate but it has at its core the linear inverse methods imbedded in a Newton–Raphson iteration scheme.

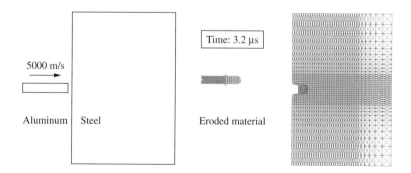

Figure 7.1: FEM simulation of erosion and penetration at very high impact velocities.

Modern Experimental Stress Analysis: completing the solution of partially specified problems. James Doyle
© 2004 John Wiley & Sons, Ltd   ISBN 0-470-86156-8

The robust solution of nonlinear problems requires more data than their linear counterparts, and so we end the chapter with a discussion of some of the issues surrounding
highly instrumented structures.

7.1   Static Inverse Method

We formulate the inverse method similar to the linear case except that we have iterations.
If data are available at a number of load steps, it can be easily adapted to be performed
at each of the loads. In that case, there are some similarities with the dynamic problem.
However, because of instantaneity of time, propagation effects need not be considered,
and hence a genuine incremental formulation can be used. The flavor of the derivation
will parallel the linear static case derived in Chapter 4.

Iterative Formulation

As shown in Chapter 1, the governing equations for a nonlinear static system can be
discretized using the finite element method as

$$[K_T]\{\Delta u\} = \{P\} - \{F\}_o = [\Phi]\{\tilde{P}\} - \{F\}_o \tag{7.1}$$

where subscript "o" refers to the current state. In this, $\{P\}$ is the $\{M_u \times 1\}$ vector of
all applied loads (some of which are zero), $\{\tilde{P}\}$ is the $\{M_p \times 1\}$ vector subset of $\{P\}$
of the nonzero applied loads, $[\Phi]$ is the $[M_u \times M_p]$ matrix that associates these loads
with the degrees of freedom, and $\{F\}$ is the $\{M_u \times 1\}$ vector of element nodal loads.
This equation is to be used iteratively until $\{\Delta u\}$ becomes zero putting $\{P\}$ and $\{F\}$ in
equilibrium. At that stage, the applied load can be incremented.

In the inverse problem of interest here, the applied loads are unknown but we know
some information about the responses. In particular, assume we have a vector of measurements $\{d\}$ of size $\{M_d \times 1\}$, which are related to the structural DoF according to

$$\{d\} \Leftrightarrow [Q]\{u\}$$

We want to find the forces $\{\tilde{P}\}$ that make the system best match the measurements.
Consider the general least squares error given by

$$E(u, \tilde{P}) = \{d - Qu\}^T \lfloor W \rfloor \{d - Qu\} + \gamma \{\tilde{P}\}^T [H]\{\tilde{P}\}$$

Our objective is to find the set of forces $\{\tilde{P}\}$ that minimize this error functional.

Consider the case when there is only one load increment and we are using Newton–Raphson iterations for convergence. The displacements are represented as

$$\{\Delta u\} = \{u\} - \{u\}_o, \qquad \{F\} = \{F\}_o$$

where both $\{u\}_o$ and $\{F\}_o$ are known from a previous iteration. The error function becomes

$$E(u, \tilde{P}) = \{d - Q\Delta u + Qu_o\}^T \lceil W \rfloor \{d - Q\Delta u + Qu_o\} + \gamma \{\tilde{P}\}^T [H]\{\tilde{P}\}$$

There is a relation between $\{\Delta u\}$ and $\{\tilde{P}\}$ due to the incremental form of the loading equation which we must now make explicit.

First decompose $[K_T]$ and then solve $[\Psi]$ as the series of back-substitutions

$$\text{solve:} \qquad [K_T][\Psi] = [U^T DU][\Psi] = [\Phi]$$

The matrix $[\Psi]$ is the collection of sensitivity responses for the perturbation loads $\{\phi\}$ applied for each of the unknown forces. We also solve for the response due to the element nodal forces

$$\text{solve:} \qquad [K_T]\{\psi_F\} = [U^T DU]\{\psi_F\} = -\{F\}_o$$

Therefore, the actual increment solution can be written as

$$\{\Delta u\} = [\Psi]\{\tilde{P}\} + \{\psi_F\}$$

The forces $\{\tilde{P}\}$ can be used as scalings on the solutions $[\Psi]$ because the loading equation for each increment is linear.

Now the error functional can be written as a function of $\{\tilde{P}\}$ only:

$$E(\tilde{P}) = \{d - Q\Psi\tilde{P} + Q\psi_F + Qu_o\}^T \lceil W \rfloor \{d - Q\Psi\tilde{P} + Q\psi_F + Qu_o\}$$

$$+\gamma \{\tilde{P}\}^T [H]\{\tilde{P}\}$$

This is minimized with respect to $\{\tilde{P}\}$ to give

$$[[Q\Psi]^T \lceil W \rfloor [Q\Psi] + \gamma [H]]\{\tilde{P}\} = [Q\Psi]^T \lceil W \rfloor \{d - Qu_o - Q\psi_F\}$$

This reduced system of equations is of size $[M_p \times M_p]$ and is relatively inexpensive to solve since the core matrix is symmetric and small in size. Once the loads are determined, the forward solution is performed; the new displacements are used to update the nodal force vector $\{F\}$, and the iteration is repeated.

This formulation of the problem is similar to the linear case except that we have iterations with the effective data (the right-hand side) needing to be updated. If we have data at a number of load steps, then it can be easily adapted to be performed at each of the loads.

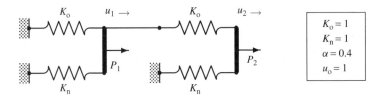

Figure 7.2: A 2-DoF nonlinear system.

A Test Problem

Consider the test problem shown in Figure 7.2. The springs K_o are linear, while the springs K_n are nonlinear and its nonlinear elastic relation is given by

$$F_i = K_n[1 + \alpha(u_i/u_o)^2]u_i$$

where K_n, α, and u_o are parameters. This arrangement allows for the easy separation of linear and nonlinear effects as well as for a straightforward assemblage. The nonlinear behavior of these springs is similar to that observed for a cantilever beam with a transverse applied load; that is, it stiffens as the load is increased.

The two equilibrium equations are

$$\mathcal{F}_1 = P_1 - F_1 = P_1 - [2K_o + K_n(1 + \alpha(u_1/u_o)^2)]u_1 + [K_o]u_2 = 0$$
$$\mathcal{F}_2 = P_2 - F_2 = P_2 + [K_o]u_1 - [K_o + K_n(1 + \alpha(u_1/u_o)^2)]u_2 = 0$$

The [2 × 2] tangent stiffness matrix is

$$[K_T] = \left[\frac{\partial F}{\partial u}\right] = \begin{bmatrix} 2K_o + K_n(1 + \beta 3\alpha(u_1/u_o)^2) & -K_o \\ -K_o & K_o + K_n(1 + \beta 3\alpha(u_2/u_o)^2) \end{bmatrix}$$

The β parameter was added so that different approximations for $[K_T]$ can be investigated. To make the inverse problem definite, consider two measurements to determine two unknown forces. Figure 7.3(a) shows the response against P_2 when $P_1 = 0$. The value of the parameters are shown in Figure 7.2.

The system matrices and vectors are

$$\{u\} = \{u_1, u_2\}^T$$

$$\{d\} = \{u_1, u_2\}^T = \begin{bmatrix} 1 & 0 \\ 0 & 1 \end{bmatrix}\{u\} = [\, Q \,]\{u\}$$

$$\{P\} = \{P_1, P_2\}^T = \begin{bmatrix} 1 & 0 \\ 0 & 1 \end{bmatrix}\begin{Bmatrix} P_1 \\ P_2 \end{Bmatrix} = [\, \Phi \,]\{\tilde{P}\}$$

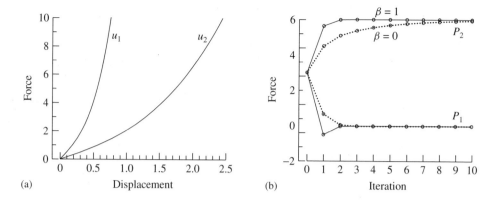

(a)   Displacement   (b)   Iteration

Figure 7.3: Results for nonlinear static case. (a) Exact force deflection behavior with $P_1 = 0$. (b) Convergence for different tangent stiffnesses.

Let the weighting arrays be simply

$$\lceil W \rfloor = \begin{bmatrix} 1 & 0 \\ 0 & 1 \end{bmatrix}, \qquad [H] = \begin{bmatrix} 1 & 0 \\ 0 & 1 \end{bmatrix}$$

In this particular instance, since the data are adequate for a unique solution, we can set $\gamma = 0.0$ and Figure 7.3(b) shows the performance against iteration.

The case of $\beta = 0$ corresponds to using the linear version of the tangent stiffness matrix, while $\beta = 1$ corresponds to using the exact tangent stiffness matrix. In both cases, we get the correct converged result, but clearly, the approximate stiffness requires more iterations.

The full results over many load increments are shown in Figure 7.4. In this case, the time variable is pseudotime.

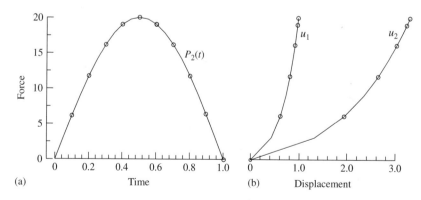

(a)   Time   (b)   Displacement

Figure 7.4: Results for determining a load history. (a) Force history with $P_1 = 0$. (b) Force/ displacement with $P_1 = 0$.

Computational Implementation

A very important issue that needs to be addressed is the assemblage of the matrices. Finite element codes for nonlinear problems require a level of sophistication well beyond that needed for linear problems. Therefore, as much as possible, the FEM portion of the inverse solution should be shifted external to the inverse programming.

To see where this can be done, consider the following abbreviated algorithm at a particular load step:

$$\textbf{sensitivity responses:} \quad [\,\Phi\,] \longrightarrow$$

$$[K_T]_0[\,\Psi\,] = [\,\Phi\,]$$

$$\longleftarrow [\,\Psi\,]$$

$$\textbf{begin iterate:}$$

$$\textbf{update:} \quad \{P\} \longrightarrow$$

$$[K_T][\Delta u] = \{P\} - \{F\}$$

$$\longleftarrow \{u\}_0, \{F\}_0$$

$$\textbf{responses:} \quad \{F\}_0 \longrightarrow$$

$$[K_T]_0\{\psi_F\} = \{F\}_0$$

$$\longleftarrow \{\psi_F\}$$

$$\textbf{least squares:} \quad \{u\}_0, [\,\Psi\,], \{\psi_F\}, \{d\} \longrightarrow$$

$$[A^T W A + \gamma H]\{\tilde{P}\} = \{B\}$$

$$\longleftarrow \{P\}$$

$$\textbf{end iterate:}$$

The updating can be done externally which is fortunate because good quality element nodal forces $\{F\}$ are required for a converged solution. The least squares solve stage is done internally. Computing the responses is problematic because the tangent stiffness matrix (associated with the nonlinear problem) is required and this is something not usually readily available external to an FEM program. Constructing this internally in the inverse program is tantamount to writing a nonlinear FEM code.

One solution to this dilemma is to realize that the Newton–Raphson iterations scheme does not require the exact tangent stiffness matrix for convergence. Thus, at the expense of the number of iterations and the radius of convergence, the sensitivity responses can be obtained from the external FEM program by using an approximate tangent stiffness matrix. An obvious choice would be to use the linear elastic stiffness matrix, another could be a linear elastic stiffness matrix that is based on an updated geometry. This issue is considered in more detail in the dynamic case.

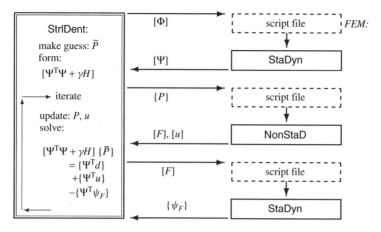

Figure 7.5: A possible distributed computing relationship between the static inverse program and the FEM programs.

With this in mind, a possible schematic for the algorithm is shown in Figure 7.5 where StaDyn is the linear FEM code and NonStaD is the nonlinear FEM code. A note about notation: the vectors $\{u\}$ and $\{F\}$ are associated with the free degrees of freedom, which necessarily depend on the internal organization of memory; the arrays $[\ u\ ]$ and $[\ F\ ]$ contain the same information but are arranged according to nodes to facilitate communication between programs. Note that the left-hand side is formed and decomposed only once, and only the right-hand side changes during the iterations.

Numerical Study: Nonlinear Deflection of a Beam

Consider the cantilever beam shown in Figure 7.6(a) under the action of a dead weight vertical load. Figure 7.6(b) shows some of the deflected shapes as the load is increased.

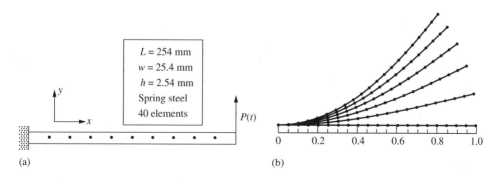

Figure 7.6: Cantilever beam. (a) Geometry and properties. (b) Some deflected shapes.

That the beam tip has a significant horizontal displacement is indicative of the problem being nonlinear. We will determine the applied load using this displacement information.

The synthetic data were generated using the nonlinear FEM program NonStaD [71]; the implicit integration option was chosen. The experimental version of this problem could have, for example, fluorescent dots painted on the side of the beam, and for each load increment, an image of the dots are digitally recorded.

Figure 7.7(a) shows some convergence results to obtain the maximum load; only the tip deflection value at the maximum load is used. The convergence is quite robust as regards initial guess. The rate of convergence is typical for the variations of the problem tested.

The u displacement results in Figure 7.7(a) are interesting because the linear model associated with the horizontal beam predicts zero horizontal displacement and therefore is not capable of determining the sensitivity responses (actually the sensitivity responses are zero). As will be discussed in greater detail in the next section dealing with structural dynamics, because the present scheme is based on Newton–Raphson equilibrium iterations, there is no requirement that the linear model (which is used to estimate the tangent stiffness) be the actual model. The algorithm as implemented in StrlDent uses two separate models: one is the actual or true model and this is used in the nonlinear updating, the other is used for estimating the tangent stiffness. When the u displacement data is used, the linear model was a cantilever beam at an angle to the horizontal such that the tip position was approximately at half of the measured tip deflection.

Figure 7.7(b) shows the convergence results when two loads are determined. The second load was taken as acting midway along the beam and it should evaluate to zero. The two-force problem is more sensitive to noise and therefore a total of four data were used—u and v displacement at each load point. The results show that it is possible to determine multiple loads. Note that in this two-force problem, the actual model was also used as the linear model.

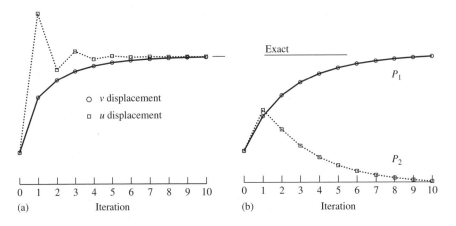

Figure 7.7: Convergence results for the cantilever beam. (a) Single force using u or v displacement with 1% noise. (b) Two forces using u and v displacements with 1% noise.

7.2   Nonlinear Structural Dynamics

In many ways, incrementing in time for dynamic problems is similar to incrementing in load for static problems. Therefore, it might seem reasonable to let, from Equation (1.44),

$$[K_T^*] \equiv \left[K_T + \frac{4}{\Delta t}C + \frac{4}{\Delta t^2}M \right]$$

$$\{F^*\} \equiv \{F\}_{t+\Delta t} - [C]\left\{ \frac{2}{\Delta t}u + \dot{u} \right\}_t - [M]\left\{ \frac{4}{\Delta t^2}u + \frac{4}{\Delta t}\dot{u} + \ddot{u} \right\}_t$$

and write the incremental solution as

$$[K_T^*]\{\Delta u\} = \{P\}_{t+\Delta t} - \{F^*\}_{t+\Delta t}$$

Since this is similar to the static relation, we can immediately write the equation for the optimized force $\{\tilde{P}\}$ as before, but obtain the sensitivity responses from

$$[K_T^*][\Psi] = [\Phi], \qquad [K_T^*]\{\Psi_F\} = \{F^*\}$$

The updating portion need only update from the previous converged state.

Unfortunately, this method will not work in general. Its main deficiency is that it attempts to relate the forces at a particular instant in time to the measurements at that time. If there is any spatial separation (as happens in wave propagation), then completely erroneous results are obtained. We must instead use fully dynamic approaches analogous to those in Chapters 5 and 6.

Iterative Dynamic Scheme

The dynamics of a general nonlinear system are described by

$$[M]\{\ddot{u}\} + [C]\{\dot{u}\} = \{P\} - \{F\} = [\Phi]\{\tilde{P}\} - \{F\}$$

The nodal forces depend on the deformations that are not known. Approximate these forces by a Taylor series expansion about a known (guessed) configuration

$$\{F\} \approx \{F\}_o + \left[\frac{\partial F}{\partial u} \right]\{\Delta u\} = \{F\}_o + [K_T]_o\{u - u_o\} = \{F\}_o - [K_T]_o\{u_o\} + [K_T]_o\{u\}$$

where the subscript 'o' means known state and $\{F\}_o - [K_T]_o\{u\}_o$ is a residual force vector due to the mismatch between the correctly reconstructed nodal forces $\{F\}$ and its linear approximation $[K_T]_o\{u\}_o$; at convergence, it is not zero. We have thus represented the nonlinear problem as a combination of known forces $\{F\}_0 - [K_T]_0\{u\}_0$ (known from the previous iteration) and a force linearly dependent on the current displacements, $[K_T]_0\{u\}$. The governing equation is rearranged to

$$[M]\{\ddot{u}\} + [C]\{\dot{u}\} + [K_T]_0\{u\} = [\Phi]\{\tilde{P}\} - \{F\}_0 + [K_T]_0\{u\}_0$$

This is a linear force identification problem with unknown $\{\tilde{P}\}$ and known $\{F\}_0$ and $[K_T]_0\{u\}_0$. Solve the three linear dynamics problems

$$[M][\ddot{\Psi}] + [C][\dot{\Psi}] + [K_T]_0[\Psi] = [\Phi]$$

$$[M]\{\ddot{\psi}_F\} + [C]\{\dot{\psi}_F\} + [K_T]_0\{\psi_F\} = -\{F\}_0$$

$$[M]\{\ddot{\psi}_u\} + [C]\{\dot{\psi}_u\} + [K_T]_0\{\psi_u\} = [K_T]_0\{u\}_0$$

Note that the left-hand operator is the same in each case. The total response can be represented as

$$\{u\} = [\Psi]\{\tilde{P}\} + \{\psi_F\} + \{\psi_u\}$$

and the data relation becomes

$$[Q]\{u\} - \{d\} = 0 \qquad \Longrightarrow \qquad [Q\Psi]\{\tilde{P}\} = \{d\} - \{Q\psi_F\} - \{Q\psi_u\} = \{d^*\}$$

The minimum principle is

$$E(\tilde{P}) = \{d^* - Q\Psi\tilde{P}\}^T\lceil W \rfloor\{d^* - Q\Psi\tilde{P}\} + \gamma\{\tilde{P}\}^T[H]\{\tilde{P}\} = \min$$

and minimizing with respect to $\{\tilde{P}\}$ then gives

$$[[Q\Psi]^T\lceil W \rfloor[Q\Psi] + \gamma[H]]\{\tilde{P}\} = [Q\Psi]^T\{d - Q\psi_F - Q\psi_u\} \qquad (7.2)$$

which is identical in form to the linear case. The core array is of symmetric size $[M_p \times M_p]$; note that it does not change during the iterations and hence the decomposition need be done only once. Further, it is of near Toeplitz [143] type just as for the linear problem, which holds the possibility of even further efficiencies in its solution; more discussion of this is given later in the chapter. The left-hand side changes during the iterations due to the updating associated with $\{\psi_F\}$ and $\{\psi_u\}$.

The basic algorithm is as follows: starting with an initial guess for the unknown force histories, the effective data in Equation (7.2) is updated, then $\tilde{P}$ is obtained from solving Equation (7.2), from which new effective data are obtained and the process is repeated until convergence. The implementation details, however, are crucial to its success and this is discussed next.

Computer Implementation

As discussed with the static algorithm, a very important issue that needs to be addressed is the formation and assemblage of the matrices associated with the above formulation and, as much as possible, to have the FEM portion of the inverse solution shifted external to the inverse programming. To see where this can be done, consider the following abbreviated algorithm where the arrows indicate the direction of parameter passing:

sensitivity responses: $[\ \Phi\ ]\longrightarrow$

$$[\ M\ ][\ \ddot{\Psi}\ ]+[\ C\ ][\ \dot{\Psi}\ ]+[\ K_{\mathrm{T}}\ ]_{\mathrm{o}}[\ \Psi\ ]=[\ \Phi\ ]$$

$$\longleftarrow[\ \Psi\ ]$$

begin iterate:

update: $\{\tilde{P}\}\longrightarrow$

$$[\ M\ ]\{\ddot{u}\}+[\ C\ ]\{\dot{u}\}=[\ \Phi\ ]\{\tilde{P}\}-\{F\}$$

$$\longleftarrow\{u\}_{\mathrm{o}},\ \{F\}_{\mathrm{o}}$$

responses: $\{F\}_{\mathrm{o}},\ \{u\}_{\mathrm{o}}\longrightarrow$

$$[\ M\ ]\{\ddot{\psi}_{F}\}+[\ C\ ]\{\dot{\psi}_{F}\}+[\ K_{\mathrm{T}}\ ]_{\mathrm{o}}\{\psi_{F}\}=-\{F\}_{\mathrm{o}}$$

$$[\ M\ ]\{\ddot{\psi}_{u}\}+[\ C\ ]\{\dot{\psi}_{u}\}+[\ K_{\mathrm{T}}\ ]_{\mathrm{o}}\{\psi_{u}\}=[\ K_{\mathrm{T}}\ ]_{\mathrm{o}}\{u\}_{\mathrm{o}}$$

$$\longleftarrow\{\psi_{F}\},\ \{\psi_{u}\}$$

least squares: $\{u\}_{\mathrm{o}},\ [\ \Psi\ ],\ \{\psi_{F}\},\ \{\psi_{u}\},\ \{d\}\longrightarrow$

$$[\Psi^{\mathrm{T}}W\Psi+\gamma\ H]\{\tilde{P}\}=\{d^{*}\}$$

$$\longleftarrow\{\tilde{P}\}$$

end iterate:

This bears a remarkable resemblance to the abbreviated algorithm of the static method and therefore it is not necessary to repeat the whole discussion.

The main conclusion again is that since the Newton–Raphson iterations scheme does not require the exact tangent stiffness matrix for convergence, then, perhaps at the expense of the number of iterations and the radius of convergence, the sensitivity responses can be obtained from the external FEM program by using an approximate tangent stiffness matrix. An obvious choice would be to use the linear elastic stiffness matrix, another could be a linear elastic stiffness matrix that is based on an updated or mean geometry. This issue is considered in more detail in the examples to follow because it also relates to how the initial guesses are obtained.

The schematic for the algorithm as implemented in the program StrIDent is shown in Figure 7.8 and superficially is almost identical to Figure 7.5 One significant difference

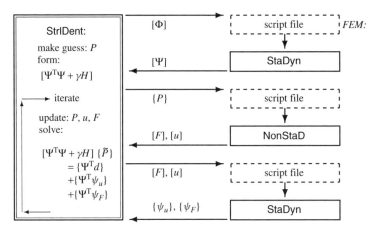

Figure 7.8: A possible distributed computing relationship between the nonlinear dynamic inverse program and the FEM programs. A possible scheme for using external FEM programs for solving the nonlinear dynamic inverse problem.

is that now [u] and [F] are histories of every DoF at every node, and therefore can require a very large storage space. Also note that the linear FEM program is expected to give the dynamic solution to the loads $\{F\}_o$ and $[K_T]_o\{u\}_o$; both are loadings where every DoF at every node is an individual history. Forming the product $[K_T]_o\{u\}_o$ where each component of $\{u\}_o$ is an individual history is not something usually available in commercial FEM packages. Requirements such as this were needed to be implemented as special services in the FEM programs available to the inverse program. It seems that for the nonlinear problem there is not a definite separation (about what must be done internally and what can be done externally) as there is for linear problems.

Approximating the Tangent Stiffness

A most crucial aspect of the foregoing algorithm is the role played by the approximate tangent stiffness [K_T]$_o$. The algorithm requires a guess of this stiffness; in the case of the nonlinear elastic deflection of a structure (such as depicted in Figure 7.9(a)), a reasonable guess is to use the linear elastic stiffness if the deflections are not too large. When the material behavior is nonlinear, especially when it is elastic/plastic (such as depicted in Figure 7.9(b) with multiple unloadings/reloadings), the choice of linear structure is not at all obvious. We now look at examples of both of these to help clarify the issues involved.

Consider the dynamics of the simple two-member truss shown in Figure 7.10 where the members have the elastic-plastic behavior shown in Figure 7.11(b). Recall that a truss member can support only axial loads.

The nonlinear FEM program NonStaD [71] was used to generate the synthetic data, and the force, velocity, and displacement are shown in Figure 7.11(a). Note that this

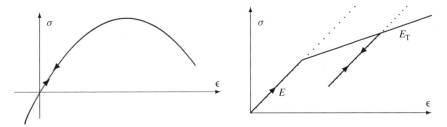

Figure 7.9: Examples of nonlinear material behavior. (a) Elastic. (b) Elastic-plastic.

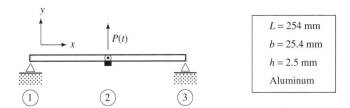

$L = 254$ mm

$b = 25.4$ mm

$h = 2.5$ mm

Aluminum

Figure 7.10: Transverse impact of a simple two-member truss.

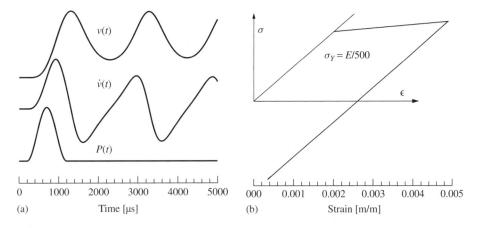

Figure 7.11: Responses for the two-member truss. (a) Force, velocity, and displacement. (b) Nonlinear stress/strain behavior with unloading.

problem is a combination of nonlinear (large) deflections and nonlinear material behavior. At the end of the impact, the truss remains in a deformed configuration. The characteristics of the long-term vibrations are that of a nonlinear nonsymmetric spring–mass system [71]; that is, they show a softening on the down swing and a stiffening on the up swing (the reverse of Figure 7.9(a)); if the amplitudes are large enough, they would show a snap-through buckling instability.

The elastic/plastic excursions of the truss are shown in Figure 7.11(b). There is unloading and reloading but the reloading never reaches the work-hardening curve; this is typical of short duration impulse loading.

Because a truss member can support only axial loads and the joints are pinned, the linear version of this problem (i.e., when the members are horizontal) does not have any structural stiffness because the truss members are not triangulated. As indicated for the static problem, the implemented algorithm uses two FEM models: one is the "true" model used in the nonlinear updating, the other is needed to obtain the (approximate) tangent stiffness used in a linear analysis. A number of possibilities present themselves for choosing this second model; for example, based on the displacement response, an A-frame truss can be constructed with a height of the average vertical excursion. This scheme actually works very well. Figure 7.12(a) instead shows convergence results for a quite different model: the flat truss is treated as a beam and the transverse stiffness then comes from the flexural behavior. The convergence was quite robust as regards the choice of flexural stiffness EI. What is important about this case is that the linear model has a different number of DoF compared to the nonlinear model—indeed, it is a different type of structure altogether.

Convergence results for a more radical linear model is shown in Figure 7.12(b). Here, the truss is treated as is initially and therefore it effectively is just that of a free floating mass. The sensitivity response for the velocity is a straight line in time. Nonetheless, the convergence is quite good.

While not a proof as such, these examples make the key point that it is the Newton–Raphson iterations on the equilibrium, and not the quality of the $[K_T]$ estimate, that produces the converged result.

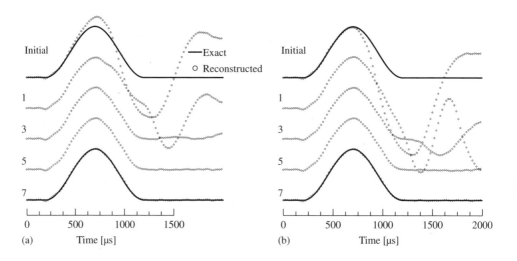

Figure 7.12: Effect of linear model on convergence behavior. (a) Linear model is a beam. (b) Linear model has no transverse stiffness. Numbers to the left are the iteration counter.

A final point to make is that in each case, the initial guess for the load history is based on the linear model. Thus, the choice of linear model affects the convergence in two ways: the initial structure about which the iterations take place, and the initial guess for the history. There are many ways to get an initial guess for the solution; the one usually available is to use the linear solution. If the nonlinearity is strong, then the linear guess may be so grossly in error that convergence problems can occur. In that case, the data can be adjusted as $(1 - \alpha^i)\{d\}$ where α^i is decreased on each iteration. This is considered in the examples.

7.3 Nonlinear Elastic Behavior

There are two types of nonlinear elastic behavior. One is where the stress/strain relation is nonlinear, and the other is where the material is linear but the deflections and rotations are large enough to cause nonlinear structural behavior. We look at an example of both.

Nonlinear Elastic Wave Propagation

One of the underlying themes of this book is the use of remote measurements to infer the unknowns of a problem. The archetypal carrier of information from and to remote regions is a stress wave. This section looks at some of the issues involved when the wave is propagating in a nonlinear elastic material.

Hyperelastic materials are elastic materials for which the current stresses are some (nonlinear) function of the current total strains; a common example is that of rubber materials. For these materials, the stresses are derivable from an elastic potential, according to Equation (1.18). We will restrict our problem to small strains and displacements.

The structure to be considered and its stress/strain behavior is shown in Figure 7.13. The material is described by the two modulus relation

$$\sigma_{xx} = E_2 \epsilon_{xx} + [E_1 - E_2]\epsilon_0[1 - f(\epsilon)], \qquad f(\epsilon) \equiv e^{-\epsilon_{xx}/\epsilon_0}$$

The strain ϵ_0 or equivalently the stress $\sigma_0 = E_1\epsilon_0$ is used to control where the bend in the stress/strain relation occurs. Depending on whether E_2 is less than or greater than E_1, the material will either soften or harden. The two extremes to be considered is shown in Figure 7.13. The tangent modulus and local wave speed are then

$$E_T = \frac{\partial \sigma_{xx}}{\partial \epsilon_{xx}} = E_2 + (E_1 - E_2)f(\epsilon), \qquad c = \sqrt{E_T/\rho}$$

As a consequence of a strain-dependent modulus, different parts of the wave will travel with different speeds; that is, the observed waves are dispersive.

The rod of Figure 7.13 is subjected to a very short duration pulse at one end and monitored at the points 0, $L/3$, $2L/3$, L; the velocity responses are shown in Figure 7.14.

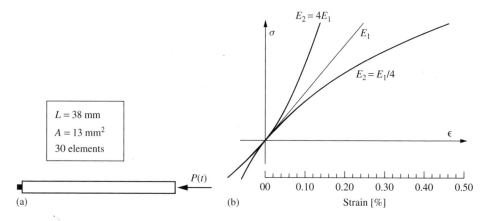

Figure 7.13: Impact of a uniaxial rod. (a) Geometry. (b) Stress/strain behavior of the materials.

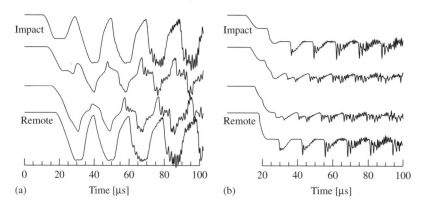

Figure 7.14: Velocity responses showing wave dispersion due to material nonlinearity. (a) Softening material. (b) Hardening material.

The force history applied is shown as the heavy line in Figure 7.15. There is a significant wave propagation effect and the dispersive nature of the waves is clear.

The character of the responses is quite different between the two materials. The softening material undergoes more strain and, therefore, the particle velocities are larger and the amplitude swings are also quite large. The hardening material undergoes less strain and, therefore, the particle velocities are smaller and the amplitude swings are also smaller.

Initial Guess

As part of the iteration process, we need to have an estimate of the tangent stiffness $[K_{To}]$. As discussed previously, there are many possible choices for this initial structure, the most

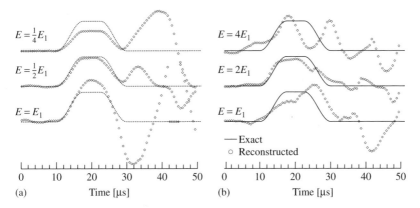

Figure 7.15: Initial guesses based on single modulus material. (a) Softening material. (b) Hardening material.

obvious being the linear version of the structure. But the guessed structure stiffness can also have another important use—it can be used to estimate the initial responses.

Figure 7.15 shows a set of initial guesses using different (single) moduli for the initial structure. For softening materials, almost any modulus will give a good estimate of the initiation of the pulse but the tail end can be grossly in error. Indeed, as seen in Figure 7.15(a) for $E = E_1$, the force goes outside the allowable search area; parenthetically, when this happens we need a rule for adjusting the estimate, as shown in the figure, for this type of problem, the load is replaced by zero.

The hardening material shows that the overall magnitude is not severely dependent on the modulus, and that with an appropriate modulus, reasonable initial behavior can be estimated.

The convergence behavior is shown in Figure 7.16 for both materials. It should be noted that, overall, the hardening material was more sensitive to the choice of problem

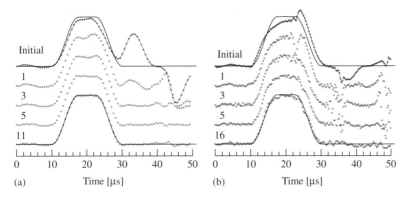

Figure 7.16: Convergence behavior over iteration with 1% noise. (a) Softening material with $E = \frac{1}{2}E_1$. (b) Hardening material with $E = \frac{3}{2}E_1$. Numbers to the left are the iteration counter.

parameters. For example, it required a smaller Δt, and convergence was robustly achieved only for $E \approx \frac{3}{2}E_1$. Even then, the quality of the converged results is quite different compared to the softening material. This could be because hardening materials are more inherently sensitive to noise (a supposition inferred from looking at Figure 7.14), or some other factors. This is something requiring more study and experience.

Transverse Impact of a Beam

We consider the large deflection of a beam as our second example of nonlinear elastic behavior. The material behavior is linear elastic, but the large deflections mean that the geometry must be updated and this gives rise to the nonlinearity. Figure 7.17(a) shows the geometry of the cantilever beam and the impact position.

Figure 7.17(b) shows the velocity responses at the four monitoring positions and their comparison to a linear analysis. It is clear that there is a significant nonlinear effect but, by the nature of the problem, it manifests itself only over time as the deflections become large. The maximum deflection at the beam tip is on the order of seven times the beam thickness—this deflection would not cause significant nonlinear effects if it were a static problem, but dynamic problems typically are more sensitive. These synthetic data were obtained using a $\Delta T = 1.0\,\mu s$, a mesh with 40 elements, and the applied force history shown as the bottom plots in Figure 7.18 with a peak value of 4.45 KN (1000 lb). These data were contaminated with 1% noise.

For the inverse problem, a time discretization of $\Delta T = 2.5\,\mu s$ was used and only time regularization was applied. Figure 7.18(a) shows the computed forces through the first six iterations. The first iteration is the linear analysis and we see the effects of mismatched oscillations in the nonzero force past $t = 400\,\mu s$. On subsequent iterations, the noticeable deviations are shifted further out in time. Note that it is possible that the estimated forces

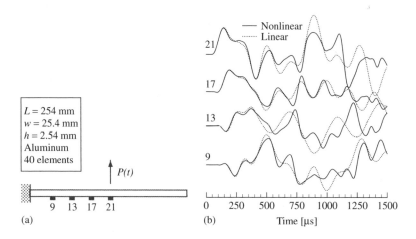

Figure 7.17: An impacted cantilever beam exhibiting nonlinear dynamic behavior. (a) Geometry and monitoring positions. (b) Velocity responses. Also shown as dotted lines is the linear response.

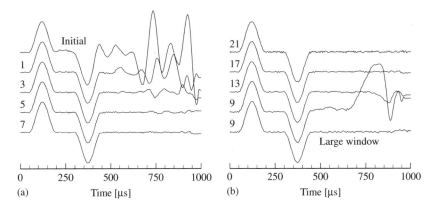

Figure 7.18: Force reconstructions for data with 1% noise. (a) Rate of convergence using data from Sensor 17. (b) Converged results with respect to sensor location.

can be large enough to cause a numerical instability and consequently, a rule must be used to suppress them. In the results presented, forces were never allowed to exceed two times the maximum force that occurred during the first iteration. There are, of course, other possibilities.

Figure 7.18(b) shows the effect of sensor location on the converged result. Sensors 13 and 9 show significant deviations at the large times. The main reason for this is that a larger time window is required when the sensor is remote from the forcing point. The bottom plot in Figure 7.18(b) is for a time window of $T = 1750\,\mu$s. In this case, the deviations are shifted out to about $t = 1400\,\mu$s.

Figure 7.19 shows two reconstructed forces using data from Sensors 17 and 9—note that neither sensor is at a force location. The second force is taken at location 11 and, of course, should reconstruct to be zero. Clearly, the method can determine multiple forces.

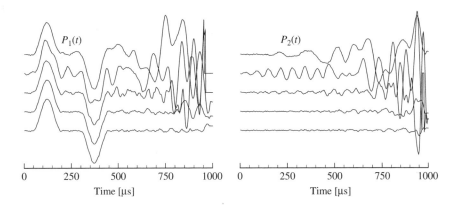

Figure 7.19: Rate of convergence for reconstructing two forces using responses from Nodes 9 and 17 with 1% noise.

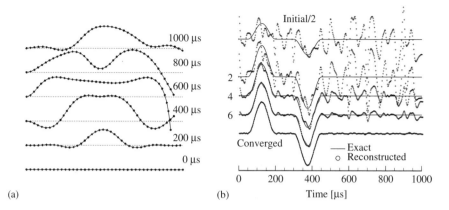

Figure 7.20: Force reconstructions using image data. (a) Sample deformed shapes (exaggerated times 15) used as image inputs. (b) Rate of convergence with 0.1% noise.

We end this section by considering the use of images as a data source for nonlinear force identification. The algorithm of Figure 7.8 is hardly affected when image data is used; the only difference is that the scheme of Figure 6.4 is used to establish the [Ψ] matrix but otherwise the nonlinear updating and evaluating the responses $\{\psi_F\}$ and $\{\psi_u\}$ follows identically.

Figure 7.20(a) shows a series of deformed shapes for the problem of Figure 7.17—the black dots indicate the data points. Although the problem is highly nonlinear, the deflections, nonetheless, are relatively small and for this reason the shapes in the figure are shown exaggerated by a factor of 15. Parenthetically, the nonlinearity is also observed through the horizontal displacement of the tip of the beam.

Figure 7.20(b) shows the force reconstructions over the iterations. The initial guess is based on the linear behavior of the beam and it bears very little relation to the exact force history. It is gratifying, however, to see the rather rapid convergence rate in spite of the poor initial guess. Nonetheless, this example shows that there is a need to develop methods to give better initial approximations.

The number of frames of data needed depends on the strength of the nonlinearity, the total duration, and the amount of noise; with no noise, the corresponding linear problem yields good reconstructions using just two frames, whereas the nonlinear problem with no noise required five frames. The results in Figure 7.20(b) required 10 frames (one every 100 µs) because of the noise. Part of the reason for needing more frames is that each frame contains so few data points (41) that the noise has fewer data points to be averaged over. Remember that 400 time points are being reconstructed. Approximately, the sensitivity of the results to noise is about the same as that illustrated in Figure 6.27.

Finally, convergence progresses with time but not as strongly as for the point sensors. This is partly explained by the uniformly poor initial guess.

7.4 Elastic-Plastic Materials

Most engineering structures are designed to sustain only elastic loads, but analysis of problems in impact, metal-forming, and fracture are based on the inelastic and plastic responses of materials. These are the situations of interest here.

Projectile Impact

One challenge of ballistic experiments is to be able to determine the force due to impact and penetration. References [36, 75] discuss two possible experimental methods; what they have in common is that they record data at the back end of the projectile from which they try to infer the applied loading at the front end. Two more recent examples [54, 32] used measurements off the side of the projectile.

Figure 7.21 shows a blunt projectile impacting a stop—the stop, as implied, does not deform. This example will be used to explore some of the issues involved in trying to use remote measurements to infer contact loads when the contact region undergoes elastic-plastic deformations.

The program NonStaD can model dynamic contact problems [40] and was used to generate the synthetic data for the moving projectile contacting the rigid stop. A blunt projectile was chosen because it gives a sharper rise time in the contact force history than the corresponding rounded projectile. Although the six contacting nodes have individual histories, only the resultant force will be identified here; that is, StrIDent has the ability to identify individual forces or resultants on groups of nodes; in the present case, just the resultant will be identified.

Figure 7.22(a) shows the velocity responses along the length of the projectile and Figure 7.22(b) shows the stress/strain excursions experienced. As can be seen, there is a substantial amount of plasticity with multiple unloadings and reloadings; however, it is interesting to note that although the head of the projectile experiences a good deal of plasticity, the rebound velocity shows only a slight loss of momentum.

Figure 7.23 shows the resultant force history as the full heavy line. For this example, the contact conditions were adjusted so as to not excite high-frequency vibrations in the elements—this would have resulted in a high-frequency oscillation in the generated force history, which in turn would have required a finer mesh to remove. Note that the

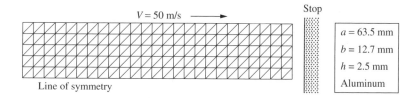

Figure 7.21: Blunt projectile impacting a stop.

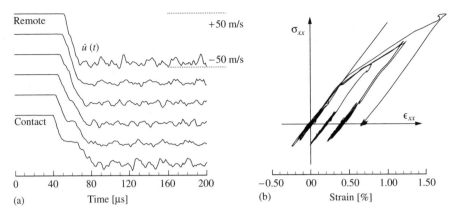

Figure 7.22: Projectile behavior. (a) Velocity responses. (b) Nonlinear $\sigma_{xx}/\epsilon_{xx}$ stress/strain response at different locations.

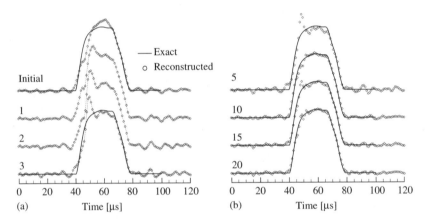

Figure 7.23: Convergence behavior with 1% noise. (a) First three iterations. (b) Samples over last twenty iterations. Numbers to the left are the iteration counter.

contact time depends primarily on the length of the projectile and not on the contact conditions.

Figure 7.23 shows the convergence behavior using data just from the tail end. The initial guess is based on the linear behavior and it looks quite good. This is not surprising because essentially the signal propagated to the monitor is dominated by the elastic response, and the measured response is insensitive to the particulars of the behavior in the vicinity of the contact. This accounts for the success of force identification methods even in the presence of localized plasticity as occurs in the impact of, say, a steel ball on aluminum.

There actually are significant differences between the linear reconstructed force and the true force. The phase delay is due to the change of modulus above yielding. The first few iterations of Figure 7.23(a) are attempting to correct for this phase shift. By the fifth iteration, this initial behavior has been corrected but in the process, the maximum force has become large. Keep in mind that this maximum force will primarily affect the long time response seen in the trailing edge.

The next twenty or so iterations of Figure 7.23(b) are concerned with reducing this peak force. Convergence is slow but, all in all, the final comparison is quite good. That the reconstructed force could have the large excursions and still return to the correct answer attests to the basic robustness of the method.

Multiple Loads on a Truss Structure

As a second elastic-plastic example, consider the multimember truss shown in Figure 7.24. This is subjected to two uncorrelated loads and their histories are shown in Figure 7.25 as the solid lines; note that the two forces are not identical.

Again, NonStaD was used to generate synthetic velocity data at each of the joints. With this particular loading, the members can experience many load/unload cycles. Three of the members yield and their stress/strain excursions are shown in Figure 7.24(b). Member 2, which is closest to P_1, shows a significant amount of plasticity.

The convergence results are shown in Figure 7.25 where the vertical velocity of the three center joints were used as data. The initial guess for P_1 is poor while that for P_2 is quite good. This can be explained by noting that P_1 on its own will cause the three members to yield whereas P_2 on its own will not cause any yielding. The forces are reconstructed exceptionally well by the tenth or so iteration. Since there is 1% noise, the fluctuations never settle down.

A variety of tests were performed on this structure during which it was noticed that all joints are not equally good as sensor locations. The impact sites, for example, seemed relatively poor. This is surprising because it is contrary to what is found for

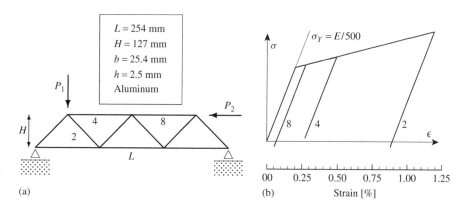

Figure 7.24: A truss with two uncorrelated loads. (a) Geometry. (b) Stress/strain behavior of members that yielded.

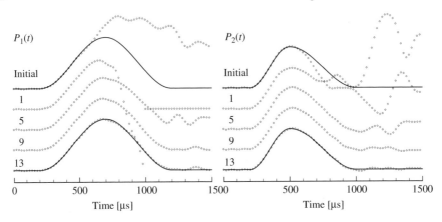

Figure 7.25: Convergence behavior with 1% noise. Numbers to the left are the iteration counter.

elastic structures. At this time, it is not clear as to why this is so and is something worth investigating.

Attempts to reconstruct the forces over a longer period of time were not especially successful. The longer period encompasses the stage when the members, especially Member 2, reverse stress sign. It seems that when the members are in the unloading/reloading stage, the responses are sensitive to the history of stress up to that point; small variations in initial history can cause large variations in this behavior.

Although not explored here, a possible solution strategy might be to reconstruct the forces with a time base expanding with the iterations. If the new force estimates are weighted in favor of retaining the already estimated early time behavior, then this might give a more robust approach to getting to the unload/reloading stage. Such a strategy would also help give better initial guesses. This is something worth investigating.

7.5   Nonlinear Parameter Identification

In previous chapters, we divided parameter identification problems into two types: those that can be parameterized as unknown forces and those that have implicit parameters. The advantages of the former is that iteration is not required and that generally a parameterized model need not be assumed *a priori*. These advantages are no longer compelling in nonlinear problems since all cases require iteration.

Consider the two cases shown in Figure 7.26 where a beam is attached to nonlinear springs. In Figure 7.26(a), the unknown spring is localized to the beam and therefore can be separated through a free-body cut. What is left is a nonlinear force identification problem of the type considered in the previous sections; only if the deflections are small (as was the case in Section 4.7) are iterations avoided. A force parameterization approach to the case of Figure 7.26(b) would require the identification of many forces and would make sense only if all the springs were different. When all the springs are the same or

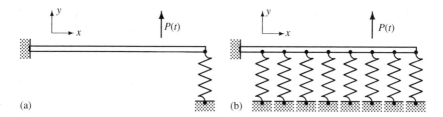

Figure 7.26: Beams attached to unknown nonlinear springs. (a) The nonlinearity of the spring is localized. (b) The nonlinearity of the springs is distributed.

can be represented by a reduced number of parameters, it makes sense to determine the spring parameters directly. We will show examples of doing both.

Determining a Localized Nonlinearity

Consider the beam shown in Figure 7.27(a); the attached nonlinear spring is governed by the equation

$$P(u) = eP_0[u/\delta]e^{-u/\delta}$$

and is shown in Figure 4.42. Note that $P(u)$ reaches a maximum value of P_0 when $u = \delta$. This spring exhibits a limit instability on the tension side—this is precisely the type of nonlinear behavior expected from unstable crack growth as we will see later.

The synthetic v velocity data were generated by the nonlinear FEM program Non-StaD [71]. This is a fully nonlinear dynamics program that uses either an explicit or implicit time integration scheme combined with a corotational formulation. In the present

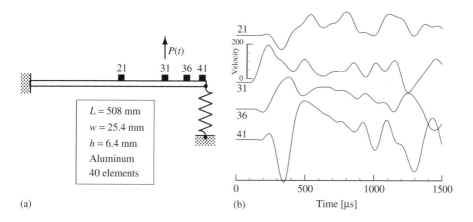

Figure 7.27: A beam with an attached nonlinear spring. (a) Geometry and properties. (b) Velocity response data.

case, the explicit option was used. It is worth pointing out that even when synthetic data are used, the FEM model must be converged for both element size and time step. This is because a different FEM solution scheme will be used for the inverse problem. The parameters for the spring were chosen as

$$P_0 = 220\,\text{N}\,(50\,\text{lb})\,, \qquad \delta = h/8$$

The velocity response data are shown in Figure 7.27(b) generated with the force history shown in Figure 7.28(a) as $P_1(t)$.

The subdomain of the inverse problem is that of the beam without the nonlinear spring, and the spring is replaced with an unknown reaction as shown in Figure 4.42(b). The reaction force, as well as the input force, are the unknowns to be determined.

The reconstructed forces using only Sensors (21) and (38) are shown in Figure 7.28(a) for two noise levels. The correspondence with the exact forces is very good except at the large time value; this deviation at large time is expected because of the propagation delay.

It is worth repeating that once the tractions have been determined, everything else about the problem (such as stress, displacements, and so on) can be determined. This includes the displacements at the nonlinear spring and consequently we will be able to determine the nonlinear spring relation.

Figure 7.28(b) shows the reconstructions plotted as force versus displacement. It is seen that the spring went past its limit point instability; this spring is elastic so that on unloading it would go back through the limit point. The agreement with the exact behavior is very good except for the large time behavior. Also note that the spring was characterized without an *a priori* assumption about the model; this can be a very useful feature in some circumstances.

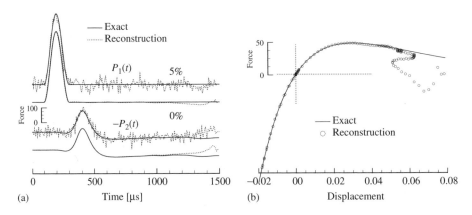

Figure 7.28: Inverse solution. (a) Force reconstructions for two levels of noise. (b) Force/displacement relation compared to the exact.

Distributed Parameters

This second example illustrates how the inverse methods can be used to identify nonlinear parameters that are distributed throughout the structure. The particular example is shown in Figure 7.26(b).

The governing equation for a nonlinear system is

$$[M]\{\ddot{u}\} + [C]\{\dot{u}\} = \{P\} - \{F\}$$

from which it is seen that not only do the unknown parameters (spring coefficients in this case) appear implicitly, but even the stiffness itself appears only implicitly (the element stiffness is used to construct $\{F\}$). In the implicit integration scheme, the structural tangent stiffness matrix $[K_T]$ is constructed but this changes throughout the loading history unless a modified Newton–Raphson scheme is adopted. We conclude that any direct manipulations of the stiffness matrix is not readily available in nonlinear problems. This is likely to be even more true if commercial FEM packages are used. Therefore, the direct differentiation method as discussed in Section 5.7 is not well suited for nonlinear problems.

We will use the sensitivity response method first discussed in Chapter 3 and also used in the static and dynamic chapters. Let us have a set of guesses for the unknown parameters, then solve the system

$$[M]_o\{\ddot{u}\}_o + [C]_o\{\dot{u}\}_o = \{P\} - \{F\}_o$$

It does not matter if an explicit or implicit integration scheme is used. In turn, change each of the parameters (one at a time) by the definite amount $a_j \rightarrow a_j + da_j$ and solve

$$[M]_j\{\ddot{u}\}_j + [C]_j\{\dot{u}\}_j = \{P\} - \{F\}_j$$

The sensitivity of the solution to the change of parameters is constructed from

$$\{\psi\}_j \equiv \left\{ \frac{u_j - u_o}{da_j} \right\} \bar{a}_j$$

where $\bar{a}_j$ is a normalizing factor. From here on, the method is identical to that for linear problems and Figure 4.36 can serve to illustrate its implementation.

The advantage of the scheme presented is that it is not necessary to have access to the system matrices. Potentially at least, any nonlinear parameter in a nonlinear setting can be determined.

By way of example, consider the beam shown in Figure 7.27(b) resting on a distributed spring. The beam and springs have the properties shown in Figure 7.27(a). The goal is to identify the parameters P_o and δ.

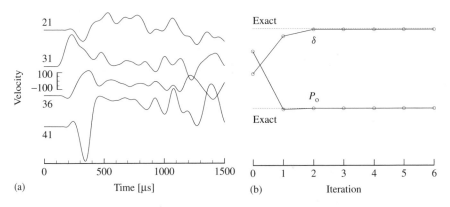

Figure 7.29: A beam with distributed springs. (a) Velocity response data. (b) Convergence results using a single sensor with 10% noise.

Figure 7.29(a) shows the recorded velocity data where the force of Figure 7.28(a) is used. The first issue to be addressed is the number of sensors required. The previous examples show that, approximately, one sensor history is required for each unknown force history. Here, however, there are just two unknowns and a single sensor has the equivalent of a large amount of data.

Figure 7.29(b) shows the convergence results using just one sensor. In spite of there being 10% noise, the single sensor correctly identified the parameters. These results were quite robust as regards initial guess and choice of sensor.

A word about testing and the appropriate level of force to apply: with reference to Figure 7.28(b), the parameter P_o governs the peak force achieved whereas the ratio P_o/δ governs the initial slope of the curve. If the force level is too low, the only real identification we can achieve is the ratio P_o/δ; to separate these parameters, it is necessary to load past the instability point. This is what was done in the test; the results correspond to all of the springs rupturing.

It is worth keeping in mind that for problems like this, it is often possible to segment the testing; that is, low force levels are first used to identify the single parameter P_o/δ. This is followed by a high force level identification of the single parameter δ. By and large, two single parameter identifications are more robust than one identification of two parameters simultaneously.

7.6 Dynamics of Cracks

There is considerable interest in the stress analysis of crack propagation and dynamic fracture in general; see Reference [79] for a comprehensive discussion of the topic. This section looks at two aspects. First is the determination of the stress intensity factor when a stress wave impinges on a crack. The second is determining the fracture properties for a propagating crack.

Consider a plate with an initial crack: under dynamic loading, there are very high stress gradients at the crack tip that change with time. If the stresses are large enough they could lead to crack propagation. The controlling quantity for this to happen is the stress intensity factor (or, equivalently, the strain energy release rate) as discussed in Sections 3.3 and 4.7. Crack propagation, as a dynamic event, is highly nonlinear and not all FEM codes are capable of analyzing it; therefore, we will first give a brief review of how it is implemented.

We will consider two problems. The first is where the strain energy release rate is not sufficient to cause crack propagation and we focus on the dynamics at the crack tip. The second is where the crack actually propagates. In both cases, we will use a fully nonlinear FEM analysis to generate the synthetic data.

Separation of Material Systems

It is the intention here to review just the main ingredients of the FEM modeling; more details can be found in Reference [39, 40]. The main feature discussed is the scheme for material separation: in this, the continuum is characterized by two constitutive relations, one that relates stress and deformation in the bulk material, the other that relates the traction and displacement jump across a cohesive surface. Consequently, all element-to-element connections are potential surfaces for separation. This formulation allows separation through cracking, tearing, rupturing, and erosion. These were implemented into the existing nonlinear code NonStaD [71].

With reference to Figures 7.30 and 7.31, each mesh is made up of separated elements that are kept together by the interfacial cohesive tractions (both normal and shear). Thus, in addition to the nodal forces arising from the body stresses, there are nodal forces associated with the cohesive stresses. This is done by introducing cohesive surfaces into the element boundary.

The governing equation is given by [71]

$$[M]\{\ddot{u}\} + [C]\{\dot{u}\} = \{P\} - \{F\} = \{P\} - \{F_b\} - \{F_c\} \tag{7.3}$$

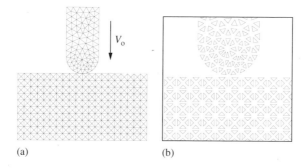

(a)   (b)

Figure 7.30: Framework for material separation. (a) Global mesh. (b) Separated elements.

Chapter 7. Nonlinear Problems

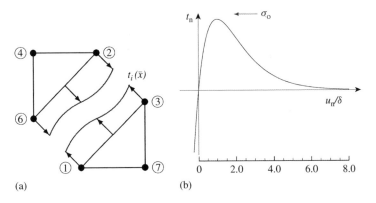

Figure 7.31: Interelement behavior. (a) Cohesive tractions between elements. (b) Normal cohesive stress behavior.

where $[\,M\,]$, $[\,C\,]$, $\{P\}$, and $\{F\}$ have their usual meanings. The element nodal forces have two contributions: $\{F_b\}$ arises from the element body stresses, and $\{F_c\}$ represents the assembled cohesive forces obtained from

$$ F_{cij} = \int_S t_i(\overline{x}) g_j(\overline{x})\, d\overline{x}\,, \qquad t_i = \frac{\partial \Phi}{\partial u_i} $$

where t_i is the traction vector on a surface in the reference configuration, S is the interface of the element, g_j are weighting functions, and Φ is an elastic potential.

The specific form of potential used is that given in Reference [175]; restricting attention to two dimensions with normal and tangential displacements u_n and u_t, respectively, this reduces to

$$ \Phi(u_n, u_t) = \Phi_n - \Phi_n e^{(-u_n/\delta)}[1 + u_n/\delta]e^{(-u_t^2/\delta^2)}\,, \qquad \Phi_n = e\sigma_o\delta $$

Here, σ_o is the cohesive surface normal strength and δ is a characteristic length. Letting $r_n = u_n/\delta$ and $r_t = u_t/\delta$, the surface tractions are obtained as

$$ t_n = e\,\sigma_{max}\, r_n e^{-r_n} e^{-r_t^2}\,, \qquad t_t = 2e\,\sigma_{max}\, r_t[1 + r_n]e^{r_n}e^{-r_t^2} \tag{7.4} $$

The variation of t_n with u_n (when $u_t = 0$) is shown in Figure 7.31(b). The maximum value of t_n is σ_o and occurs when $u_n = \delta_n$. On the tension side, when u_n exceeds δ decohesion begins to occur. When complete decohesion occurs ($u_n > 6\delta$), the surfaces are specified as "postrupture"; that is, the cohesive forces are zero unless the surfaces come into contact with each other or some other surfaces. On the compression side, the surfaces experience an increasing resistance to interpenetration. The tangential behavior is different in that it can decohere for both positive and negative u_t.

The specific values of the cohesive parameters are important for brittle materials especially when there is fast crack propagation whereas for ductile materials, the bulk stress is not likely to be anywhere near σ_0 and therefore decohering as such will not occur; but rupture through tearing and erosion can occur. The cohesive parameters chosen are such that at low stress levels, the cohesive modeling recreates the modeling using perfectly bonded elements [40].

The corotational scheme [23, 46, 71] as discussed in Chapter 1 was adopted as the underlying solution method. In this, a local coordinate system moves with each element, and constitutive relations and the like are written with respect to this coordinate system. This scheme is especially convenient when materials fragment because then the local coordinate system moves with each fragment. Lastly, the explicit time integration scheme is used to time integrate the resulting equations.

Stationary Crack

Consider the plate of Figure 7.32 with an initial crack and subjected to symmetric forces with the history shown. Reference [41] used this model to study crack propagation; here the crack will remain stationary. Parenthetically, the relatively short duration of pulse combined with the small model were imposed by the constraints of the FEM program [40] used to generate the crack propagation data.

The mesh for the top half of the cracked plate is shown in Figure 7.33. The dynamic event lasts a very short time, about 6 μs, but includes multiple reflections. Hence, a small ΔT is needed for StrIDent to give accurate reconstructions. To do so, a converged mesh must be used—the mesh shown is converged with $\Delta T = 0.05$ μs.

The synthetic $v(x, y, t)$ displacement data were generated by NonStaD and the Moiré fringe patterns are shown in Figure 7.33. Because of symmetry, the reaction forces of the interface are in the vertical direction only. Thus, as shown in Figure 7.34(a), the applied force and 21 reactions constitute the total set of unknowns. This data will be used to reconstruct these forces.

The program StrIDent will be used to reconstruct the forces on the assumption that the subdomain problem is linear; hence the first task is to assess this assumption. Subset problems were constructed, as shown in Figures 7.34(b, c), that involved the applied

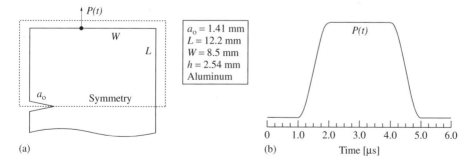

Figure 7.32: Plate with an initial crack. (a) Geometry and properties. (b) Applied load.

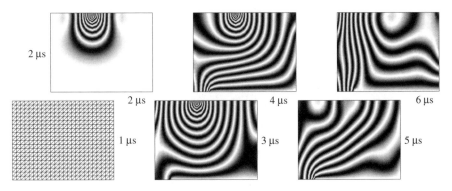

Figure 7.33: Moiré $v(x, y)$ fringe patterns. The frames are 1 μs apart.

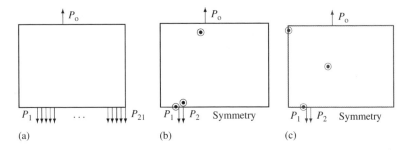

Figure 7.34: Possible exposed unknown forces and sensor locations. (a) All forces are exposed. (b) Two exposed forces with sensors close to loads. (c) Two exposed forces with sensors remote to loads.

load plus the reactions at the first two crack nodes. These forces were identified using two sensor arrangements and the results are shown in Figure 7.35. The near sensor arrangement essentially gives the exact solution in each case.

There is a definite difference between the mismatching of the models. The differences would be exacerbated if even more forces were identified. For problems like this, there is a space–time connection between inferences drawn. The mismatched model means that sensor signals cannot be propagated over large distances and hence must be placed close to the unknown forces; for whole-field image data, it means that the frames must be recorded at a finer frame rate. For example, the linear–linear analysis gave very good results using frames every 3 μs, whereas the nonlinear–linear analysis required frames every 0.5 μs.

The 22 unknown forces were identified by StrIDent using frame data (with noise of 0.1% added) at every 0.5 μs. The reconstructed input force history is almost identical to that shown in Figure 7.32(b). The reconstructed force histories along the edge of the model are shown in Figure 7.36(a). The edge has a complicated traction history with a traction distribution changing in both space and time but the inverse method is capable of determining it. The arrival of the various waves is discernible.

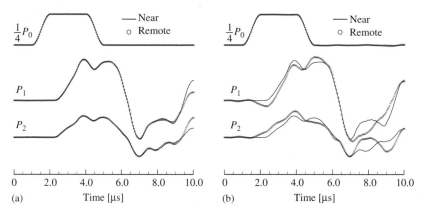

Figure 7.35: Effect of sensor position on the reconstructed force histories using different modeling. (a) StaDyn data, StaDyn reconstruction. (b) NonStaD data, StaDyn reconstruction.

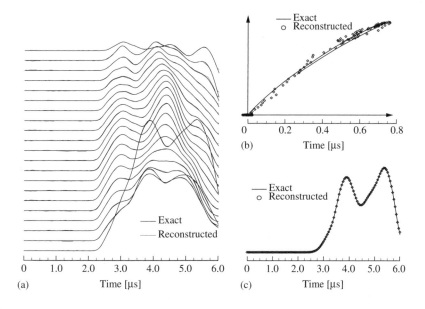

Figure 7.36: Reconstructions for the complete line of symmetry. (a) Cohesive force histories. (b) Cohesive behavior at crack tip. (c) Strain energy release rate.

Since all the tractions are known, it is possible to obtain the displacements and, in particular, the crack tip opening displacement. The elastic behavior of the crack tip is similar to that of the nonlinear spring except that the loads are not high enough to reach the limit load point.

The strain energy release rate and stress intensity factor were computed by the virtual crack closure technique (VCCT) [151] as discussed in Section 4.7. This is shown in

Figures 7.36(c). The VCCT was also applied to the linear version of this problem (where the mesh is a standard FEM mesh) and is also compared in Figure 7.36(c), the agreement is very good. Strictly speaking, the mesh used does not meet the criterion for an accurate application of VCCT [151] because the elements are too large relative to the crack length.

Discussion of Stationary Crack Results

The theoretical framing rate for this problem is 1.25 μs but it was necessary to use a rate of 0.5 μs to get good results. The reasons for this are not entirely clear; the worst-case scenario used for estimating the frame rate (as discussed in Chapter 6) was for a single force on one side of the model. A force on both sides of the model would obviously require a faster frame rate by a factor of two. The question here is whether the applied load is just one more load along with the 21 from the crack, or if it is different because it is on the other side of the model. This will need further investigation to be answered. There is the possibility of segmenting the inverse problem so that the applied force is determined separately from the others. In the particular case of the stationary crack, it can be determined without using a subdomain such that there is only one unknown force.

Reference [41] discusses the same model and loading as shown in Figure 7.32 but with the load scaling increased so as to cause the crack to propagate. In comparison to the stationary crack, this problem required an yet faster frame rate of 0.3 μs in order to get good reconstructions.

Thus, nearly 20 frames would be required to reconstruct the histories over the 6 μs time window; this is getting close to the sampling rate associated with point sensors. We conclude that reconstructing cohesive force histories directly from frame data is not feasible. In the next series of studies, we parameterize the problem in terms of the cohesive properties as a way of reducing the number of unknowns.

Dynamic Crack Propagation

A theme of this book is the central role played by the model in stress analysis. We illustrate here how it can be used in the design of a complicated experiment; that is, it is generally too expensive in terms of both time and money to design experiments by trial and error and a well-constructed model can be used to narrow the design parameters.

To make the scenario explicit, consider a double cantilever beam (DCB) specimen made from two aluminum segments bonded together as shown in Figure 7.37. The crack (or delamination) is constrained to grow along the interface and the interface properties of the bond line are the basic unknowns of the experiment. The FEM model takes advantage of the symmetry of the DCB specimen.

In order to get accurate estimates of the interface properties, it is necessary to record the data as the crack passes through the frame of view. Typical frames are shown in Figure 7.38 along with the exaggerated deformed shape. The applied load, shown in

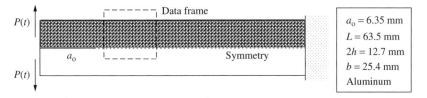

Figure 7.37: Double cantilevered beam. Geometry, properties, and mesh.

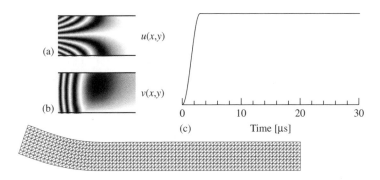

Figure 7.38: Double cantilever beam data. (a) Moiré $u(x, y)$ and $v(x, y)$ fringes at 10.6 μs. (b) Deformed shape (exaggerated by 10). (c) Applied load history.

Figure 7.38(c), is a rectangular pulse of long duration, and for identification purposes is taken as known.

This is a problem with a wide range of scales. As seen from Figure 7.38(a), the tip deflection is on the order of $h/10$ but h is on the order of $\delta \times 10^5$. At rupture, the deflections along the line of symmetry are $1\delta \sim 10\delta$, which is up to four orders of magnitude different. To increase the accuracy of the data, the frame came from just the 12 mm × 12 mm window indicated in Figure 7.37(a). In previous studies using synthetic data, the added noise was based on the average of the data; here the standard deviation was specified at $0.2 \times \delta$. With a setup sensitivity of 100 line/mm this works out at about 0.2% of a fringe.

The convergence behavior is shown in Figure 7.39, in which a single $v(x, y)$ frame of data at 10.6 μs was used; both parameters have converged by the tenth iteration. The results are robust within reasonable initial guesses of the parameters. In the present tests, the allowable search windows spanned 10 to 200% of the exact values with the initial guesses being 50%.

Keep in mind that during the iteration process, at each iteration, a complete crack propagation problem is solved three times (once for the nominal parameters, and once each for the variation of the parameters). Thus, the computational cost is quite significant.

Once the parameters are determined, reconstructions of any quantity can be obtained. Figure 7.40(a) shows, for example, the interface force histories at every other node. These histories are almost identical to the exact histories (which is not surprising because of the

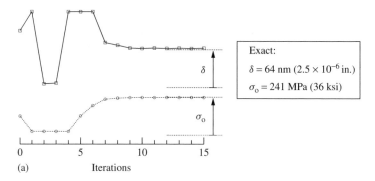

Figure 7.39: Convergence behavior using a single frame of data with 0.2% noise.

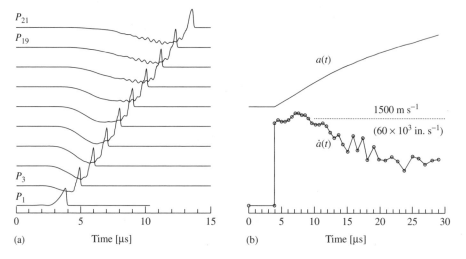

Figure 7.40: Reconstruction of the DCB specimen behavior. (a) Cohesive force history. (b) Crack tip position and speed.

goodness of the parameter estimates). The quality of these reconstructions is far superior to those achieved in Reference [41] using many more frames; this underlines the value of a reparameterization of a problem.

The position of the crack tip is quite apparent in Figure 7.40(a); using the peak value of the force as the indicator, Figure 7.40(b) shows crack tip position and its speed. The speed is not constant because of the decrease in beam stiffness as the crack propagates.

Discussion of Propagating Crack Results

The stress history in a crack propagation problem is very complicated, being a combination of dynamic effects arising from the incident waves plus waves emanating from the propagating crack. Hence, the quality of frame data is very important.

Reparameterization of the problem in terms of the cohesive properties reduces the problem to a couple of unknowns and therefore considerably relieves the burden on the experimental data. It is important to realize, however, that all this was achievable only because the FEM formulation already has these parameters.

It is possible to use other parameterizations. For example, Reference [109] discusses how measured crack tip position versus time data is used to drive an FEM code to extract various dynamic variables during the event. This could be modified so that fracture toughness (slowly varying over the history) is the basic unknown parameter.

7.7 Highly Instrumented Structures

A feature to emerge from the foregoing sections and chapters is that realistic situations involving real structures entail dealing with a large number of unknowns. Depending on our parameterization of the problem, this means either a large number of unknown forces or a large number of unknown parameters. In either case, it is essential that we have a corresponding large number of sensors to provide the required independent information.

Emerging technologies are reducing the size, weight, and cost of sensors, and also improving their spatial and time resolutions. The reason these developments are so exciting is that, for the first time, it is economically feasible to contemplate using a large number of sensors. In the future, structures are expected to have a large number of embedded sensors [7, 80, 81] for time data (e.g., laser scanning, in-fiber velocimeter), but spatial information (e.g., thermographic imaging interferometry, infrared imaging) is also expected to be available to provide additional information.

It is, therefore, not unreasonable to contemplate determining many unknowns, and as a consequence, this refined spatial information can be used to advantage in structural monitoring, damage detection, and operational reliability. Realizing this potential will require the development of innovative algorithms and schemes for processing this large amount of data. The objective of this section is to explore some of the implications that may be realized in the future for highly instrumented structures. We explore this question within the context of modern experimental stress analysis.

Some Possibilities

We ask the open-ended question: if we have a great number of sensors, what new things can we do with them? The immediate answer, of course, is that we can determine many forces and many unknowns. But the deeper answer revolves around how we might rephrase or reorganize the identification problem. We show some possibilities here.

I: Subdomain Inverse Concept

Often in experimental situations, it is impossible, inefficient, or impractical to model the entire structure. Also, in many cases, data can only be collected from a limited region. For example, a Moiré pattern cannot be obtained for a whole structure when the geometry is complex as was seen in Chapter 6 for the cylinder. Another example is when the data resolution from certain parts of the structure is not clear enough as happens around notches using photoelasticity. Conversely, increasing the resolution at the notch region would require just that small region to be focused on at the expense of the rest of the structure. Furthermore, we may only be interested in a subset of the unknowns; for example, we may want to know if a flaw is present and not particularly care about the actual boundary conditions.

Additionally, the probability of achieving a good inverse solution is enhanced if the number of unknowns can be reduced. This would be possible if we could just isolate the subregion or subdomain of interest (the aircraft wing detached from the fuselage, the loose joint separate from the beams, the center span independent of the rest of the bridge). The basic idea is shown schematically in Figure 7.41; the subdomain has, in addition to the primary unknowns of interest, a set of additional unknown tractions that are equivalent to the remainder of the structure. As we can see, this introduces a new set of unknowns associated with the attachment boundaries; that is, while in each of the examples discussed above and in the previous chapters, the data come from a limited region the whole model is actually used. The subdomain inverse method is a completely different concept: here, only a small portion of the complete structure is actually modeled.

The significant advantage of analyzing a subdomain with unknown tractions over analyzing the complete domain is the opportunity to remove nonlinearities from the analysis. For example, a sloshing fuel tank in an aircraft or a sunken foundation pile would simply not be included in the subdomain. It may also be that the totality of unknowns in the complete system is itself unknown, again justifying a subdomain. A further advantage is that it gives us a formal way of scaling our analysis; that is, global analysis as well as detailed analysis is possible within the same framework.

II: Structural Monitoring

In general, structural health monitoring can be understood as [6] "the measurement of the operating and loading environment and the critical responses of a structure in order to

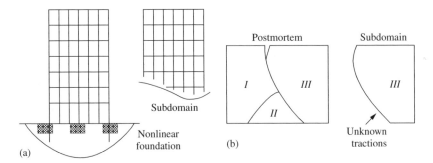

Figure 7.41: Application of the subdomain concept. (a) A building structure removed from its non-linear foundation. (b) A fragment of a brittle fracture.

track and evaluate the symptoms of operational anomalies and/or deterioration or damage that may impact service or safety reliability". Modern civil and aeronautic structures require advanced health-monitoring systems to increase safety and operational reliability not only because of their increasing size and complexity but also because they are becoming essential parts of sophisticated engineering integrated systems (such as transport and communication systems) [6, 7]. Integrated systems will rely on field information gathered by different types of sensors and therefore the processing and analysis of the information will also be essential. Continuous and remote monitoring [140] is increasingly being used and different evaluation strategies are needed in order to handle appropriately large amounts of time and spatial information.

One of the most important applications of structural monitoring is damage detection and evaluation. The response of a complex structure to a stress wave is governed by the equations

$$[M]\{\ddot{u}\} + [C]\{\dot{u}\} = \{P\} - \{F(u)\} \qquad \text{or} \qquad [M]\{\ddot{u}\} + [C]\{\dot{u}\} + [K]\{u\} = \{P\}$$

for nonlinear and linear systems, respectively. The changed condition due to damage is represented as a perturbation of the stiffness and mass matrices from the original state as

$$\{F_D\} = \{F\} - \{\Delta F\}, \qquad [K_D] = [K] - [\Delta K], \qquad [M_D] = [M] - [\Delta M]$$

Substitute into the governing equations and rearrange for the linear system as

$$[M]\{\ddot{u}\} + [C]\{\dot{u}\} + [K]\{u\} = \{P\} + [\Delta K]\{u\} + [\Delta M]\{\ddot{u}\} = \{P\} + \{D\}$$

and for the nonlinear system as

$$[M]\{\ddot{u}\} + [C]\{\dot{u}\} = \{P\} - \{F\} + \{\Delta F\} + [\Delta M]\{\ddot{u}\} = \{P\} - \{F\} + \{D\}$$

In both cases, the vector $\{D\}$ clearly has information about the changed state; if we can somehow determine $\{D\}$, then it should be a direct process to extract the desired information about the changed state. Indeed, in some of the earlier chapters, we showed some examples of doing this.

The key issue in the monitoring scheme is the ability to identify the vector $\{D\}$. The components of this vector are associated with the nodes of the FEM model and therefore constitute a large set of unknowns. Note that it appears above as a force vector similar to $\{P\}$; our problem therefore reduces to one of determining a large set of unknown applied forces.

III: Monitoring by Partitions

Monitoring involves determining if an event such as an impact or a rapid change of properties has occurred. What makes it difficult is that the whole structure must be monitored simultaneously. The task would be easier if subregions could be monitored separate from other regions.

The idea of partitioning is closely related to the concept of the subdomain. Here, the sensors are arranged to partition the structure into regions such that only sensors directly related to a changed region show a response.

To motivate the approach, consider the beam of Figure 7.42. As per the stipulation of this section, we assume that we have many sensors distributed throughout the structure. The idea we want to develop is that if the location of the force is unknown, how can the sensor information to delimit the region of the force be used.

If we make a free-body cut to the left of the applied load as in Figure 7.42(b) and use the sensors to identify the exposed forces, then these forces are equivalent to the removed material plus the applied load. This is the subdomain concept.

Suppose that, instead, we retain the full model but identify forces at a location to the left of the applied forces as in Figure 7.42(c). (Note that the identified forces must be the same number as exposed by a free-body cut.) Then these force are equivalent to the applied load. If, in addition to these two forces, we also identify a third force to their left, then this third force will be zero. This third force acts as a monitor to show that there are no unaccounted forces in the left domain.

Figure 7.42(d) shows the force reconstructions (using four sensors) when the forces are to the left of the actual load. Noise of 1.0% was used. The force at the monitoring site, P_3, shows only the noise.

This position for P_2 and T_2 can be moved arbitrarily and therefore, the monitoring force can unambiguously determine in which region the applied force is not located. Note that in each case the same recorded data is used and the same model is used.

Complex Structures

A complex structure such as an aircraft wing or highway bridge is formed as a combination of structural elements that can broadly be modeled as frame, plate, or solid entities

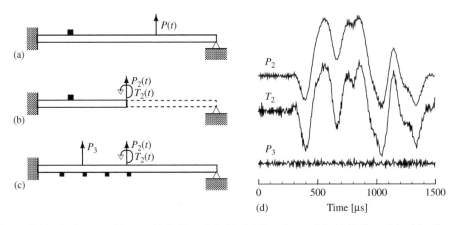

Figure 7.42: An impacted beam. (a) Full model. (b) Subdomain model. (c) Full model with relocated forces. (d) Reconstructed forces for impacted beam with 1% noise.

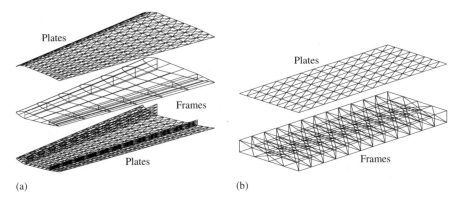

Figure 7.43: Complex structures are formed as combinations of frame, plate, and solid elements. (a) Aircraft wing. (b) Highway bridge.

as shown in Figure 7.43. Different parts of these structures will be accessible in different ways. It is therefore clear that a general monitoring scheme will require the simultaneous use of different types of sensors. The synergistic use of different sensor types is an issue that must be addressed when dealing with complex structures.

Consider the aircraft wing, for example. Extended surfaces such as the skin could be monitored using some whole-field data scheme, but the interior regions would have to be monitored using discrete sensors since it is not accessible visually. The requirement then is to be able to use both data sets simultaneously.

As developed in the previous chapters, we envision many types of sensors being available; discrete (in space) sensors that give time-based signals and are recorded using digital acquisition systems, and whole-field (continuous in space) sensors that are recorded with some sort of optical system at a limited time rate.

A very useful attribute of the developed sensitivity response method is that the complete set of unknowns are reduced to the scalings $\tilde{P}_m$. Mixing data of different types is then a matter of appending more rows to the system of equations; that is, using the subscripts 1 and 2 to refer to two different sensor types (possibly time sensors and space sensors respectively or maybe strain gages and accelerometers respectively), the separate system of equations are written as

$$[\,A_1\,]\{\tilde{P}\} = \{B_1\}, \qquad [\,A_2\,]\{\tilde{P}\} = \{B_2\}$$

These are combined as

$$\begin{bmatrix} [\,A_1\,] \\ [\,A_2\,] \end{bmatrix} \{\tilde{P}\} = \left\{ \begin{array}{c} \{B_1\} \\ \{B_2\} \end{array} \right\}$$

The least squares minimizing leads to

$$[[\,A_1\,]^T \lceil W_1 \rfloor [\,A_1\,] + [\,A_2\,]^T \lceil W_2 \rfloor [\,A_2\,]]\{\tilde{P}\} = \{[\,A_1\,]^T \lceil W_1 \rfloor \{B_1\} + [\,A_2\,]^T \lceil W_2 \rfloor \{B_2\}\}$$

Because [A_1] and [A_2] could be of different magnitudes, this form of the least squares would give inadvertent weighting. Therefore, it is necessary to normalize the equations. A simple normalization is to use the average diagonal value, that is,

$$\mu_i^2 = \text{Tr}[[A_i]^\text{T} \lceil W_i \rfloor [A_i]]/N_i$$

The regularized form of the system of equations becomes

$$\left[\sum_i^{N_s} \frac{1}{\mu_i^2}[A_i]^\text{T} \lceil W_i \rfloor [A_i] + \gamma [H] \right] \{\tilde{P}\} = \left\{ \sum_i^{N_s} \frac{1}{\mu_i^2}[A_i]^\text{T} \lceil W_i \rfloor \{B_i\} \right\}$$

where N_s is the number of sensor types. A more explicit form for the regularization could be

$$\gamma [H] \Longrightarrow \gamma_\text{t}[H_\text{t}] + \gamma_\text{s}[H_\text{s}]$$

if both isolated forces and traction distribution are to be determined.

This feature of combining different sensor types is also very useful for problems using nonlinear data since in that case one or two strain gage rosettes could be used in conjunction with the optical method to give a very robust inverse solution.

Scalability Issues

The practical application of the methods suggested here will need very efficient algorithms that scale adequately to the large number of unknowns and large number of sensors. The primary cost factors for the least squares–based sensitivity response method (SRM) are forming the least squares system and then solving it. The costs of these are

$$\text{LST cost:} \quad N^3 N_p^2 N_d , \qquad \text{Solve cost:} \quad N^3 N_p^3$$

where N is the number of time points, N_d is the number of sensors, and N_p is the number of forces. Both parts of the cost are on the order of $O(N^3)$; however, to make the scheme scalable for many forces, effort must be expended on reducing the N_p^3 dependence of the solution part. We will attempt to make the scheme more efficient by taking advantage of the special forms of the matrices. But first, we show how the least squares part can be reduced significantly.

I: Reducing the Least Squares Cost

The N_p unknown force histories are related to the N_d sensor histories by

$$\begin{Bmatrix} \{d\}_1 \\ \{d\}_2 \\ \vdots \\ \{d\}_{N_d} \end{Bmatrix} = \begin{bmatrix} [\Psi]_{11} & [\Psi]_{12} & \cdots & [\Psi]_{1N_g} \\ [\Psi]_{21} & [\Psi]_{22} & \cdots & [\Psi]_{2N_g} \\ \vdots & \vdots & \ddots & \vdots \\ [\Psi]_{N_d 1} & [\Psi]_{N_d 1} & \cdots & [\Psi]_{N_d N_g} \end{bmatrix} \begin{Bmatrix} \{\tilde{P}\}_1 \\ \{\tilde{P}\}_2 \\ \vdots \\ \{\tilde{P}\}_{N_g} \end{Bmatrix} \qquad (7.5)$$

We also write this as

$$\{d\}_i = \sum_j [\ \Psi\ ]_{ij}\{\tilde{P}\}_j \qquad \text{or} \qquad \{\bar{d}\} = [\ \overline{\Psi}\ ]\{\overline{P}\}$$

The $[M_p \times M_p]$ array $[\ \Psi\ ]_{ij}$ relates the i location sensor history to the j location applied force. It contains the response $\psi_m(t)$ arranged as a column vector with each column shifted down one; that is, the columns are the same except shifted down by one to yield a matrix that is lower triangular as shown in Figure 7.44(a).

The least squares system matrix is formed as the product $[\ A\ ] = [\ \overline{\Psi}\ ]^T[\ \overline{\Psi}\ ]$. A typical submatrix is

$$[\ A\ ]_{ij} = \sum_k [\ \psi\ ]_{ik}^T [\ \psi\ ]_{kj}$$

The individual products $[\ A\ ]_{ij}^k = [\ \psi\ ]_{ik}^T [\ \psi\ ]_{kj}$ can be accomplished efficiently by noting that $[A_{11}]_{ij}^k$ is the same as $[A_{22}]_{ij}^k$ but with one extra multiplication, that $[A_{22}]_{ij}^k$ is the same as $[A_{33}]_{ij}^k$ but with one extra multiplication, and so on. Thus, the diagonal terms are computed in reverse order as

$$[A_{N,N}]_{ij}^n = \sum_k^1 [\psi_{1k}]_{in}^T [\psi_{1k}]_{nj}$$

$$[A_{N-1,N-1}]_{ij}^n = \sum_k^2 [\psi_{1k}]_{in}^T [\psi_{1k}]_{nj} = [A_{N,N}]_{ij}^n + [\psi_{12}]_{in}^T [\psi_{12}]_{nj}$$

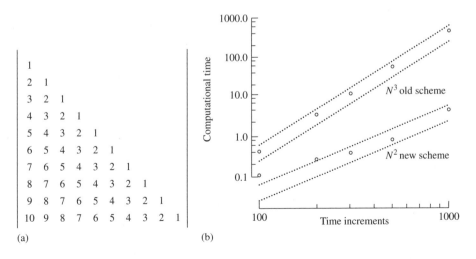

(a) (b)

Figure 7.44: Computational cost of the least squares scheme. (a) Structure of the matrix $[\psi]_{ij}$. (b) Performance of the new assemblage.

all the way up to the first term

$$[A_{1,1},]^n_{ij} = \sum_k^N [\psi_{1k}]^T_{in}[\psi_{1k}]_{nj} = [A_{2,2},]^n_{ij} + [\psi_{1N}]^T_{in}[\psi_{1N}]_{nj}$$

In this manner, the complete diagonal terms are computed in N products. The off-diagonal terms are done similarly with one product less for each column. Thus the complete product is formed in N^2 products and the total least squares procedure is reduced to

$$\text{LST cost:} \quad N^2 N_p^2 N_d$$

Since $N \approx 1000$, the least squares has been reduced to an insignificant factor of the solution cost.

Figure 7.44(b) shows how the cost varies with the number of time increments. The results shown are for one force and four sensors. There seems to be a computational overhead in the new assemblage procedure, but even so, there is a significant reduction in the cost.

Simultaneously with reducing the assemblage cost, the new scheme also avoids the necessity of establishing the full matrix [Ψ]; this saves memory of the size [$(N_d \times N) \times (N_p \times N)$], which can be quite substantial when the number of sensors and forces increase. For example, a modest 10 forces and 10 sensors with 1000 time steps requires a RAM memory of 800 Mbyte.

II: Utilizing the Near Toeplitz Form of Matrices

A Toeplitz matrix is such that each row is the same as the one above it but shifted laterally by one space; that is,

$$\begin{bmatrix} a_0 & a_{-1} & a_{-2} & \cdots & a_{-(N-1)} \\ a_1 & a_0 & a_{-1} & \cdots & a_{-(N-2)} \\ a_2 & a_1 & a_0 & \cdots & a_{-(N-3)} \\ \vdots & \vdots & \ddots & \vdots & \vdots \\ a_{N-1} & a_{N-2} & a_{N-3} & \cdots & a_0 \end{bmatrix}$$

The Toeplitz matrix is symmetric if $a_i = a_{-i}$. Considerations of the Toeplitz nature of the matrices are important because Toeplitz solvers have a computational cost of only $O(N^2)$ while the superfast Toeplitz algorithms [10, 11, 37, 98, 136, 157] can have a computational cost as low as $O(N \log^2 N)$; these are to be compared to the $O(N^3)$ costs of the decomposition solvers. Additionally, the memory requirements are minimal since only a vector of size $\{N_p \times N\}$ would need to be stored.

After application of the least squares procedure to time trace problems, the system matrix is symmetric but only of near Toeplitz form. Sample rows from the matrix [A]

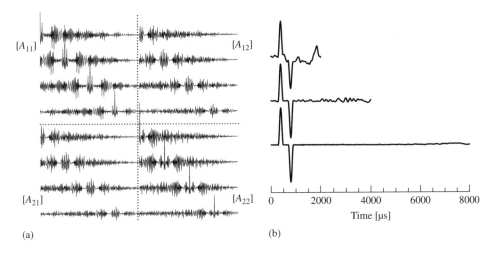

Figure 7.45: The near Toeplitz structure of the system array. (a) Sample rows of the system array. (b) Force reconstructions using different time windows.

(for two forces and four sensors) are shown in Figure 7.45(a). Note that $[A_{11}]$ and $[A_{22}]$ are symmetric but only near Toeplitz (the average amplitude of each row diminishes), while $[A_{12}]$ and $[A_{21}]$ are not symmetric but still near Toeplitz. The reason the submatrix products $[\psi]_{ik}^T[\psi]_{kj}$ (and hence their summations) are only near Toeplitz is because the columns of $[\psi]_{ij}$ are finite. As shown earlier, the diagonal terms of $[A_{11}]$, for example, are similar but with one product less.

The Toeplitz form of these matrices can be enhanced if the sensitivity responses are such as to die down near zero within the time window. The situations in which this can occur are as follows: the structure is large but the force and sensor locations are in proximity to each other, there is significant damping, the damping is mild but the time window is very large.

Figure 7.45(b) shows reconstructions for one force for the cantilever beam with mild damping. The reconstructions treat the matrix $[A]$ as being Toeplitz. It is clear that as the time window is increased, a more accurate reconstruction is obtained. Actually, we can write the system array as

$$[A] = [A_{\text{Toep}}] + [\Delta A]$$

and the system of equations to be solved then becomes

$$[A_{\text{Toep}}]\{\tilde{P}\} = \{b\} + [A - A_{\text{Toep}}]\{\tilde{P}\}$$

This is solved iteratively where the right-hand side is updated for the force components. This iterative scheme works best when the initial force estimate is good.

III: Possible Algorithm for Multiple Forces

For the multiple force case, the submatrices $[A_{ij}]$ are near Toeplitz but the system matrix $[A]$ is definitely not Toeplitz. Hence to take advantage of the Toeplitz solvers, an algorithm that works with the submatrices is required. We now present a possible algorithm.

For ease of notation, let A_{ij} represent the submatrix $[A_{ij}]$ and d_i represent the right-hand side. For concreteness, we will consider a system of $[3 \times 3]$ submatrices. Let the matrix have the [UDU] decomposition [67]

$$\begin{bmatrix} A_{11} & A_{12} & A_{13} \\ A_{21} & A_{22} & A_{23} \\ A_{31} & A_{32} & A_{33} \end{bmatrix} = \begin{bmatrix} 1 & 0 & 0 \\ U_{21} & 1 & 0 \\ U_{31} & U_{32} & 1 \end{bmatrix} \begin{bmatrix} D_{11} & 0 & 0 \\ 0 & D_{22} & 0 \\ 0 & 0 & D_{33} \end{bmatrix} \begin{bmatrix} 1 & U_{12} & U_{13} \\ 0 & 1 & U_{23} \\ 0 & 0 & 1 \end{bmatrix}$$

The elements of the decomposition are obtained by the sequence of calculations

$$D_{11} = A_{11}, \qquad\qquad U_{12} = D_{11}^{-1} A_{12}, \qquad\qquad U_{13} = D_{11}^{-1} A_{13}$$

$$D_{22} = A_{22} - U_{21} A_{12}, \qquad\qquad U_{23} = D_{22}^{-1}(A_{23} - U_{21} A_{13})$$

$$D_{33} = A_{33} - U_{31} A_{13} - U_{32}(A_{23} - U_{31} A_{13})$$

Clearly, many of the intermediate computations can be reused. Since the diagonal terms A_{ii} are near Toeplitz, then the diagonal terms D_{ii} are also near Toeplitz and can be inverted using a fast algorithm.

The solution is obtained in three steps. The system is first written as

$$\begin{bmatrix} 1 & 0 & 0 \\ U_{21} & 1 & 0 \\ U_{31} & U_{32} & 1 \end{bmatrix} \begin{Bmatrix} z_1 \\ z_2 \\ z_3 \end{Bmatrix} = \begin{Bmatrix} d_1 \\ d_2 \\ d_3 \end{Bmatrix}$$

giving the solution

$$z_1 = d_1, \qquad z_2 = d_2 - U_{21} z_1, \qquad z_3 = d_3 - U_{31} z_1 - U_{32} z_2$$

Then

$$\begin{bmatrix} D_{11} & 0 & 0 \\ 0 & D_{22} & 0 \\ 0 & 0 & D_{33} \end{bmatrix} \begin{Bmatrix} y_1 \\ y_2 \\ y_3 \end{Bmatrix} = \begin{Bmatrix} z_1 \\ z_2 \\ z_3 \end{Bmatrix}$$

giving

$$y_1 = D_{11}^{-1} z_1, \qquad y_2 = D_{22}^{-1} z_2, \qquad y_3 = D_{33}^{-1} z_3$$

Finally, the forces are found from the back-substitutions from

$$\begin{bmatrix} 1 & U_{12} & U_{13} \\ 0 & 1 & U_{23} \\ 0 & 0 & 1 \end{bmatrix} \begin{Bmatrix} P_1 \\ P_2 \\ P_3 \end{Bmatrix} = \begin{Bmatrix} y_1 \\ y_2 \\ y_3 \end{Bmatrix}$$

to give

$$P_3 = y_3, \qquad P_2 = y_2 - U_{23} P_3, \qquad P_1 = y_1 - U_{12} P_2 - U_{13} P_3$$

The equations are easily generalized to a system on the order of N_p; analogous coding is given in Reference [67].

For this scheme to reap the benefits of the near Toeplitz form, it is essential that all products take advantage of the special forms of A_{ij}. Thus, extra long time sequences should be used. If necessary, an iterative correction can be incorporated.

All matrix products and the inversion of D_{ii} are at most $O(N^2)$. There are $O(N_p^3)$ matrix products in the decomposition, hence the total cost is on the order of

$$\text{solve cost:} \quad N^2 N_p^3$$

The scheme is still cubic in the number of forces.

Some Implications for Highly Instrumented Structures

What the foregoing have in common is that they require the determination of a great number of forces. We assume that we have enough sensors for the task but if we are to determine a great number of forces, the algorithms for doing this must be able to scale appropriately. Therefore, the first issue to be addressed is that of scalability.

It is not clear how the data from future sensors will be recorded, whether it will be digitized time traces or some whole-field recording method, but it does seem clear that data of different types will be used. Therefore, the algorithms must provide for the mixing of data. It also seems reasonable to conjecture that the sensors (being inexpensive) do not have linear data relations nor good time fidelity (the sampling rate does not meet the Shannon criterion, for example). This must also be taken into consideration.

Finally, the issue of excitation must be addressed. All analysis is based on responses. In some cases, the excitation is part of the unknown event, but in terms of diagnostics, a designed excitation can be used. When a large number of unknowns are being determined, it is important to insure that the responses have information about these unknowns. Appropriate design of the excitation is an important consideration.

Implementing the improved least squares scheme is straightforward but much research will need to be done to implement the improved solver to take advantage of the near

Toeplitz form of the matrices. Furthermore, fast and superfast Toeplitz solvers can be unstable and several modifications and techniques are required in order to have adequate solutions [17, 173].

Both improvements bring the cost down to $O(N^2)$ in the number of time steps but leave it as $O(N_p^3)$ in the number of forces. There still remains, therefore, a potential bottleneck in the implementation of the SRM to large systems. An intriguing possibility is to use the nonlinear algorithm (even for linear problems) and take advantage of the second model to predetermine [Ψ] and the system decomposition; the cost would then be just that of the forward FEM solutions. Along the same lines, a sliding time window of fixed duration could be used thus making N (and hence [Ψ]) small. In both cases, the cost with respect to n is decreased in exchange for doing more iterations.

7.8 Discussion

The direct handling of nonlinearities can be computationally expensive. Furthermore, because of the iterations, they are prone to failure because of nonconvergence or convergence to the wrong solution.

The example problems discussed make it clear that first and foremost, it is necessary to have very robust identification schemes for linear problems because these become the core algorithmic piece of the nonlinear scheme. Good initial guesses can be crucial to the success of an iterative scheme, but how to obtain these is usually problem dependent. Again, this is a case where the linear methods can be very useful.

Two additional points are worth commenting on. First, the inverse part of the solution does not actually solve a nonlinear problem—the specifics of the mechanics of the problem is handled entirely by the FEM program. Consequently, the results show that potentially all types of material nonlinearities (elastic/viscoplastic, for example) can be handled as long as the FEM program can handle it. Second, the sensitivity responses and the nonlinear updating are computed using different FEM programs and therefore possibly different time integration algorithms; indeed, in the examples discussed, the linear analysis used an implicit scheme while the updating sometimes used an implicit scheme and sometimes used an explicit scheme. The implication is that a very large variety of nonlinear problems is amenable to solution by the present methods, again, as long as the FEM nonlinear program can handle the corresponding forward problem.

The second point is about convergence of these types of problems. The radius of convergence depends on the total time window over which the force is reconstructed. Since convergence usually proceeds along time, the implementation of a scheme that proceeds incrementally in time (increasing or sliding the time window on each iteration) could be very beneficial. Such a scheme could also be used to adjust the initial structure $[K_{\text{To}}]$ and thereby determine an even more appropriate set of sensitivity responses [Ψ]. It may well be that constructing a sophisticated (adaptive) algorithm for the initial guess will become a very important part of the successful application of the method to complex problems.

Afterword

... to solve an inverse problem means to discover the cause of a known result. Hence, all problems of the interpretation of observed data are actually inverse [problems].

A.N. TIKHONOV and A.V. GONCHARSKY [166]

It is important, however, to realize that most physical inversion problems are ambiguous — they do not possess a unique solution, and the selection of a preferred unique solution from the infinity of possible solutions is an imposed additional condition.

S. TWOMEY [172]

We take this opportunity to peer a little into the future and assess some of the future needs for full implementation of the ideas expressed in this book. As a backdrop, we remind ourselves that the book deals with computational methods and algorithms directed toward the active integration of FEM modeling and experiment.

For experimental stress analysts to fully utilize the methods presented, it is necessary that they be adept at both experimental methods and FEM analysis. The potential power of the sensitivity response methods (SRM) is that they can leverage the versatility and convenience of commercial FEM programs by using them in a distributed computing manner. That is, the structural analysis aspects of the inverse problem can be handled by their favorite FEM programs identically to that of the forward problem, thereby retaining their investment in these programs. However, to take advantage of this is no trivial matter; indeed, all the examples discussed used the StaDyn/NonStaD/GenMesh FEM package [67, 71] because access to the source code was available, allowing special features to be programmed as needed by the inverse methods. Therefore, a very important next step is to explore the writing of a program that interfaces between the inverse methods and commercial programs such as ABAQUS [178] and ANSYS [179]. This should be done in such a way as to make it convenient to switch between forward and inverse problems. Ideally, the program would identify a sufficient number of "generics" such that it would be easy to switch from one commercial package to another. In all probability, cooperation will be needed from the commercial people to provide hooks into their codes.

Modern Experimental Stress Analysis: completing the solution of partially specified problems. James Doyle
© 2004 John Wiley & Sons, Ltd ISBN 0-470-86156-8

The most challenging force identification algorithm presented is that associated with elastic/plastic loading/unloading because of the nonuniqueness of the load state at a particular instant in time—the uniqueness is established by tracking the history of the loading. Since convergence usually proceeds along time, the development of a solution strategy that also proceeds along time (either increasing or sliding the time window on each iteration) would be very beneficial. Indeed, it could become a core algorithm irrespective of the nature of the problem because it would also mitigate the computational costs associated with large unknown time histories.

A surprisingly difficult class of problems to solve is that of finding a location (either of a force or a parameter change). These are global problems that appear to require a good number of sensors plus a high computational cost for their solution. With the advent of the new sensor technologies, this problem now seems ripe for a new and robust general solution. Such a solution would also have enormous implications in the general area of global monitoring.

There is room for making some of the algorithms faster (such as implementing variations on the superfast solvers for Toeplitz systems), but it would be more to the point to place them inside some new global schemes. For example, all direct parameter identification schemes and all nonlinear problems required the use of iteration; this could be an opportunity to implement new solution strategies that are feasible only when iteration is done. Nonlinear problems use two separate structural models, the true model and the one used for incremental refinement. Iteration raises the intriguing prospect of using two separate models *ab initio* for all problems (including linear problems), where convergence to the correct solution is forced through the Newton–Raphson iterations on the true model, but the sensitivity responses are computed from the simpler model. This would make the recursive method of Chapter 5 feasible because only the FEM formulation of the simpler model need be coded.

The form of regularization implemented in all the examples discussed is based on levels of smoothness between neighboring parameters. But regularization, as a concept, is much more general—it is a mechanism to incorporate additional information into the solution process. The key is identifying the additional information for particular classes of problems: it might be as simple as nonnegativity for intensity problems or as complex as using multiple modelings. This is an area that needs a good deal more of research.

In conclusion, this book has only begun to touch the issues involved in the active integration of FEM modeling and experiment. There are many more challenges yet to be overcome.

References

[1] Aben, H.K., "Optical Phenomena in Photoelastic Models by the Rotation of Principal Axes", *Experimental Mechanics*, **6**(1), 13–22, 1966.

[2] Aben, H.K., *Integrated Photoelasticity*, McGraw-Hill, New York, 1980.

[3] Adams, R.A. and Doyle, J.F., "Force Identification in Complex Structures", *Recent Advances in Structural Dynamics*, pp. 225–236, ISVR, Southampton, 2000.

[4] Adams, R.A. and Doyle, J.F., "Multiple Force Identification for Complex Structures", *Experimental Mechanics*, **42**(1), 25–36, 2002.

[5] Adams, R.A., *Force Identification in Complex Structures*, M.S. Thesis, Purdue University, 1999.

[6] Aktan, A.E. and Grimmelsman, K.A., "The Role of NDE in Bridge Health Monitoring", *Proceedings of the SPIE Conference on Nondestructive Evaluation of Bridges and Highways II*, SPIE 3587, Newport Beach, CA, 1999.

[7] Aktan, A.E., Helmicki, A.J. and Hunt, V.J., "Issues in Health Monitoring for Intelligent Infrastructure", *Journal of Smart Materials and Structures*, **7**(5), 674–692, 1998.

[8] Alifanov, O.M., "Methods of Solving Ill-Posed Inverse Problems", *Journal of Engineering Physics*, **45**(5), 1237–1245, 1983.

[9] Allman, D.J., "Evaluation of the Constant Strain Triangle with Drilling Rotations", *International Journal for Numerical Methods in Engineering*, **26**, 2645–2655, 1988.

[10] Ammar, G. and Gragg, W., "Superfast Solution of Real Positive Definite Toeplitz Systems", *SIAM Journal of Matrix Analysis and Applications*, **9**(1), 111–116, 1988.

[11] Ammar, G. and Gragg, W., "Numerical Experience with a Superfast Real Toeplitz Solver", *Linear Algebra and Its Applications*, **121**, 185–206, 1989.

[12] Aprahamian, R., Evensen, D.A., Mixson, J.S. and Jacoby, J.L., "Holographic Study of Propagating Transverse Waves in Plates", *Experimental Mechanics*, **11**, 357–362, 1971.

Modern Experimental Stress Analysis: completing the solution of partially specified problems. James Doyle
© 2004 John Wiley & Sons, Ltd ISBN 0-470-86156-8

[13] Aprahamian, R., Evensen, D.A., Mixson, J.S. and Wright, J.E., "Application of Pulsed Holographic Interferometry to the Measurement of Propagating Transverse Waves in Beams", *Experimental Mechanics*, **11**, 309–314, 1971.

[14] Arsenin, V.Ya., "Problems on Computer Diagnostics in Medicine", In *Ill-Posed Problems in the Natural Sciences*, A.N. Tikhonov and A.V. Goncharsky, editors, pp. 206–219, Mir Publishers, Moscow, 1987.

[15] Asundi, A. and Sajan, M.R., "Digital Dynamic Photoelasticity—Recent Advances", *Proceedings SEM*, **28**, 212–217, 1994.

[16] Atluri, S.N., "Energetic Approaches and Path-Independent Integrals in Fracture Mechanics", *Computational Methods in the Mechanics of Fracture*, pp. 121–165, North-Holland, New York, 1986.

[17] Van Barel and Kravanja, P., "A Stabilized Superfast Solver for Indefinite Hankel Systems", *Linear Algebra and its Applications*, **284**, 335–355, 1998.

[18] Bathe, K.-J., *Finite Element Procedures in Engineering Analysis*, 2nd ed., Prentice Hall, Englewood Cliffs, NJ, 1982, 1995.

[19] Batoz, J.-L., Bathe, K.-J. and Ho, L.-W., "A Study of Three-Node Triangular Plate Bending Elements", *International Journal for Numerical Methods in Engineering*, **15**, 1771–1812, 1980.

[20] Batoz, J.-L., "An Explicit Formulation for an Efficient Triangular Plate-Bending Element", *International Journal for Numerical Methods in Engineering*, **18**, 1077–1089, 1982.

[21] Bellman, R. and Kalaba, R., *Dynamic Programming and Modern Control Theory*, Academic Press, New York, 1965.

[22] Belytschko, T., Schwer, L. and Klein, M.J., "Large Displacement, Transient Analysis of Space Frames", *International Journal for Numerical Methods in Engineering*, **11**, 65–84, 1977.

[23] Belytschko, T. and Hsieh, B.J., "Non-Linear Transient Finite Element Analysis with Convected Coordinates", *International Journal for Numerical Methods in Engineering*, **7**, 255–271, 1973.

[24] Belytschko, T., "An Overview of Semidiscretization and Time Integration Procedures", In *Computational Methods for Transient Analysis*, T. Belytschko and T.J.R. Hughes, editors, pp. 1–65, Elsevier, Amsterdam, 1993.

[25] Bergan, P.G. and Felippa, C.A., "A Triangular Membrane Element with Rotational Degrees of Freedom", *Computer Methods in Applied Mechanics and Engineering*, **50**, 25–69, 1985.

[26] Bergan, P.G. and Felippa, C.A., "Efficient Implementation of a Triangular Membrane Element with Drilling Freedoms", *FEM Methods*, pp. 128–152, CRC Press, New York, 1987.

[27] Bickle, L.W., "The Response of Strain Gages to Longitudinally Sweeping Strain Pulses", *Experimental Mechanics*, **10**, 333–337, 1970.

[28] Biemond, J., Lagendik, R.L. and Mersereau, R.M., "Iterative Methods for Image Deblurring", *Proceedings of IEEE*, **78**(5), 856–883, 1990.

[29] Blackman, R.B. and Tukey, J.W., *The Measurement of Power Spectra*, Dover Publications, New York, 1958.

[30] Boley, B.A. and Weiner, J.H., *Theory of Thermal Stress*, Wiley & Sons, New York, 1960.

[31] Brigham, E.O., *The Fast Fourier Transform*, Prentice Hall, Englewood Cliffs, NJ, 1973.

[32] Bruck, H.A., Casem, D., Williamson, R.L. and Epstein, J.S., "Characterization of Short Duration Stress Pulses Generated by Impacting Laminated Carbon-Fiber/Epoxy Composites with Magnetic Flyer Plates", *Experimental Mechanics*, **42**(3), 279–287, 2002.

[33] Busby, H.R. and Trujillo, D.M., "Numerical Solution to a Two-Dimensional Inverse Heat Conduction Problem", *International Journal for Numerical Methods in Engineering*, **21**, 349–359, 1985.

[34] Busby, H.R. and Trujillo, D.M., "Solution of an Inverse Dynamics Problem using an Eigenvalue Reduction Technique", *Computers & Structures*, **25**(1), 109–117, 1987.

[35] Chandrupatla, T.R. and Belegundu, A.D., *Introduction to Finite Elements in Engineering*, 3rd ed., Prentice Hall, Englewood Cliffs, NJ, 1991, 2002.

[36] Chang, A.L. and Rajendran, A.M., "Novel In-situ Ballistic Measurements for Validation of Ceramic Constitutive Models", In *14th US Army Symposium on Solid Mechanics*, K.R. Iyer and S.-C. Chou, editors, pp. 99–110, Batelle Press, Columbus, OH, October, 1996.

[37] Chan, R. and Ng, M., "Conjugate Gradient Methods for Toeplitz Systems", *SIAM Review*, **38**(3), 427–482, 1996.

[38] Chatfield, C., *The Analysis of Time Series: An Introduction*, Chapman & Hall, London, 1984.

[39] Choi, S.-W. and Doyle, J.F., "A Computational Framework for the Separation of Material Systems", *Recent Advances in Structural Dynamics*, ISVR, Southampton, CD #23, 2003.

[40] Choi, S.-W., *Impact Damage of Layered Material Systems*, Ph.D. Thesis, Purdue University, August, 2002.

[41] Cho, S.-M., *A Sub-Domain Inverse Method for Dynamic Crack Propagation Problems*, M.S. Thesis, Purdue University, December, 2000.

[42] Cloud, G.L., "Practical Speckle Interferometry for Measuring In-plane deformation", *Applied Optics*, **14**(4), 878–884, 1975.

[43] Cloud, G.L., *Optical Methods of Engineering Analysis*, Cambridge University Press, Cambridge, UK, 1995.

[44] Cole, R.H., *Underwater Explosions*, Princeton University Press, NJ, 1948.

[45] Cook, R.D., Malkus, D.S. and Plesha, M.E., *Concepts and Applications of Finite Element Analysis*, 3rd ed., Wiley & Sons, New York, 1989.

[46] Crisfield, M.A., *Nonlinear Finite Element Analysis of Solids and Structures, Vol 2: Advanced Topics*, Wiley & Sons, New York, 1997.

[47] Dally, J.W. and Read, D.T., "Electron Beam Moire", *Experimental Mechanics*, **33**, 270–277, 1993.

[48] Dally, J.W. and Riley, W.F., *Experimental Stress Analysis*, 3rd ed., McGraw-Hill, New York, 1991.

[49] Dally, J.W. and Sanford, R.J., "Multiple Ruby Laser System for High Speed Photography", *Optical Engineering*, **21**, 704–708, 1982.

[50] Dally, J.W., "Dynamic Photoelastic Studies of Dynamic Fracture", *Experimental Mechanics*, **19**, 349–361, 1979.

[51] Dally, J.W., "An Introduction to Dynamic Photoelasticity", *Experimental Mechanics*, **20**, 409–416, 1980.

[52] Dally, J.W., Riley, W.F. and McConnell, K.G., *Instrumentation for Engineering Measurements*, 2nd ed., Wiley, New York, 1993.

[53] Danial, A.N. and Doyle, J.F., "A Massively Parallel Implementation of the Spectral Element Method for Impact Problems in Plate Structures", *Computing Systems in Engineering*, **5**, 375–388, 1994.

[54] Degrieck, J. and Verleysen, P., "Determination of Impact Parameters by Optical Measurement of the Impactor Displacement", *Experimental Mechanics*, **42**(3), 298–302, 2002.

[55] Doyle, J.F. and Danyluk, H.T., "Integrated Photoelasticity for Axisymmetric Problems", *Experimental Mechanics*, **18**(6), 215–220, 1978.

[56] Doyle, J.F. and Sun, C.-T., *Theory of Elasticity*, A&AE 553 Class Notes, Purdue University, 1999.

[57] Doyle, J.F., "A Unified Approach to Optical Constitutive Relations", *Experimental Mechanics*, **18**, 416–420, 1978.

[58] Doyle, J.F., "Mechanical and Optical Characterization of a Photoviscoplastic Material", *Journal of Applied Polymer Science*, **24**, 1295–1304, 1979.

[59] Doyle, J.F., "Constitutive Relations in Photomechanics", *International Journal of Mechanical Sciences*, **22**, 1–8, 1980.

[60] Doyle, J.F., "An Interpretation of Photoviscoelastic/ Plastic Data", *Experimental Mechanics*, **20**, 65–67, 1980.

[61] Doyle, J.F., "On a Nonlinearity in Flow Birefringence", *Experimental Mechanics*, **22**, 37–38, 1982.

[62] Doyle, J.F., "Further Developments in Determining the Dynamic Contact Law", *Experimental Mechanics*, **24**, 265–270, 1984.

[63] Doyle, J.F., "Determining the Contact Force during the Transverse Impact of Plates", *Experimental Mechanics*, **27**, 68–72, 1987.

[64] Doyle, J.F., "Experimentally Determining the Contact Force During the Transverse Impact of Orthotropic Plates", *Journal of Sound and Vibration*, **128**, 441–448, 1987.

[65] Doyle, J.F., "Impact and Longitudinal Wave Propagation", *Experimental Techniques*, **2**, 29–31, 1988.

[66] Doyle, J.F., "Toward In-Situ Testing of the Mechanical Properties of Composite Panels", *Journal of Composite Materials*, **22**, 416–426, 1988.

[67] Doyle, J.F., *Static and Dynamic Analysis of Structures*, Kluwer, The Netherlands, 1991.

[68] Doyle, J.F., "Force Identification from Dynamic Responses of a Bi-Material Beam", *Experimental Mechanics*, **33**, 64–69, 1993.

[69] Doyle, J.F., "A Wavelet Deconvolution Method for Impact Force Identification", *Experimental Mechanics*, **37**, 404–408, 1997.

[70] Doyle, J.F., *Wave Propagation in Structures*, 2nd ed., Springer-Verlag, New York, 1997.

[71] Doyle, J.F., *Nonlinear Analysis of Thin-Walled Structures: Statics, Dynamics, and Stability*, Springer-Verlag, New York, 2001.

[72] Doyle, J.F., "Reconstructing Dynamic Events from Time-Limited Spatially Distributed Data", *International Journal for Numerical Methods in Engineering*, **53**, 2721–2734, 2002.

[73] Doyle, J.F., "Force Identification Using Dynamic Photoelasticity", *International Journal for Numerical Methods in Engineering*, 2004, in press.

[74] Doyle, J.F., Kamle, S. and Takezaki, J., "Error Analysis of Photoelasticity in Fracture Mechanics", *Experimental Mechanics*, **17**, 429–435, 1981.

[75] Espinosa, H.D., Xu, Y. and Lu, H.-C., "A Novel Technique for Penetrator Velocity Measurements and Damage Identification in Ballistic Penetration Experiments", In *14th US Army Symposium on Solid Mechanics*, K.R. Iyer and S.-C. Chou, editors, pp. 111–120, Batelle Press, Columbus, OH, October, 1996.

[76] Ewins, D.J., *Modal Testing: Theory and Practice*, Wiley & Sons, New York, 1984.

[77] Fienup, J.R. and Wackerman, C.C., "Phase-Retrieval Stagnation Problems and Solutions", *Journal of the Optical Society of America*, **3**(11), 1897–1907, 1986.

[78] Fienup, J.R., "Phase Retrieval Algorithms: A Comparison", *Applied Optics*, **21**(15), 2758–2769, 1982.

[79] Freund, L.B., *Dynamic Fracture Mechanics*, Cambridge University Press, Cambridge, 1990.

[80] Friebele, E.J., Askins, C.G., Kersey, A.D., Patrick, H.J., Pogue, W.R., Simon, W.R., Tasker, F.A., Vincent, W.S., Vohra, S.T., "Optical Fiber Sensors for Spacecraft Applications", *Journal of Smart Materials and Structures*, **8**(6), 813–838, 1999.

[81] Friswell, M.I. and Inman, D.J., "Sensor Validation for Smart Structures", *IMAC-XVIII: Conference on Structural Dynamics*, San Antonio, pp. 483–489, 2000.

[82] Friswell, M.I. and Mottershead, J.E., *Finite Element Model Updating in Structural Dynamics*, Kluwer, The Netherlands, 1995.

[83] Frocht, M.M., *Photoelasticity*, Vol. I & II, Wiley & Sons, New York, 1948.

[84] Galvanetto, U. and Crisfield, M.A., "An Energy-Conserving Co-Rotational Procedure for the Dynamics of Planar Beam Structures", *International Journal for Numerical Methods in Engineering*, **39**, 2265–2282, 1992.

[85] Gerchberg, R.W. and Saxon, W.O., "A Practical Algorithm for the Determination of Phase from Image and Diffraction Plane Pictures", *Optik*, **35**(2), 237–246, 1972.

[86] Gordon, J.E., *Structures, or Why Things Don't Fall Down*, Penguin, London, 1978.

[87] Graff, K.F., *Wave Motion in Elastic Solids*, Ohio State University Press, Columbus, 1975.

[88] Gupta, A.D., "Dynamic Elasto-Plastic Response of a Generic Vehicle Floor Model to Coupled Transient Loads", In *14th US Army Symposium on Solid Mechanics*, K.R. Iyer and S.-C. Chou, editors, pp. 507–520, Batelle Press, Columbus, OH, October, 1996.

[89] Hamilton, W.R., *The Mathematical Papers of Sir W.R. Hamilton*, Cambridge University Press, Cambridge, 1940.

[90] Hecht, E., *Theory and Problems of Optics*, Schaum's Outline Series, McGraw-Hill, New York, 1975.

[91] Hollandsworth, P.E. and Busby, H.R., "Impact Force Identification using the General Inverse Technique", *International Journal of Impact Engineering*, **8**, 315–322, 1989.

[92] Holman, J.P., *Experimental Methods for Engineers*, 5th ed., McGraw-Hill, New York, 1989.

[93] Hung, Y.Y., "Shearography: A New Optical Method for Strain Measurement and Non-Destructive Testing", *Optical Engineering*, **21**, 391–395, 1982.

[94] Ince, E.L., *Ordinary Differential Equations*, Dover Publications, New York, 1956.

[95] Jazwinski, A.H., *Stochastic Processes and Filtering Theory*, Academic Press, New York, 1970.

[96] Jeyachandrabose, C., Kirkhope, J. and Babu, C.R., "An Alternative Explicit Formulation for the DKT Plate-Bending Element", *International Journal for Numerical Methods in Engineering*, **21**, 1289–1293, 1985.

[97] Jih, C.J. and Sun, C.T., "Evaluation of a Finite Element Based Crack-Closure Method for Calculating Static and Dynamic Strain Energy Release Rates", *Engineering Fracture Mechanics*, **37**, 313–322, 1990.

[98] Jin, X., "Fast iterative Solvers for Symmetric Toeplitz Systems - A Survey and an Extension", *Journal of Computational and Applied Mathematics*, **66**, 315–321, 1996.

[99] Johnson, W. and Mellor, P.B., *Plasticity for Mechanical Engineers*, Van Nostrand, London, 1962.

[100] Jones, R.M., *Mechanics of Composite Materials*, McGraw-Hill, New York, 1975.

[101] Kachanov, L.M., *Fundamentals of the Theory of Plasticity*, Mir Publishers, Moscow, 1974.

[102] Kailath, T., "Some New Algorithms for Recursive Estimation in Constant, Linear, Discrete-Time Systems", *IEEE Transactions on Information Theory*, **AC-19**(4), 83–92, 1973.

[103] Kalman, R.E., "A New Approach to Linear Filtering and Prediction Problems", *ASME Journal of Basic Engineering*, **82D**, 35–45, 1960.

[104] Kalman, R.E., "New Methods in Wiener Filtering Theory", *IEEE*, **82D**, 270–388, 1962.

[105] Kleiber, M., *Parameter Sensitivity in Nonlinear Mechanics*, Wiley & Sons, Chichester, 1997.

[106] Kobayashi, A.S., "Hybrid Experimental-Numerical Stress Analysis", *Experimental Mechanics*, **23**, 338–347, 1983.

[107] Kobayashi, A.S., *Handbook on Experimental Mechanics*, VCH Publishers, New York, 1993.

[108] Kobayashi, A.S., "Hybrid Experimental-Numerical Stress Analysis", In *Handbook on Experimental Mechanics*, A.S. Kobayashi, editor, pp. 751–783, VCH Publishers, New York, 1993.

[109] Kobayashi, A.S., Emery, A.G. and Liaw, B.-M., "Dynamic Fracture Toughness of Reaction Bonded Silicone Nitride", *Journal of the American Ceramics Society*, **66**(2), 151–155, 1983.

[110] Kolsky, H., *Stress Waves in Solids*, Dover Publications, New York, 1963.

[111] Kuhl, D. and Crisfield, M.A., "Energy-Conserving and Decaying Algorithms in Non-Linear Structural Dynamics", *International Journal for Numerical Methods in Engineering*, **45**, 569–599, 1999.

[112] Laermann, K.-H., "Recent Developments and Further Aspects of Experimental Stress Analysis in the Federal Republic of Germany and Western Europe", *Experimental Mechanics*, **21**(1), 49–57, 1981.

[113] Lapidus, L. and Luus, R., *Optimal Control of Engineering Processes*, Blaisdell Publishing Company, Waltham, MA, 1967.

[114] Liu, T., Guille, M. and Sullivan, J.P., "Accuracy of Pressure Sensitive Paint", *AIAA Journal*, **39**(1), 103–112, 2001.

[115] Larmore, L., *Introduction to Photographic Principles*, Dover Publications, New York, 1965.

[116] Martinez, Y. and Dinth, A., "A Generalization of Tikhonov's Regularizations of Zero and First Order", *Computers and Mathematics with Applications*, **12B**(5/6), 1203–1208, 1986.

[117] Martin, L.P. and Ju, F.D., "The Moire Method for Measuring Large Plane Deformations", *Journal of Applied Mechanics*, **36**(4), 385–391, 1969.

[118] Martin, M.T. and Doyle, J.F., "Impact Force Identification from Wave Propagation Responses", *International Journal of Impact Engineering*, **18**, 65–77, 1996.

[119] Martin, M.T. and Doyle, J.F., "Impact Force Location in Frame Structures", *International Journal of Impact Engineering*, **18**, 79–97, 1996.

[120] McConnell, K.G. and Riley, W.F., "Force-Pressure-Motion Measuring Transducers", In *Handbook on Experimental Mechanics*, A.S. Kobayashi, editor, pp. 79–118, VCH Publishers, New York, 1993.

[121] McCoy, R.W., *Dynamic Analysis of a Thick-Section Composite Cylinder Subjected to Underwater Blast Loading*, Ph.D. Thesis, Purdue University, August, 1996.

[122] Michaels, J.E. and Pao, Y.-H., "The Inverse Source Problem for an Oblique Force on an Elastic Plate", *Journal of Acoustical Society of America*, **77**(6), 2005–2011, 1985.

[123] Miller, K., "Least Squares Methods for Ill-Posed Problems with a Prescribed Bound", *SIAM Journal of Mathematical Analysis*, **1**(1), 52–74, 1970.

[124] Morf, M., Sidhu, G. and Kailath, T., "Some New Algorithms for Recursive Estimation in Constant, Linear, Discrete-Time Systems", *IEEE Transactions on Automatic Control*, **AC-19**(4), 315–323, 1974.

[125] Morimoto, Y., "Digital Image Processing", In *Handbook on Experimental Mechanics*, A.S. Kobayashi, editor, pp. 969–1030, VCH Publishers, New York, 1993.

[126] Mottershead, J.E. and Friswell, M.I., "Model Updating in Structural Dynamics: A Survey", *Journal of Sound & Vibration*, **162**(2), 347–375, 1993.

[127] Neumaier, A., "Solving Ill-Conditioned and Singular Linear Systems: A Tutorial on Regularization", *Society for Industrial and Applied Mathematics*, **40**(3), 636–666, 1998.

[128] Niblack, w., *An Introduction to Digital Image Processing*, Prentice Hall, Englewood Cliffs, NJ, 1986.

[129] Nishioka, T., "An Intelligent Hybrid Method to Automatically Detect and Eliminate Experimental Measurement Errors for Linear Elastic Deformation Fields", *Experimental Mechanics*, **40**(2), 170–179, 2000.

[130] Nour-Omid, B. and Rankin, C.C., "Finite Rotation Analysis and Consistent Linearization using Projectors", *Computer Methods in Applied Mechanics and Engineering*, **93**(1), 353–384, 1991.

[131] Oi, K., "Transient Response of Bonded Strain Gages", *Experimental Mechanics*, **6**, 463–469, 1966.

[132] Oliver, D.E., "Stress Pattern Analysis by Thermal Emission", In Chapter 14, *Handbook on Experimental Mechanics*, 1st ed., A.S. Kobayashi, editor, Prentice Hall, Englewood Cliffs, NJ, 1986.

[133] Ostrovsky, Yu.I., *Holography and Its Applications*, Mir Publishers, Moscow, 1977.

[134] Pacoste, C. and Eriksson, A., "Beam Elements in Instability Problems", *Computer Methods in Applied Mechanics and Engineering*, **144**, 163–197, 1997.

[135] Pacoste, C., "Co-Rotational Flat Facet Triangular Elements for Shell Instability Analyses", *Computer Methods in Applied Mechanics and Engineering*, **156**, 75–110, 1998.

[136] Pan, V., Zheng, A., Dias, O. and Huang, X., "A Fast, Preconditioned Conjugate Gradient Toeplitz and Toeplitz-Like Solvers", *Computers and Mathematical Applications*, **30**(8), 57–63, 1995.

[137] Patterson, E.A., "Digital Photoelasticity: Principles, Practice, and Potential", *Strain*, **38**(1), 27–39, 2002.

[138] Pavlids, T., *Algorithms for Graphics and Image Processing*, Computer Science Press, Rockville, 1982.

[139] Phillips, D.L., "A Technique for the Numerical Solution of Certain Integral Equations of the First Kind", *Journal of Association for Computing Machinery*, **9**(1), 84–97, 1962.

[140] Pines, D.J. and Lovell, P.A., "Conceptual Framework of a Remote Wireless Health Monitoring System for Large Civil Structures", *Journal of Smart Materials and Structures*, **7**(5), 627–636, 1998.

[141] Pirodda, L., "Shadow and Projection Moiré Techniques for Absolute or Relative Mapping of Surface Shapes", *Optical Engineering*, **21**(4), 640–649, 1982.

[142] Post, D., Han, B. and Ifju, P., *High Sensitivity Moiré*, Springer-Verlag, New York, 1994.

[143] Press, W.H., Flannery, B.P., Teukolsky, S.A. and Vetterling, W.T., *Numerical Recipes*, 2nd ed., Cambridge University Press, Cambridge, 1992.

[144] Ramesh, K., *Digital Photoelasticity—Advanced Techniques and Applications*, Springer, New York, 2000.

[145] Ranson, W.F., Sutton, M.A. and Peters, W.H., "Holographic and Laser Speckle Interferometry", In *Handbook on Experimental Mechanics*, A.S. Kobayashi, editor, pp. 365–406, VCH Publishers, New York, 1993.

[146] Rauch, B.J. and Rowlands, R.E., "Thermoelastic Stress Analysis", In Chapter 14, *Handbook on Experimental Mechanics*, 2nd ed., A.S. Kobayashi, editor, pp. 581–599, VCH Publishers, New York, 1993.

[147] Rizzi, S.A. and Doyle, J.F., "Spectral Analysis of Wave Motion in Plane Solids with Boundaries", *Journal of Vibration and Acoustics*, **114**, 133–140, 1992.

[148] Rizzi, S.A. and Doyle, J.F., "A Spectral Element Approach to Wave Motion in Layered Solids", *Journal of Vibration and Acoustics*, **114**, 569–577, 1992.

[149] Rizzi, S.A. and Doyle, J.F., "A Simple High-Frequency Force Transducer for Impact", *Experimental Techniques*, **4**, 45–48, 1990.

[150] Rizzi, S.A., *A Spectral Analysis Approach to Wave Propagation in Layered Solids*, Ph.D. Thesis, Purdue University, 1989.

[151] Rybicki, E.F. and Kanninen, M.F., "A Finite Element Calculation of Stress Intensity Factors by a Modified Crack-Closure Integral", *Engineering Fracture Mechanics*, **9**, 931–938, 1977.

[152] Sakaue, H. and Sullivan, J.P., "Time Response of Anodized Aluminum Pressure Sensitive Paint", *AIAA Journal*, **39**(10), 1944–1949, 2001.

[153] Schell, A.C., *Programs for Digital Signal Processing*, IEEE Press, New York, 1979.

[154] Sciammarella, C.A., "The Moire Method–A Review", *Experimental Mechanics*, **22**(11), 418–443, 1982.

[155] Scurria, N.V. and Doyle, J.F., "Photoelastic Analysis of Contact Stresses in the Presence of Machining Irregularities", *Experimental Mechanics*, **22**, 342–347, 1997.

[156] Scurria, N.V., *Three-Dimensional Analysis of the Shrink-Fit Problem by Stress Freezing*, M.S. Thesis, Purdue University, 1980.

[157] Serra, S., "Superlinear PCG Methods for Symmetric Toeplitz Systems", *Mathematics of Computation*, **68**, 22–26, 1999.

[158] Smith, C.W., McGowan, J.J. and Jolles, M., "Effects of Artificial Cracks and Poisson's Ratio upon Photoelastic Stress Intensity Determination", *Experimental Mechanics*, **16**(5), 188–193, 1976.

[159] Sneddon, I.N., *Fourier Transforms*, McGraw-Hill, New York, 1951.

[160] Speigel, M.R., *Fourier Analysis with Applications to Boundary Value Problems*, McGraw-Hill, New York, 1974.

[161] Stevens, K.K., "Force Identification Problems: An Overview", *Proceedings of SEM Spring Meeting*, Houston, pp. 838–844, 1987.

[162] Stricklin, J.A, Haisler, E.E., Tisdale, P.R. and Gunderson, R., "A Rapidly Converging Triangular Plate Element", *AAIA Journal*, **7**(1), 180–181, 1969.

[163] Taylor, C.E., "Holography", In *Manual on Experimental Stress Analysis*, 5th ed., J.F. Doyle and J.W. Phillips, editors, pp. 136–149, Society for Experimental Mechanics, Bridgeport, Conn, 1989.

[164] Tedesco, J.W., McDougal, W.G. and Allen Ross, C., *Structural Dynamics: Theory and Applications*, Addison-Wesley, Menlo Park, CA, 1999.

[165] Tikhonov, A.N. and Arsenin, V.Y., *Solutions of Ill-Posed Problems*, Wiley & Sons, New York, 1977.

[166] Tikhonov, A.N. and Goncharsky, A.V., *Ill-Posed Problems in the Natural Sciences*, MIR Publishers, Moscow, 1987.

[167] Timoshenko, S.P. and Goodier, J.N., *Theory of Elasticity*, McGraw-Hill, New York, 1970.

[168] Trujillo, D.M. and Busby, H.R., "Optimal Regularization of the Inverse Heat-Conduction Problem", *Journal of Thermophysics*, **3**(4), 423–427, 1989.

[169] Trujillo, D.M. and Busby, H.R., *Practical Inverse Analysis in Engineering*, CRC Press, New York, 1997.

[170] Trujillo, D.M., "Application of Dynamic Programming to the General Inverse Problem", *International Journal for Numerical Methods in Engineering*, **12**, 613–624, 1978.

[171] Twomey, S., "On the Numerical Solution of Fredholm Integral Equations of the First Kind by the Inversion of the Linear System Produced by Quadratures", *Journal of Association for Computing Machinery*, **10**, 97–101, 1963.

[172] Twomey, S., *Introduction to the Mathematics of Inversion in Remote Sensing and Indirect Measurements*, Elsevier, Amsterdam, 1977.

[173] Van Barel, M., Heinig, G. and Kravanja, P., *A Stabilized Superfast Solver for Nonsymmetric Toeplitz Systems*, Report TW293, Department of Computer Science, Katholieke Universiteit Leuven, Belgium, 1999.

[174] Williams, J.G., "On the Calculation of Energy Release Rates for Cracked Laminates", *International Journal of Fracture*, **45**, 101–119, 1988.

[175] Xu, X.-P. and Needleman, A., "Numerical Simulations of Fast Crack Growth in Brittle Solids", *Journal of Mechanics and Physics of Solids*, **42**(9), 1397–1434, 1994.

[176] Yourgrau, W. and Mandelstam, S., *Variational Principles in Dynamics and Quantum Theory*, Dover Publications, New York, 1979.

[177] "Instructions for Using Photoelastic Plastics Liquid PLM-4 and Pre-Cast Block PLM-4B", Photoelastic Inc., Bulletin I-206, 1974.

[178] ABAQUS, Hibbitt, Karlsson and Sorenson, Inc., Pawtucket, RI.

[179] ANSYS, Swanson Analysis Systems, Inc., Houston, PA.

[180] DASH-18, Astro-Med, Inc., West Warwick, RI.

[181] Norland 3001, Norland Corp., Fort Atkinson, WI.

[182] PCB#, PCB Piezotronics, Depew, NY.

Index

Modern Experimental Stress Analysis: completing the solution of partially specified problems. James Doyle
© 2004 John Wiley & Sons, Ltd ISBN 0-470-86156-8